AQUATIC PREDATORS AND THEIR PREY

Aquatic Predators
and
their Prey

Edited by

Simon P.R. Greenstreet

and

Mark L. Tasker

Fishing News Books

Copyright © 1996 Fishing News Books
A division of Blackwell Science Ltd
Editorial Offices:
Osney Mead, Oxford OX2 0EL
25 John Street, London WC1N 2BL
23 Ainslie Place, Edinburgh EH3 6AJ
238 Main Street, Cambridge,
    Massachusetts 02142, USA
54 University Street, Carlton,
    Victoria 3053, Australia

Other Editorial Offices:
Arnette Blackwell SA
    1, rue de Lille, 75007 Paris
    France

Blackwell Wissenschafts-Verlag GmbH
    Kurfürstendamm 57
    10707 Berlin, Germany

Zehetnergasse 6, A-1140 Wien
Austria

First published 1996

Set in Ehrhardt 9/10.5 pt
by Best-set Typesetter Ltd. Hong Kong
Printed and bound in Great Britain by
The University Press, Cambridge.

DISTRIBUTORS

Marston Book Services Ltd
PO Box 87
Oxford OX2 0DT
(*Orders:*   Tel: 01865 206206
            Fax: 01865 721205
            Telex: 83355 MEDBOK G)
USA
Blackwell Science, Inc.
238 Main Street
Cambridge, MA 02142
(*Orders*:   Tel: 800 215-1000
            617 876-7000
            Fax: 617 492-5263)

Canada
Oxford University Press
70 Wynford Drive
Don Mills
Ontario M3C 1J9
(*Orders:*   Tel: 416 441 2941)

Australia
Blackwell Science Pty Ltd
54 University Street
Carlton, Victoria 3053
(*Orders:*   Tel: 03 9347-0300
            Fax: 03 9349 3016)

A catalogue record for this book is
available from the British Library

ISBN 0-85238-230-8

Contents

Contributors

ADAMS C.E. *Fish Behaviour and Ecology Group, University Field Station, Glasgow University, Rowardennan, Glasgow G63 0AW, UK*

BRAND A.R. *Port Erin Marine Laboratory, University of Liverpool, Port Erin, Isle of Man IM9 6JA, UK*

CALDOW R.W.G. *Applied Ornithology Unit, Division of Evolutionary and Environmental Biology, Institute of Biomedical and Life Sciences, University of Glasgow, Glasgow G12 8QQ, UK*
Present address: Institute of Terrestrial Ecology, Furzebrook Research Station, Wareham, Dorset BH20 5AS, UK

CARSS D.N. *Institute of Terrestrial Ecology, Hill of Brathens, Banchory, Kincardineshire AB31 4BY, UK*

CORPE H.M. *Department of Zoology, University of Aberdeen, Lighthouse Field Station, Cromarty IV11 8YJ, UK*

DAVIES J.M. *School of Biological and Earth Sciences, Liverpool John Moores University, Byrom Street, Liverpool L3 3AF, UK*

DECKER M.B. *Department of Ecology and Evolutionary Biology, University of California, Irvine, CA 92717, USA*

DEROOS A. *Department of Pure and Applied Ecology, University of Amsterdam, Kruislaan 320, 1098 SM Amsterdam, Netherlands*

des CLERS S. *Renewable Resources Assessment Group, Centre for Environmental Technology, Imperial College of Technology, Science and Medicine, 8 Prince's Gardens, London SW7 1NA, UK*

DUNNET G.M. *Department of Zoology, University of Aberdeen, Tillydrone Avenue, Aberdeen AB9 2TN, UK*

ESTES J.A. *National Biological Service, Institute of Marine Sciences, University of California, Santa Cruz, CA 95064, USA*

FELTHAM M.J. *School of Biological and Earth Sciences, Liverpool John Moores University, Byrom Street, Liverpool L3 3AF, UK*

FURNESS R.W. *Applied Ornithology Unit, Division of Evolutionary and Environmental Biology, Institute of Biomedical and Life Sciences, University of Glasgow, Glasgow G12 8QQ, UK*

GOSS-CUSTARD J.D. *Institute of Terrestrial Ecology, Furzebrook Research Station, Wareham, Dorset BH20 5AS, UK*

GREENSTREET S.P.R. *Scottish Office Agriculture, Environment and Fisheries Department, Marine Laboratory, PO Box 101, Victoria Road, Aberdeen AB9 8DB, UK*

GURNEY W.S.C. *Department of Statistics and Modelling Science, University of Strathclyde, Glasgow G1 1XH, UK*

HAWKINS S.J. *Port Erin Marine Laboratory, University of Liverpool, Port Erin, Isle of Man IM9 6JA, UK*

HILL A.S. *Port Erin Marine Laboratory, University of Liverpool, Port Erin, Isle of Man IM9 6JA, UK*

HUNT G.L. Jr *Department of Ecology and Evolutionary Biology, University of California, Irvine, CA 92717, USA*

KAISER M.J. *Ministry of Agriculture, Fisheries and Food, Directorate of Fisheries Research, Fisheries Laboratory, Benarth Road, Conwy, Gwynedd LL32 8UB, UK*

KITAYSKY A. *Department of Ecology and Evolutionary Biology, University of California, Irvine, CA 92717, USA*

KRUUK H. *Institute of Terrestrial Ecology, Hill of Brathens, Banchory, Kincardineshire AB31 4BY, UK*

LYNDON A.R. *Department of Biological Sciences, University of Exeter, Hatherly Laboratories, Prince of Wales Road, Exeter EX4 4PS, UK*
Present address: Department of Biological Sciences, Heriot-Watt University, Edinburgh EH 14 4AS, UK

McCAULEY E. *Department of Biological Sciences, University of Calgary, Calgary, Alberta T2N 1N4, Canada*

MACKAY A. *Department of Zoology, University of Aberdeen, Lighthouse Field Station, Cromarty IV11 8YJ, UK*

MARQUISS M. *Institute of Terrestrial Ecology, Hill of Brathens, Banchory, Kincardineshire AB31 4BY, UK*

MIDDLETON D.A.J. *Department of Statistics and Modelling Science, University of Strathclyde, Glasgow G1 1XH, UK*

MURDOCH W.W. *Department of Biology, University of California at Santa Barbara, Santa Barbara, CA 93106, USA*

NISBET R.M. *Department of Biology, University of California at Santa Barbara, Santa Barbara, CA 93106, USA*

ORMEROD S.J. *Catchment Research Group, School of Pure and Applied Biology, University of Wales College of Cardiff, PO Box 915, Cardiff CF1 3TL, UK*

PERROW M.R. *ECON, School of Biological Sciences, University of East Anglia, Norwich NR4 7TJ, UK*

PHILLIPS G.L. *National Rivers Authority (Anglian Region), Cobham Road, Ipswich IP3 9JE, UK*

PHILLIPS R.A. *Applied Ornithology Unit, Division of Evolutionary and Environmental Biology, Institute of Biomedical and Life Sciences, University of Glasgow, Glasgow G12 8QQ, UK*

PIERCE G.J. *Department of Zoology, University of Aberdeen, Tillydrone Avenue, Aberdeen AB9 2TN, UK*

PRIME J. *Birch Croft, Kinloch, Ledaig, Oban PA37 1QU, UK*

ROSS A.H. *National Institute of Water and Atmospheric Research, PO Box 8602, Christchurch, New Zealand*

ROTHE M.J. *NCCOSC-NRaD, Code 511, San Diego, CA 92152, USA*

SANTOS M.B. *Department of Zoology, University of Aberdeen, Tillydrone Avenue, Aberdeen AB9 2TN, UK*

SHIPPEE S.F. *NCCOSC-NRaD, Code 511, San Diego, CA 92152, USA*

SPENCER B.E. *Ministry of Agriculture, Fisheries and Food, Directorate of Fisheries Research, Fisheries Laboratory, Benarth Road, Conwy, Gwynedd LL32 8UB, UK*

STANSFIELD J. *National Rivers Authority (Anglian Region), Cobham Road, Ipswich IP3 9JE, UK*

TASKER M.L. *JNCC, Seabirds and Cetaceans Branch, Wynne-Edward House, 17 Rubislaw Terrace, Aberdeen AB1 1XE, UK*

THOMPSON P.M. *Department of Zoology, University of Aberdeen, Lighthouse Field Station, Cromarty IV11 8YJ, UK*

TOLLIT D.J. *Department of Zoology, University of Aberdeen, Lighthouse Field Station, Cromarty IV11 8YJ, UK*

VEALE L.O. *Port Erin Marine Laboratory, University of Liverpool, Port Erin, Isle of Man IM9 6JA, UK*

WILLIAMS T.M. *Department of Biology, University of California, Santa Cruz, CA 95064, USA*

WILLOWS R.I. *Plymouth Marine Laboratory, Prospect Place, West Hoe, Plymouth PL1 3DH, UK*

WILSON U.A.W. *Port Erin Marine Laboratory, University of Liverpool, Port Erin, Isle of Man IM9 6JA, UK*

WRIGHT P.J. *Scottish Office Agriculture, Environment and Fisheries Department, Marine Laboratory, PO Box 101, Victoria Road, Aberdeen AB9 8DB, UK*

Acknowledgements

A great many people were involved in planning, organizing and running the conference, and in the preparation of this proceedings volume. We are very grateful to them all.

The conference organizing committee included: Professor F.T. Last, Professor F.W. Robertson and Miss S. McDougall from The Royal Society of Edinburgh; Professor A.D. Hawkins and Dr S.P.R. Greenstreet from The Scottish Office Agriculture and Fisheries Department Marine Laboratory, Aberdeen; M.L. Tasker from the Joint Nature Conservation Committee support unit, Professor J.B.L. Matthews from the Scottish Association for Marine Science; and Dr R.N. Gibson from the Dunstaffnage Marine Laboratory.

The conference was sponsored by: The European Commission (Programme AIR); The Scottish Office Agriculture and Fisheries Department; the Scottish Association for Marine Science; the Atlantic Salmon Trust; Scottish Natural Heritage; the National Rivers Authority; the Salmon and Trout Association; and the World Wide Fund for Nature.

Conference delegates were welcomed to the City of Aberdeen by Councillor M. Savidge, on behalf of the Lord Provost, at a reception given by the City of Aberdeen District Council.

Sandra McDougall and Paula Couts at The Royal Society of Edinburgh were responsible for the conference administration. Conference facilities, accommodation and catering were provided by Ian Pirie and the conference support staff at the University of Aberdeen.

The conference was attended by 126 delegates; 27 oral and 21 poster papers were presented. Invited speakers were: Dr H. Kruuk, Mrs T. Similä, Dr J.D. Goss-Custard (paper presented by Dr R.I. Willows), Professor J.A. Estes, Professor W.S.C. Gurney, Dr S. des Clers and Dr R.W. Furness. Session chairmen were: Professor P. Racey, Dr J. Harwood, M.L. Tasker, Dr F.A. Huntingford, Dr S.J. Hall, Dr S.P.R. Greenstreet, Professor A.D. Hawkins. Delegates were welcomed by Dr T.L. Johnston, President of The Royal Society of Edinburgh, and the introductory talk was given by Professor A.D. Hawkins. Professor G.M. Dunnet summed up the conference for us and gave us his 'take home messages'. We thank everyone for coming all the way to Aberdeen and making the conference a worthwhile event, particularly those people who gave presentations. We are especially grateful to those participants who submitted manuscripts for consideration in this proceedings volume.

Each submitted manuscript was examined by two referees to whom we are very grateful for all the time and trouble they expended in helping us to put this collection of papers together. Our referees were: C.E Adams, J. Armstrong, N. Bailey, C. Bannister, D.J. Basford, N. Broekhuizen, A.D. Bryant, C.J. Camphuysen, H.Q.P. Crick, G.M. Dunnet, R.W. Furness, M.L. Gorman, J.D. Goss-Custard, W.S.C. Gurney, S.J. Hall, M.P. Harris, J. Harwood, A.S. Hill, J.R.G. Hislop, F.A. Huntingford, M.J. Kaiser, P.A. Kunzlik, B.J. McConnell, K. MacKenzie, S.P. Northridge, J.G. Ollason, I.J. Patterson, G.L. Phillips, G.J. Pierce, A. Pike, D. Raffaelli, J.W. Smith, C.J. Spray, P.M. Thompson, G. Turner, P.M. Walsh, S. Wanless and P.J. Wright.

Finally, thanks to Jan Roberts and Richard Miles at Blackwell Science for their help in bringing this project to fruition.

CHAPTER 1

Aquatic predators and their prey: an introduction

S.P.R. Greenstreet and M.L. Tasker

Predator–prey interactions are among the most influential in determining the structure and dynamics of multi-species communities. Predators have direct impact on their prey populations, and conversely, the availability of prey can directly limit predator populations. In addition, less obvious, indirect interactions also occur. The outcome of competitive interactions between prey populations can be affected by variation in predation pressure; when predation loadings are high, maintaining low population levels of a particular prey species, the populations of its competitors will be more able to expand. Consequently, altering the strength of just one predator–prey interaction can affect the structure of the whole community of species occupying a particular habitat because of knock-on effects up and down food chains throughout the food web.

Man enters this complex web of interactions as a particularly effective, and often highly selective, predator. His effect on communities, both through direct impacts on target species populations, and through cascading indirect effects up, down and sideways through food webs, can be enormous. Nowhere is this more obvious than in aquatic ecosystems where the direct effects of man's exploitation have included huge and sometimes catastrophic changes in commercially important fish stocks. In addition, the largely unpredictable outcome, at current levels of knowledge, of his indirect effects on other components of the system are also giving rise to increasing concern (e.g. Furness, this volume). That exploited aquatic ecosystems require careful management in order for sustainable harvesting to be maintained has become widely recognized.

Early fisheries' science concentrated almost exclusively on the biology of the commercially important target fish species. More recently the importance of indirect effects of fishing has become clear; in particular, how changes in the abundance and distribution of one species, brought about by fishing, affects the stocks of other species through knock-on impacts

on shared prey resources. This has been reflected in the development of multi-species approaches to fish stock assessments; for example, multi-species Virtual Population Analysis. However, even these modern fisheries' management tools remain focused on fish–fish interactions. Only cursory account has been taken of the impact of other top predators, such as seals, otters, cetaceans and birds, on some (the commercially important) fish species. Very little attention has been directed to the opposite interaction, the effect of fishing practices on other (non-fish) components of aquatic ecosystems.

Public interest in these components of aquatic ecosystems has increased over the last decade. In many parts of the world for example, marine mammals and seabirds are increasingly regarded as another resource to be exploited as major tourist attractions. Mass mortalities, or breeding failures, among aquatic top predators are regularly blamed on fishing activities. In many cases this may not be entirely fair (e.g. Wright, this volume), but it does highlight the sometimes conflicting nature of these two ways in which man can exploit aquatic ecosystems. As ways of utilizing aquatic ecosystems proliferate, the requirement for more 'global' aquatic habitat management plans will increase; plans which promote the best compromise between the conflicting demands of different users. The alternative to this may be permanently altered aquatic ecosystems, often changed to the detriment of one, if not all, users.

Developing such an approach to the management of aquatic ecosystems requires a greater and more detailed understanding as to how these systems work than current knowledge allows. Considerable effort has been expended in understanding terrestrial multi-species systems, and how predator and prey populations within such systems interact. But how applicable is this knowledge to aquatic ecosystems? Clearly the principal goal of both terrestrial and aquatic predators is the same: namely, to find sufficient food to survive and reproduce. However, the solutions found by predators to this common problem may differ between the two environments.

Foraging theory has been developed mainly from studies of warm-blooded terrestrial predators; however, a high proportion of the predators in aquatic ecosystems are cold-blooded, and this is especially so in marine environments. Since maintenance energy requirements are much lower in cold-blooded predators, the consequences of not finding food on a regular basis are very much less drastic than would be the case for their warm-blooded counterparts. Birds and mammals in terrestrial systems are therefore generally constrained to pursue strategies which maximize energy intake, but this is not necessarily the case for fish predators in aquatic systems. Herring and sprat in the North Sea hardly feed during the winter, when their zooplankton prey are least abundant, and instead utilize their energy reserves to survive and sustain their activity levels. Among warm-blooded terrestrial predators, similar strategies can only be followed by animals which have the capability to hibernate, or to enter a state of torpor, and so allow their body temperatures to drop to near ambient levels. Warm-blooded predators that remain active have to continue to feed efficiently and generally only have energy reserves sufficient to sustain their maintenance energy demands for only a few days of low, or no food intake.

Since predators generally find food most readily in areas where the density of their prey is high, this tends to lead, particularly in terrestrial systems, to a high degree of spatial coherence between the distributions of predators and their prey. In aquatic systems, however, there are many examples where this is not the case. Clearly predators that are not particularly interested in feeding have little cause to track the distributions of their prey, but other factors also come into play. Sight is a far more useful sense in the medium of air than it is in water. Thus a predator, such as a redshank *Tringa totanus* using visual cues to detect its prey on a mud-flat, is capable of assessing the merit of different potential feeding sites much more readily than, for example, a guillemot *Uria aalge* using sight to search for pelagic fish prey from the surface of the sea. Wading bird distributions are frequently closely correlated with the density of their invertebrate prey, while seabird distributions often show no correlation with the density of their pelagic fish prey, and where they do, these correlations are usually weak. Visually searching predators in terrestrial systems can more easily detect, and so track, variations in the density of their prey, whereas in aquatic systems visually searching predators more often have to adopt a strategy of occupying a site, where prior knowledge suggests that the chances of

encountering prey within a reasonable period of time are acceptable, and then simply waiting for their prey to turn up. This sort of strategy need not only apply to warm-blooded aquatic predators. A predatory fish, on encountering a shoal of its prey, may fill its stomach at one go and then take several days to digest the meal. Clearly continuing to sample the habitat, detecting prey, and moving to track mobile prey during this digestion process would be wasteful of energy.

Aquatic predators may also find their prey using different predation avoidance strategies to those encountered by terrestrial predators. Classic fish schooling behaviour may only have its terrestrial equivalents in bird flocking and herd formation among grazing animals; in the latter example, an intended prey animal can only escape in two dimensions rather than three.

The examples discussed above suggest that, at least under certain circumstances and in some situations, interactions between aquatic predators and their prey may differ markedly from equivalent interactions in terrestrial environments. This strengthens the argument that, if aquatic ecosystems are to be managed effectively, the predator–prey relationships that make up aquatic food webs need to be understood in their own right. The wealth of knowledge and theory developed from the many studies of terrestrial systems over several decades cannot simply be extrapolated to the aquatic habitat. If, as seems possible, the extent to which predatory forces control community structure (top-down control) differs between the two environments, predictions regarding the impact of man's activities on aquatic ecosystems, derived from models developed principally from the study of terrestrial systems, may be inaccurate and lead to the adoption of inappropriate management policies. Indeed, determining how and why models derived from terrestrial studies fail to simulate aquatic situations may take us a long way down the road to actually understanding how aquatic ecosystems function.

Terrestrially derived predator–prey models may fail to predict aquatic systems because of the difficulty of studying interactions in the aquatic environment at the relevant temporal and spatial scales. Large terrestrial study areas can be sampled with a frequency and at a spatial resolution normally not possible in aquatic habitats. This is especially so in marine habitats where sampling intensity might rarely exceed one 1 h trawl per $2500\,km^{-2}$ four times per year. Increasing our ability to understand interactions at a variety of scales may provide enormous insight into the mechanics of aquatic predator–prey interactions.

The Royal Society of Edinburgh conference on 'Aquatic Predators and Their Prey', held at the University of Aberdeen in August 1994, was intended to bring together biologists studying all aspects of aquatic predator–prey interactions. The resulting proceedings might therefore go some way towards serving as a statement of our current state of knowledge regarding the subject, not only describing what we do know, but also indicating areas where necessary information is not available and highlighting problems requiring urgent research. Papers were presented to the conference which addressed three main themes: the behavioural strategies of predators and prey; the role of predators in ecosystem structure; and predators, prey and man. Thus this conference proceedings volume starts by describing the behavioural tactics used by predators to capture prey and the evasion strategies employed by prey in avoiding capture, the factors which affect these behaviour patterns and their evolution. It goes on to examine how predators affect prey populations directly, and vice versa, before exploring how predation forces influence other species indirectly throughout the food web, thereby structuring entire communities. Finally the impact of man on aquatic systems, as an active harvester, through the alteration of food resource supply, and as a potential system manager is addressed.

Strategies for reducing foraging costs in dolphins

T.M. Williams, S.F. Shippee and M.J. Rothe

SUMMARY

(1) The major energetic costs faced by an aquatic endotherm include those associated with maintaining basal metabolism, thermoregulation, locomotion, and digestion.

(2) Wave-riding, intermittent modes of swimming, and limiting transit swimming to maximum range speeds reduce the cost of locomotion and increase the time available for foraging in the diving dolphin.

(3) Energetic costs associated with high speed ascent and descents, and thermoregulation (if not compensated for by locomotor thermogenesis) result in significant decreases in the time available for foraging by diving dolphins.

(4) Behaviours that dolphins use to reduce locomotor and thermoregulatory costs can benefit the foraging animal by conserving limited oxygen reserves during a dive. As a result, a greater proportion of the reserve will be available for predator–prey interactions.

Key-words: aquatic predation, bottlenose dolphin, diving, energetics, *Tursiops truncatus*

INTRODUCTION

Many conventional foraging models for aquatic predators maximize the net rate of energy intake over the period of foraging (Stephens & Krebs 1986, Kramer 1988). Important elements of these models include:

(1) energetic costs defined as the amount of energy expended to perform a particular behaviour,
(2) energetic benefits defined as the amount of energy gained by performing the behaviour, and
(3) the time required for these behaviours.

The latter element is especially critical for aquatic mammals in which the duration of a foraging dive is dictated by the parsimonious utilization of oxygen stores. For these animals the costs and benefits of foraging must be balanced against limited oxygen reserves in the muscles, blood and lungs (Dunstone & O'Connor 1979, Williams *et al.* 1993b).

There has been considerable effort in detailing the energetic benefits of foraging in aquatic mammals. For example, prey consumption and diets have been described for sea otters (Riedman & Estes 1990), pinnipeds (Condit & Le Boeuf 1984, Gentry *et al.* 1986), and dolphins (Barros & Odell 1990, Cockcroft & Ross 1990, Hanson & Defran 1993). In contrast, comparatively little is known about the various energetic costs associated with foraging by aquatic mammals. Often these costs are averaged in a field metabolic rate obtained from the doubly-labelled water method (Costa 1991).

The major energetic costs faced by foraging endothermic swimmers include the energy expended for basic biological functions, thermoregulation, locomotion (swimming and diving), and digestion (heat increment of feeding and the cost of heating ingested prey). These costs are incorporated in the travel, search, pursuit and handling costs typically used in ecological models of foraging (see Charnov 1976, Stephens & Krebs 1986). Our focus is on the physiological rather than behavioural capabilities of the animal. Furthermore, in our model individual energetic costs may not always be mutually exclusive, whereas in the ecological models they necessarily are. For example, locomotor activity may offset thermoregulatory demands (see below). The non-exclusive nature of these costs means that the combined effects of the various energetic costs must be considered when assessing the total cost of foraging by an animal.

During the past five years, we have conducted a series of physiological studies to determine energetic costs in the bottlenose dolphin *Tursiops truncatus*. This paper is our first attempt to apply these measurements to foraging theory. We examine the individual energetic costs associated with a simulated foraging dive and estimate their effect on oxygen reserves, and hence foraging time in these animals.

MATERIALS AND METHODS

Adult dolphins (average body weight = 145 kg) were trained to rest in a metabolic chamber (n = 3 males and 3 females, Williams *et al.* 1992b), swim next to moving boat (n = 1 male and 1 female, Williams *et al.* 1992a), and dive to submerged targets (n = 1 male and 2 females, Williams *et al.* 1993b). Ascent and descent rates, heart rate and behaviour during diving were monitored with microprocessors (Wildlife Computers, Inc.) placed on a pectoral fin strap or on a body harness. Measurements of oxygen consumption, heart rate and blood lactate for resting and active dolphins were used to determine the energetic cost of each activity. These costs were then used to compare the energetic consequences of basal metabolism, locomotor mode and thermoregulation during a dive. By defining the various energetic components of diving, we were able to identify behaviours that would enable the aquatic mammal to reduce costs associated with travelling between prey patches or between the water surface and depth.

RESULTS AND DISCUSSION

Maintenance and thermoregulatory costs

Maintenance costs and thermoregulatory costs were determined from the metabolic rate of bottlenose dolphins resting in water ranging from 3.6°C to 17.3°C (Williams *et al.* 1992b). The experiments were conducted on adult animals acclimated to ambient water temperatures of 15°C in San Diego (California) or 25°C in Kaneohe Bay (Hawaii). The results of the study showed that basal metabolism and lower critical temperature of the dolphins depended on acclimation temperature. Animals acclimated to 15°C had a basal metabolic rate that was 1.4 times the value measured for dolphins acclimated to 25°C. Lower critical temperature was less than 6°C for the cold acclimated dolphins and 11–16°C for the animals living in warm water. Thus, both acute and chronic changes in water temperature altered total energetic costs in the dolphin.

Several behaviours will enable a foraging dolphin to maintain low maintenance and thermoregulatory costs. These include limiting foraging excursions to areas where water temperatures are within the thermal neutral zone, and recycling heat generated by locomotor movements. Thermogenic muscular activity has been demonstrated for other aquatic and semi-aquatic mammals, and may be important for small cetaceans that live in cold seas. Swimming minks show a complex balance of locomotor heat production, body temperature and heat storage. Increased heat production during high speed swimming can offset thermoregulatory demands and stabilize the core body temperature of these small mammals (Williams 1986). Likewise, active sea otters (Costa & Kooyman 1984) and California sea lions (T. Williams, unpublished data) can compensate for thermoregulatory costs when moving or swimming at speeds as low as $1 \text{ m} \cdot \text{s}^{-1}$. Further studies are needed to determine the magnitude of thermogenic muscular activity in cetaceans and its effect on total energetic costs.

Locomotor costs

Swimming and diving represent major energetic expenditures in aquatic mammals. In wild bottlenose dolphins, locomotor activities comprise more than 82% of the animal's daytime activity budget (Hanson & Defran 1993). The energetic costs of these activities have been examined in ocean-trained bottlenose dolphins. Williams *et al.* (1992a) investigated the swimming energetics of two adult dolphins trained to match their swimming speed with that of a 22 ft Boston whaler. Acoustic signals were used to position the dolphins 5–15 m amidships from the boat out of the wake zone. The results from this study demonstrated that the transport costs for adult dolphins swimming at various speeds (Fig. 2.1) showed a U-shaped curve typical of other swimming mammals (Williams 1983, Davis *et al.* 1985). No significant difference

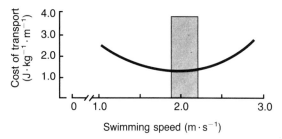

Fig. 2.1. Cost of transport in relation to transit speed for swimming bottlenose dolphins. The curve was calculated from values of oxygen consumption determined for two adult dolphins swimming next to a moving boat. The shaded area denotes the routine cruising speeds of coastal dolphins (Würsig & Würsig 1979) and corresponds to the predicted speeds for maximum range (redrawn from Williams *et al.* 1993a).

in heart rate, blood lactate or swimming energetics were observed for similarly sized male and female dolphins. The minimum cost of transport was $1.29 \pm 0.05 \, J \cdot kg^{-1} \cdot min^{-1}$ and occurred at a swimming speed of $2.1 \, m \cdot s^{-1}$. The theoretical maximum range speeds, those swimming speeds that allow the animal to achieve the greatest distance per unit power input, were 1.7 to $2.3 \, m \cdot s^{-1}$ for these dolphins.

Two behavioural strategies, wave-riding and utilization of maximum range speeds, enable dolphins to reduce the energetic cost of transit swimming. Transport costs for wave-riding dolphins moving at $3.8 \, m \cdot s^{-1}$ are similar to those of freely swimming dolphins moving at $2.1 \, m \cdot s^{-1}$. This behaviour enables the dolphin to achieve comparatively high transit speeds at a relatively low cost. In the absence of wave-riding, dolphins can maintain low transport costs by selecting swimming speeds between 1.7 and $2.3 \, m \cdot s^{-1}$, that is, the maximum range speeds (Fig. 2.1).

Diving dolphins may also use burst-and-glide swimming to reduce locomotor costs. We designed an experiment in which bottlenose dolphins were trained to dive in a straight line path to submerged targets. The targets were placed at 54 m depth to simulate a relatively easy dive, and at 206 m to test a depth at which oxygen stores may become limiting. Total oxygen reserves had been determined previously and were $33 \, mlO_2 \cdot kg^{-1}$ body weight for adult bottlenose dolphins (Williams *et al.* 1993b).

Rather than swim at the maximum range speeds, the dolphins descended slower and ascended faster than predicted for transit swimming. Total energetic cost for a 206 m dive with a mean descent rate of $1.5 \, m \cdot s^{-1}$ and mean ascent of $2.5 \, m \cdot s^{-1}$ was $42.2 \, mlO_2 \cdot kg^{-1}$ as calculated from the swimming costs in Figure 2.1.

These calculations assumed that the diving dolphin used constant propulsion. The result was a total energetic cost for diving that exceeded the animal's oxygen stores by nearly 30%. Subsequent video sequences of the dives showed that the dolphins switched from constant propulsion to an intermittent mode of swimming during the ascent from longer dives. More than 70% of the ascent period was spent gliding. If we account for the lower metabolic cost of the glide sequences, then the energetic cost of the 206 m dive is reduced by $6.2 \, mlO_2 \cdot kg^{-1}$ and remains within 10% of the animal's total oxygen reserve.

Digestive costs

The energetic costs associated with ingesting food have not been examined in bottlenose dolphins. However, the heat increment of feeding and the warming of ingested prey represent important energetic expenditures in a variety of aquatic predators. In the smallest species of marine mammal (sea otter, *Enhydra lutris*) and species of penguin (little blue, *Eudyptula minor*), the calorigenic effect of digestion plays an important role in the thermal balance of the resting animal. The metabolic rate of sea otters resting in water may increase by 54% following feeding (Costa & Kooyman 1984). Baudinette *et al.* (1986) found that the rates of oxygen consumption of the fed little blue penguin resting in air was 1.9 times the mean value of the post-absorptive bird. These elevated metabolic rates associated with digestion and termed the heat increment of feeding or specific dynamic action, coincided with a decrease in thermogenic muscular activity. In addition, the cost of heating ingested prey led to significant increases in the resting metabolism of marine birds (Wilson & Culik 1991) and mammals (Davis & Williams 1992). Food-induced thermogenesis may also enable small seabird species to forage in northern temperate and arctic oceans (Croll & McLaren 1993).

Strategies for reducing energetic costs during diving

Small cetaceans display a wide variety of foraging methods that vary in energetic cost. In relatively shallow water (<60 m) with sandy substrates, bottlenose dolphins will hold station quietly, head down, presumably echolocating on buried prey (D. Herzing, personal communication; T. Williams, unpublished data). Detected prey items are subsequently captured after digging through the sand with their rostrum (Fig. 2.2). Several fish may be taken on a single dive. Dolphins

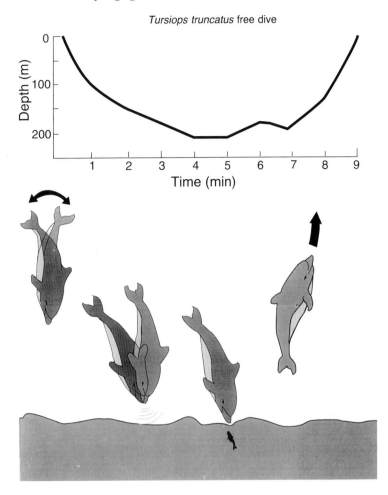

Fig. 2.2. Foraging behaviour in the bottlenose dolphin. The upper portion of the figure illustrates a typical dive profile (depth in relation to time) from a foraging dolphin carrying a time/depth microprocessor. The lower portion of the figure shows one type of foraging behaviour in which dolphins use echolocation to detect buried prey items.

will also herd fish into shallow areas (Shane 1990), or as seen with spinner *Stenella longirostris* and spotted dolphins *S. plagiodon*, allow fishing operations to herd the fish for them (Allen 1985, Pryor *et al.* 1990). For coastal populations of bottlenose dolphins, chases involving prey items are often of short duration and are associated with aggregations of fish or physical features (reefs, shallow water) that limit fish movements (Cockcroft & Ross 1990, Shane 1990). Feeding by coastal bottlenose dolphins in San Diego has been characterized by frequent steep and rapid dives (Hanson & Defran 1993).

We can begin to understand how individual energetic costs may affect foraging in bottlenose dolphins by examining the search dive. These dives consist of descent, bottom and ascent phases (Fig. 2.3). For comparative purposes, we will assume that the dolphin is relatively quiescent during the bottom phase, as

may occur with slow transits between prey patches or predictable, non-mobile prey. This phase will necessarily be shorter if the animal is active while at depth, but will not be addressed here. We also assume that the animal attempts to maximize bottom time which in turn allows it to increase the time spent in locating, pursuing, capturing and handling prey. Maximum dive time is dictated by a reserve of $33\,mlO_2 \cdot kg^{-1}$ body weight for dolphins that must be replenished when the animal surfaces to breathe.

The various energetic costs deplete the oxygen reserve in accordance with metabolic demands. This places limits on the amount of oxygen available for supporting the bottom phase. For instance, if a dolphin uses the predicted maximum range speed of $2\,m \cdot s^{-1}$ for ascent and descent on a 100 m dive, its metabolic rate is $8\,mlO_2 \cdot kg^{-1} \cdot min^{-1}$ (Williams *et al.* 1992a) during transit phases lasting 50 s each. Total cost for the

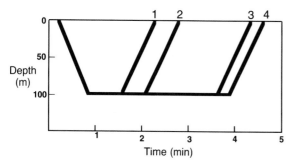

Dive strategy	Foraging time (min)
1 High speed ascent/descent	0.8
2 Winter metabolism	1.3
3 Optimum swim speed	2.6
4 Burst and glide ascent	2.8

Fig. 2.3. The effect of diving strategy on bottom (foraging) time in bottlenose dolphins. Upper portion of the figure shows a schematic profile for a 100 m search dive. In this model, foraging time is equivalent to bottom time (horizontal portion of the dive profile). Listed numbers for each strategy correspond to the timing of the ascent phase of the dive and the coincident bottom time.

ascent and descent phases is $13.3\,mlO_2\cdot kg^{-1}$; the remaining oxygen store will support a quiescent bottom time of 2.6 min. If the dolphin uses burst-and-glide swimming on the ascent, metabolic rate during the ascent is reduced by approximately 20% and bottom time concomitantly increases to 2.8 min.

High speed ascent and descents, and elevated thermoregulatory costs, if not compensated for by locomotor thermogenesis, will reduce bottom time in diving dolphins. Bottom time is less than 60 s if the animal swims at $3.0\,m\cdot s^{-1}$ when travelling between the water surface and depth. This reduction is due to the prohibitively high transport costs associated with high speed swimming (Fig. 2.1). Thus, to move quickly between resources (i.e. oxygen at the water surface and submerged prey) the animal must bear a marked reduction in foraging time. Although an energetically costly strategy when searching for prey, high speed movements may be advantageous once prey that is patchy in distribution and ephemeral in time has been identified.

Maintaining an elevated metabolic rate to compensate for cold water temperatures or warming ingested prey also limit bottom, and therefore foraging, time for the dolphin (Fig. 2.3). Such limitations may be avoided by thermogenic muscular activity, but require further research in cetaceans. It is likely that the relationships between locomotor heat production, thermoregulatory

demands, digestive costs and prey availability will be important factors in determining the range of foraging in wild populations of dolphins (Wells *et al.* 1990).

In view of the above discussion, behaviours that dolphins use to reduce locomotor and thermoregulatory costs can benefit the foraging animal by conserving limited oxygen reserves during a dive. As a result, a greater proportion of the reserve will be available for potential predator–prey interactions. An unknown component of our energetic model is the net rate of energy gained when using these different behavioural strategies. This becomes evident over a series of dives rather than the single event examined here, and awaits investigation.

ACKNOWLEDGEMENTS

The authors thank Naval Ocean Systems Center (Kaneohe, Hawaii), the Office of Naval Research, and the University of California (Santa Cruz) for supporting this study and the preparation of this manuscript. We also thank D. Croll and J. Estes for reviewing various drafts of the manuscript and for providing insightful discussions. Suggestions from two anonymous reviewers and the editors of this volume are gratefully acknowledged.

REFERENCES

Allen, R.L. (1985) Dolphins and the purse-seine fishery for yellowfin tuna. *Marine Mammals and Fisheries* (eds J.R. Beddington, R.J.H. Beverton & D.M. Lavigne), pp. 238–252. George Allen & Unwin, London.

Barros, N.B. & Odell, D.K. (1990) Food habits of bottlenose dolphins in the southeastern United States. *The Bottlenose Dolphin* (eds S. Leatherwood & R.R. Reeves), pp. 309–328. Academic Press, Inc., San Diego, CA.

Baudinette, R.V., Gill, P. & O'Driscall, M. (1986) Energetics of the Little Penguin, *Eudyptula minor*: temperature regulation, the calorigenic effect of food, and moulting. *Australian Journal of Zoology*, **34**, 35–45.

Charnov, E.L. (1976) Optimal foraging, the marginal value theorem. *Theoretical Population Biology*, **9**, 129–136.

Cockcroft, V.G. & Ross, G.J.B. (1990) Food and feeding of the Indian bottlenose dolphin off Southern Natal, South Africa. *The Bottlenose Dolphin* (eds S. Leatherwood & R.R. Reeves), pp. 295–308. Academic Press, Inc., San Diego, CA.

Condit, R. & Le Boeuf, B.J. (1984) Feeding habits and feeding grounds of the Northern elephant seal. *Journal of Mammalogy*, **65**(2), 281–290.

Costa, D.P. (1991) Reproductive and foraging energetics of high latitude penguins, albatrosses and pinnipeds: implications for life history patterns. *American Zoologist*, **31**, 111–130.

Costa, D.P. & Kooyman, G.L. (1984) Contribution of specific dynamic action to heat balance and thermoregulation in the sea otter, *Enhydra lutris. Physiological Zoology*, **57**, 199–203.

Croll, D.A. & McLaren, E. (1993) Diving metabolism and thermoregulation in common and thick-billed murres. *Journal of Comparative Physiology B*, **163**, 160–166.

Davis, R.W. & Williams, T.M. (1992) The effect of water temperature on the swimming energetics of sea lions. *The Physiologist*, **35**(4), 176.

Davis, R.W., Williams, T.M. & Kooyman, G.L. (1985) Swimming metabolism of yearling and adult harbor seals, *Phoca vitulina. Physiological Zoology*, **58**, 590–596.

Dunstone, N. & O'Connor, R.J. (1979) Optimal foraging in an amphibious mammal. I. The aqualung effect. *Animal Behaviour*, **27**, 1182–1194.

Gentry, R.L., Kooyman, G.L. & Goebel, M.E. (1986) Feeding and diving behaviour of Northern fur seals. *Fur Seals: Maternal Strategies on Land and at Sea* (eds R.L. Gentry & G.L. Kooyman), pp. 61–78. Princeton University Press, Princeton, NJ.

Hanson, M.T. & Defran, R.H. (1993) The behaviour and feeding ecology of the Pacific coast bottlenose dolphin, *Tursiops truncatus. Aquatic Mammals*, **19**(3), 127–142.

Kramer, D.L. (1988) The behavioural ecology of air breathing by aquatic animals. *Canadian Journal of Zoology*, **66**, 89–94.

Pryor, K., Lindbergh, J., Lindbergh, S. & Milano, R. (1990) A dolphin–human fishing cooperative in Brazil. *Marine Mammal Science*, **6**(1), 77–82.

Riedman, M.L. & Estes, J.A. (1990) The sea otter (*Enhydra lutris*): Behaviour, ecology, and natural history. *Fish and Wildlife Service Biological Report*, **90**(14), 1–126.

Shane, S.H. (1990) Behaviour and ecology of the bottlenose dolphin at Sanibel Island, Florida. *The Bottlenose Dolphin* (eds S. Leatherwood & R.R. Reeves), pp. 245–266. Academic Press, Inc., San Diego, CA.

Stephens, D.W. & Krebs, J.R. (1986) *Foraging Theory*. Princeton University Press, Princeton, NJ. 247 pp.

Wells, R.S., Hansen, L.J., Baldridge, A., Dohl, T.P., Kelly, D.L. & Defran, R.H. (1990) Northward extension of the range of bottlenose dolphins along the California coast. *The Bottlenose Dolphin* (eds S. Leatherwood & R.R. Reeves), pp. 421–434. Academic Press, Inc., San Diego, CA.

Williams, T.M. (1983) Locomotion in the North American mink, a semi-aquatic mammal. I. Swimming energetics and body drag. *Journal of Experimental Biology*, **103**, 155–168.

Williams, T.M. (1986) Thermoregulation of the North American mink during rest and activity in the aquatic environment. *Physiological Zoology*, **59**(3), 293–305.

Williams, T.M., Friedl, W.A., Fong, M.L., Yamada, R.M., Sedivy, P. & Haun, J.E. (1992a) Travel at low energetic cost by swimming and wave-riding bottlenose dolphins. *Nature*, **355**, 821–823.

Williams, T.M., Haun, J.E., Friedl, W.A., Hall, R.W. & Bivens, L.W. (1992b) Assessing the thermal limits of bottlenose dolphins: a cooperative study between trainers, scientists, and animals. *IMATA Soundings*, **4**, 16–17.

Williams, T.M., Friedl, W.A. & Haun, J.E. (1993a) The physiology of bottlenose dolphins (*Tursiops truncatus*): heart rate, metabolic rate, and plasma lactate concentration during exercise. *Journal of Experimental Biology*, **179**, 31–46.

Williams, T.M., Friedl, W.A., Haun, J.E. & Chun, N.K. (1993b) Balancing power and speed in bottlenose dolphins (*Tursiops truncatus*). *Symposium Zoological Society, London*, **66**, 383–394.

Wilson, R.P. & Culik, B.M. (1991) The cost of a hot meal: facultative specific dynamic action may ensure temperature homeostasis in post-ingestive endotherms. *Comparative Biochemistry and Physiology*, **100A**(1), 151–154.

Würsig, B. & Würsig, M. (1979) Behaviour and ecology of the bottlenose dolphin, *Tursiops truncatus*, in the south Atlantic. *Fisheries Bulletin of the U.S. Fish and Wildlife Service*, **77**, 399–412.

CHAPTER 3

Costs and benefits of fishing by a semi-aquatic carnivore, the otter *Lutra lutra*

H. Kruuk and D.N. Carss

SUMMARY

(1) A model is described for the general relationship between prey capture rates, predator energy expenditure and time budget, as an aid to evaluating habitat suitability. As an example, otter predation on eels in NE Scotland is discussed.

(2) Data on diet and evidence for prey as a limiting factor in populations of otters are reviewed. Otters eat small fish; salmonids and eels are prominent prey species. In some study areas there is a positive correlation between otter use and fish density. Otters can take a high proportion of fish production and there is evidence for seasonal food shortage and high otter mortality at times of low fish numbers.

(3) Experiments which measured the oxygen consumption of swimming captive otters are reviewed. Energy expenditure during foraging was high because of heat loss, determined largely by water temperature. Resting metabolic rate was also estimated.

(4) Prey intake rates were estimated in the field, from combined direct observations of otters and faecal analyses, over a range of ambient temperatures.

(5) A provisional model is constructed for the relationship between otter prey capture rates, water temperature, and foraging time needed to meet daily energy expenditure.

(6) As prey capture rates increased (e.g. with increasing prey availability) the predator could spend less time foraging. This was an inverse linear relationship. At a critical level of prey availability, a relatively small reduction in net energy intake would make hunting unprofitable in a given area. This was particularly obvious in cases where hunting was relatively energy-expensive (e.g. for otters foraging in cold water). Similar conditions are also likely to occur in several other species of predator.

Key-words: energetics, foraging, *Lutra lutra*, otter, predation

INTRODUCTION

The availability of potential prey is an essential factor in the assessment of habitat quality for all species of predator. Estimates of prey density and productivity provide important information regarding prey availability, but this is only part of the picture. The costs to predators in obtaining prey have also to be assessed.

Such assessment is particularly important in situations where there is evidence that predator populations are food limited. In these cases, the relationship between prey availability and feeding costs is likely to have a direct effect on survival. Net energy gained from prey caught can be assessed against foraging effort to provide a measure of success, but it is important that effort is measured appropriately. In certain situations, the number or rate of individual predation attempts has been defended as a measure of foraging effort (Ostfelt 1991). However, in most circumstances, this is unlikely to be the case because of differences in the amount of time and energy expended per predation attempt (Mech 1966, Kruuk 1972, Kruuk *et al.* 1990).

Here we develop a model in which both foraging effort and returns are estimated in energetic terms. Analysis of the energetic costs of foraging for semi-aquatic mammals is especially interesting, because these costs are likely to be high due to the energy demands of swimming, and the high thermoconductivity of water (Williams 1983, Schmidt-Nielsen 1983). Suggestions are made concerning how the model could be used for assessing prey availability and for evaluating risks to the predator entailed in potential fluctuations in that availability. As an example, the interactions between a semi-aquatic predator, the Eurasian otter *Lutra lutra* L., and one of its main fish prey species, the eel *Anguilla anguilla* L., are described. In our example we review evidence for the hypothesis that numbers of the Eurasian otter are food limited in at least part of its range. A general model for the relationship between foraging time, and both energetic expenditure and intake in a hunting predator, is developed and then applied to previous information and new data in the specific case of the otter. This model of otter foraging enables an estimate to be made of the potential consequences to the predator of changes in fish availability.

OTTERS AND THEIR PREY: A REVIEW

Diet

There are many studies of otter diet, usually involving spraint analysis, but sometimes direct observations. Data on prey species taken by otters are summarized by Mason & Macdonald (1986); since then, several further studies have been made (e.g. Kyne *et al.* 1989, Libois & Rosoux 1989, Carss *et al.* 1990, Kruuk & Moorhouse 1990, Watt 1991, Brzezinski *et al.* 1993, Kruuk *et al.* 1993). The diet is dominated by fish; species taken are often bottom-living, and it has been suggested that other species are taken especially when inactive and close to the bottom (e.g. Kruuk & Moorhouse 1990; Carss, in press). Most prey are relatively small, with mean weights of fresh water prey between 20 and 22 g (Wise *et al.* 1981), less than 20 g (Brzezinski *et al.* 1993), or median weights of 25 g (Libois & Rosoux 1989), 15 g (salmonids) and 29 g (eels) (Jenkins & Harper 1980). In the sea around Shetland median prey weight was 28 g (Kruuk & Moorhouse 1990). In some areas prey may be somewhat larger, e.g. in Ireland,

where otters fed on eels with estimated median weights of 72 g and salmonids of 49 g (Kyne *et al.* 1989).

Prey quality

The energetic value of otter food has been estimated for a number of different prey species. The mean value for eight marine species of fish in Shetland was $4.30 \, kJ^{-1} \, g$ wet weight (± 0.06 SE) (Nolet & Kruuk 1989); there was little variation between species. However, energetic content of eel is higher, at $6.08 \, kJ^{-1} \, g$ (Norman 1963), due to higher lipid values. Elliott (1976) recorded variation for brown trout *Salmo trutta* between 4.1 and $7.2 \, kJ^{-1} \, g$, again dependent on the lipid content of the fish.

Consumption

Consumption of fish by otters in captivity has been variously reported as 'about 1 kg per day' (approximately 15% of body weight) (Stephens 1957), and 12.2% and 12.8% of body weight per day (Wayre 1979). One 9.5 kg adult male maintained body mass in an open-air enclosure over 89 days on 1.13 kg fish per day (± 0.24 kg SD), or 11.9% of his body mass. Over the same period a neighbouring group of two males (8 and 10.5 kg) and one female (6.5 kg) similarly maintained body mass by eating 3.15 kg per day (± 0.29 kg), or 12.6% of their combined body mass (H. Kruuk & D.N. Carss, unpublished observations).

These estimates made in captivity are possibly lower than the values for animals in the wild. Consumption of large adult salmon *Salmo salar* by an otter in the wild was estimated for one 8 kg male, which ate 12.2% of his body mass daily. This was an underestimate of his total consumption as the animal also took some other small prey which could not be assessed as accurately (Carss *et al.* 1990). In Shetland a 5.4 kg lactating female was estimated to eat 1.5 kg of fish per day (± 0.2 kg) over a 12 day period, 28% of her body weight per day (Nolet & Kruuk 1994). Sea otters *Enhydra lutra*, which have higher energy requirements than *Lutra* spp., consume an amount of food equivalent to 23–33% of their body weight each day (Costa 1978).

Otter densities

To evaluate data on diet and food intake on a larger scale, for otter populations, numbers of otters and prey productivity should be known. However, otters use large ranges, up to 40 km (Green *et al.* 1984) or even

80 km (Durbin 1993) of river and stream, or 14 km of coast (Kruuk & Moorhouse 1991), with varying degrees of overlap between individual home-ranges. Along coasts there is the additional problem that different depths of water are fished with varying intensity, otters concentrating on the shallows (Nolet *et al.* 1993). It was estimated that there were some 0.8 otters per km of coast in Shetland (Watson 1978, Kruuk *et al.* 1989). It was shown that 98% of their dives were within 80 m of the shore (Kruuk & Moorhouse 1991), and if that distance is taken as the width of their range, this works out at about one otter per 10 ha of water. In rivers and streams in north-east Scotland a density of one otter per 3–83 km (mean about 15 km) was estimated, or one otter per 2–50 ha of water (Kruuk *et al.* 1993).

Within these very long otter home-ranges there are large differences in fish density, which cause wide confidence limits to fish population estimates. To overcome this problem for the purpose of estimating food intake by otters in relation to fish biomass, utilization of streams has been expressed in terms of otter time (Kruuk *et al.* 1993). First, the proportion of time radio-tracked otters spent in tributaries, or particular sections of stream or river was assessed. These animals were also labelled with a radio-isotope, so their presence could be related to time spent there by other otters in the same area, using the ratio of labelled to unlabelled spraints. Thus, the total otter presence could then be related to fish biomass as estimated by electrofishing, in relatively small sections of water. There was a significant positive correlation between otter usage of an area and fish biomass (Kruuk *et al.* 1993).

Are otters food limited?

It was estimated that in north-east Scottish streams for which information was available, otters took 53–67% of the annual productivity of salmonid fishes, their main food (Kruuk *et al.* 1993). This high rate of predation on their staple prey would be expected if otter numbers there were food limited. Similarly consistent with a food-limitation hypothesis was the observation that otters ate more of their 'less preferred' food (e.g. mammals and carrion) during spring (Kruuk *et al.* 1993, Durbin 1993), the time of lowest fish abundance (e.g. Egglishaw 1970). Furthermore, the otters' natural mortality (i.e. not caused by traffic) was considerably higher in spring, with 60% of natural mortality on Shetland occurring between March and June, and on the Scottish mainland 42% during the month of April alone (Kruuk & Conroy 1991, Kruuk *et al.* 1993). Recruitment to otter populations was also related to fish biomass (Kruuk *et al.* 1991). Thus, various observations

support the hypothesis that otter populations in Scottish streams and rivers are food limited.

METHODS

The foraging behaviour of otters was observed in Lochs Davan (42 ha) and Kinord (80 ha) at Dinnet in north-east Scotland (OS grid refs NJ442018 and NO441995), from January 1992 to May 1994. Both lochs are used by several otters (Jenkins & Harper 1980). Water temperatures were measured daily at 20 cm below the surface, and from April 1988 until December 1991 they fluctuated between 0.7 and 23.5°C, with an annual mean of 9.4°C (unpublished observations).

The animals were observed mostly during early morning, at all times of year, and data were collected on any otter seen. The animals were timed from entering the water until leaving it; prey capture was scored, and whenever possible prey species and an estimate of size were recorded. Detailed information on the size of ingested prey, particularly eels, was derived from faecal analysis. Spraints were collected monthly from the shores of both lochs from March until May 1994; they included spraints from the observed animals as well as others. The undigested remains of fish were identified (Webb 1976); the length of throracic vertebrae of eels was measured and from that the length of eels eaten by the otters was calculated, taking into account the size-related differential recovery of these bones (Carss & Elston, in press). From this the weight of eels taken by otters was then calculated by the relationship $W = 0.000859L^{3.18}$, with W in g, L in cm (Carss & Elston, in press).

GENERAL MODEL OF PREDATION EFFORT

To model the basic energetic requirements which a predator must meet by fishing, it was assumed that the animal is always either foraging or resting. This is clearly an over-simplification but it provides a useful minimum figure: energetic requirements for all other activities (e.g. social behaviour) will be additional.

Daily energy expenditure can then be expressed as:

$$E_s \cdot H_d + E_r \cdot (24 - H_d) \qquad (1)$$

in which E_s is energy expenditure when foraging in kilojoules per hour ($kJ \cdot h^{-1}$), E_r is energy expenditure when resting in $kJ \cdot h^{-1}$, and H_d is hours per day spent foraging. The energy acquired per day (food intake) equals:

$$H_d \cdot x \cdot k \qquad (2)$$

in which x is energy in kJ captured per hour (prey capture) and k is utilization efficiency, i.e. proportion of energy content of prey which is utilized. To break even, daily energy expenditure will equal acquisition, i.e. (1) = (2):

$$E_s \cdot H_d + E_r \cdot (24 - H_d) = H_d \cdot x \cdot k \qquad (3)$$

This equation may be resolved to estimate the foraging time per day required to break even:

$$H_d = 24 E_r / (k \cdot x + E_r - E_s) \qquad (4)$$

COSTS AND BENEFITS OF FEEDING IN OTTERS

The parameters of Equation (4) can be estimated for otters from observations of animals in captivity and in different habitats in the wild. It was demonstrated that the resting metabolic rate (E_r), in what were probably thermoneutral conditions, was $11.4 \, \text{kJ} \cdot \text{kg}^{-1} \cdot \text{h}^{-1}$ (Kruuk *et al.* 1994), slightly higher than that of most other mustelids, but below that of the sea otter (13.3–$14.4 \, \text{kJ} \cdot \text{kg}^{-1} \cdot \text{h}^{-1}$) (Morrison *et al.* 1974).

In captivity, the metabolic rate of swimming otters was dependent mostly on water temperature T_w. Differences in behaviour were unimportant whilst the animals were swimming quietly or floating still: only when otters were swimming fast, or showed other high-intensity behaviour, did this significantly affect their metabolism. However, such differences in their aquatic behaviour were themselves temperature-dependent. In general when swimming, the energy required by the otters (in $\text{kJ} \cdot \text{kg}^{-1} \cdot \text{h}^{-1}$, T_w in °C) could be calculated as

$$E = 43.6 - 1.44 T_w$$

which explained 55% of the variation in their oxygen uptake. In converting oxygen consumption to energy requirements, a calorific equivalent of $20.1 \, \text{J} \cdot \text{mlO}_2^{-1}$ was used (Bartholomew 1977, Williams 1986). If aquatic behaviour differences (proportion of high-intensity swimming and substrate scraping) were also taken into account, a further 13% of the variation could be explained (Kruuk *et al.* 1994). For our present purposes these smaller effects caused by differences in aquatic behaviour have been ignored.

Otters bring all fish prey to the water surface before it is consumed (Kruuk & Moorhouse 1990, Watt 1991), so prey capture rates were determined by direct observation. The identity and size of prey were estimated by spraint analysis taking into account the size related differential recovery of hard parts (Carss & Elston,

in press). Thus, knowing the energy contents of the various prey (see above), the rate of energy intake can be estimated. However, as yet there are no studies which have quantified the assimilation efficiency for *Lutra lutra*. The assimilation efficiency is defined as the proportion k of the energy content of prey which is utilized by the predator (Loudon and Racey 1987). For northern fur seals *Callorhinus ursinus* feeding on fish the digestive efficiency (Loudon and Racey 1987) was 91% (Miller 1978 (in Costa 1978)), for sea otters 82% (Costa 1978), for weasels *Mustela nivalis* 90% (Moors 1977). In addition, the loss of energy in urine is estimated as 10% for sea otters (Costa 1978), and 10.5% for weasels (Moors 1977). By averaging these figures and assuming that they will be comparable in *Lutra*, k was estimated to be 77%.

Using the above estimates for energy consumption and utilization efficiency in Equation (4), the relationship between food intake and daily foraging time can be modelled (Fig. 3.1), for otters hunting in waters of different temperatures. This model describes the amount of time otters need to spend hunting each day, in order to make up for energy lost during hunting and resting. The figure shows that daily foraging time must be increased sharply when prey capture rates fall below about $100 \, \text{kJ} \cdot \text{kg}^{-1} \cdot \text{h}^{-1}$.

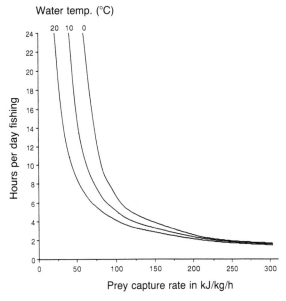

Fig. 3.1. Model of required time per day foraging for otters, to compensate for the energetic costs of foraging itself, plus the costs of the remaining time resting. Foraging time is modelled against different prey capture rates, and at different water temperatures.

The energy uptake during foraging, estimated in studies of otters both in the sea and in fresh water, can be compared with the model (Fig. 3.1). In Shetland during the summer, when there is a glut of fish (Kruuk *et al.* 1987), Nolet & Kruuk (1989) estimated that otters catch $1966 \, kJ \cdot h^{-1}$ fishing, or $317 \, kJ \cdot kg^{-1} \cdot h^{-1}$ for an average weight adult (6.2 kg in Shetland) (Kruuk 1995). Equation (4) and Figure 3.1 show that under these conditions and at $T_w = 10°C$ otters need to fish only about $1.2 \, h \cdot day^{-1}$ to meet basic requirements. It will be interesting to collect comparable observations for the late winter and early spring period, when both water temperature and fish densities are much lower (Kruuk *et al.* 1987). Watt (1991) found that adult otters on the Scottish west coast took, on average throughout the year, $286 \, g \cdot h^{-1}$ of bottom-living fish. He observed mostly female otters, and it was estimated that their intake would be the equivalent of $176 \, kJ \cdot kg^{-1} \cdot h^{-1}$ for an animal of 7.0 kg (the mean weight of mainland females, Chanin 1991). With reference to Figure 3.1, Watt's observations suggest that $2.6 \, h \cdot day^{-1}$ was necessary to meet basic energy requirements.

In the Dinnet freshwater lochs in north-east Scotland, otters were seen to spend a mean of 14.5 min (±10.7 min SD) swimming and diving per foraging trip (Fig. 3.2). Of 119 foraging trips, 83 (69.7%) were successful; thus, the mean number of prey caught per hour fishing was 2.88. Faecal analysis suggested that the mean size of eels taken was 79.9 g (Fig. 3.3); these would have an energy content of $6.08 \, kJ \cdot g^{-1}$ (Norman 1963). Thus, otters obtained $1399 \, kJ \cdot h^{-1}$ from their

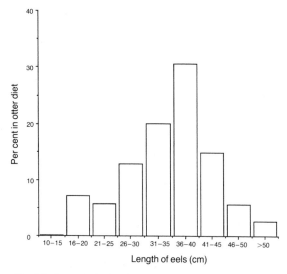

Fig. 3.3. Length of eels taken by otters at Dinnet, NE Scotland, March–May 1994, as calculated from vertebrae found in spraints, and corrected for size-related differential recovery (see text, and Carss & Elston, in press). Number of vertebrae measured = 434.

hunting in the lochs. With a mean body mass of otters captured at the lochs of 7.2 kg (n = 13) (H. Kruuk & D.N. Carss, unpublished observations), the animals caught $194 \, kJ \cdot kg^{-1} \cdot h^{-1}$ when fishing, a figure fairly close to that for otters on the Scottish west coast. They would need about $2.3 \, h \cdot day^{-1}$ fishing to meet their basic energy costs of fishing and resting metabolism.

DISCUSSION

The model suggests that in a declining prey population, at a relatively narrow range of densities, foraging by an aquatic predator such as the otter may suddenly become unprofitable. This prediction is based on a number of assumptions, particularly as extrapolations were made from the behaviour of captive animals. Measurements in captivity suggested however, that the costs of aquatic foraging were associated largely with thermoregulation (Kruuk *et al.* 1994), and that over a range of common aquatic behaviour patterns, differences in activity were far less important than variation in water temperature. This strengthens the validity of the extrapolation from captivity observations. However, it should be borne in mind that at high levels of aquatic activity, e.g. very fast swimming which is itself correlated negatively with water temperature, energy expenditure increased (Kruuk *et al.* 1994).

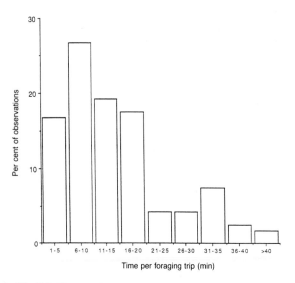

Fig. 3.2. Timing of foraging trips (time spent in water) of otters at Dinnet, NE Scotland, 1990–1994. N = 120.

The pattern of costs and benefits of foraging as described here relates to the minimum level of energy consumption, i.e. only the energy necessary to maintain daily resting metabolic rates plus the costs of foraging itself. The costs of all other activities and needs will be additional, and although these may also be high, they are likely to affect the level, but not the overall shape of, the curve in Figure 3.1. Such other energetic needs could be associated with social behaviour, and particularly with reproduction: for instance Oftedal & Gittleman (1989) estimate that lactating carnivores need to increase their energy intake by 150–300% to meet the requirements of increased milk production.

The calculations suggest that in the Dinnet lochs, otters could meet daily energy requirements of foraging and resting by 2.3 h of foraging for eels in winter, at $T_w = 0°C$. However, to put this in some perspective, if the energy gain of prey capture were to be reduced by half, otters would need $6.4 \, h \cdot day^{-1}$ fishing at $T_w = 0°C$, and along the Scottish west coast the figure would be $7.7 \, h \cdot day^{-1}$ just to break even on energy consumption of foraging plus resting. Otters would probably be able to do this, although it is likely that additional energy expenditure such as reproduction would be impossible to sustain at that level of prey availability. Time budget observations in our own studies and elsewhere suggest that otters are not usually active for more than $5–6 \, h \cdot day^{-1}$ (Durbin 1993, Green *et al.* 1984). Like most carnivores they appear to need long resting periods each day.

A long-term reduction in a fish population, to half of its mean biomass over large areas, is not likely to occur often. For instance, Elliott (1984) found that in only one year out of 17 does a population of brown trout fall to about 50% of its mean annual density. However, it would require only a relatively small reduction in total eel biomass, for instance by fisheries taking the larger sized eels, to bring about a considerable decline in profitability for foraging otters. At Dinnet, for instance, removing the largest 16% of the eels from the diet of otters (Fig. 3.3) would reduce food intake (by weight) by 46%. Our model suggests that this would be a large enough reduction to make foraging for eels unprofitable, and insufficient to sustain an otter population. These relationships are being studied at present.

The relatively large effect of a small reduction in prey availability could possibly be applicable to other predators. It is especially likely to occur in species where hunting is relatively costly, such as in other species of otter, or the African wild dog *Lycaon pictus* (Fanshawe & Fitzgibbon 1993), and the American badger *Taxidea taxus* (Lampe 1979). For such species, when habitat suitability is evaluated, prey availability estimates should take account of the energetic aspects of hunting.

ACKNOWLEDGEMENTS

We are grateful to G. Olsthoorn and K. Nelson for the faecal analyses, and to Drs P. Bacon, M. Gorman and M. Marquiss for comments on the manuscript.

REFERENCES

Bartholomew, G.A. (1977) Energy metabolism. *Animal Physiology: Principles and Adaptations* (ed. M.S. Gordon), pp. 57–110. Macmillan, New York.

Brzezinski, M., Jedrzejewski, W. & Jedrzejewska, B. (1993) Diet of otters (*Lutra lutra*) inhabiting small rivers in the Bialowieza National Park, eastern Poland. *Journal of Zoology, London*, **230**, 495–501.

Carss, D.N. (in press) Foraging behaviour and feeding ecology of the otter *Lutra lutra*: a selective review. *Hystrix*.

Carss, D.N. & Elston, D.A. (in press) Errors associated with otter *Lutra lutra* faecal analysis. II Estimating prey size distribution from bones recovered in spraints. *Journal of Zoology, London*.

Carss, D.N., Kruuk, H. & Conroy, J.W.H. (1990) Predation on adult Atlantic salmon, *Salmo salar*, by otters *Lutra lutra*, within the River Dee system, Aberdeenshire, Scotland. *Journal of Fish Biology*, **37**, 935–944.

Chanin, P. (1991) Otter *Lutra lutra*. The Handbook of British Mammals (eds G.B. Corbet and S. Harris), pp. 424–431. Blackwell, Oxford.

Costa, D.P. (1978) *The Ecological Energetics, Water and Electrolyte Balance of the California Sea Otter, Enhydra lutris*. PhD Thesis, University of California, Santa Cruz.

Durbin, L. (1993) *Food and Habitat Utilization of Otters* (Lutra lutra L.) *in a Riparian Habitat – the River Don in North-East Scotland*. PhD Thesis, University of Aberdeen.

Egglishaw, H.W. (1970) Production of salmon and trout in a stream in Scotland. *Journal of Fish Biology*, **2**, 117–136.

Elliott, J.M. (1976) Body composition of brown trout (*Salmo trutta* L.) in relation to temperature and ration size. *Journal of Animal Ecology*, **45**, 273–289.

Elliott, J.M. (1984) Numerical changes and population regulation in young migratory trout *Salmo trutta* in a Lake District stream. *Journal of Animal Ecology*, **53**, 327–350.

Fanshawe, J.H. & Fitzgibbon, C.D. (1993) Factors influencing the hunting success of an African wild dog pack. *Animal Behaviour*, **45**, 479–490.

Green, J., Green, R. & Jefferies, D.J. (1984) A radio-tracking survey of otters *Lutra lutra* on a Perthshire river system. *Lutra*, **27**, 85–145.

Jenkins, D. & Harper, R.J. (1980) Ecology of otters in northern Scotland: II. Analyses of otter (*Lutra lutra*) and mink (*Mustela vison*) faeces from Deeside, N.E. Scotland in 1977–1978. *Journal of Animal Ecology*, **49**, 737–754.

Kruuk, H. (1972) *The Spotted Hyena*. University of Chicago Press, Chicago.

Kruuk, H. (1995) *Wild Otters: Predation and Populations*. Oxford University Press, Oxford.

Kruuk, H. & Conroy, J.W.H. (1991) Mortality of otters *Lutra lutra* in Shetland. *Journal of Applied Ecology*, **28**, 83–94.

Kruuk, H. & Moorhouse, A. (1990) Seasonal and spatial differences in food selection by otters *Lutra lutra* in Shetland. *Journal of Zoololology, London*, **221**, 621–637.

Kruuk, H. & Moorhouse, A. (1991) The spatial organization of otters (*Lutra lutra* L.) in Shetland. *Journal of Zoology, London*, **224**, 41–57.

Kruuk, H., Conroy, J.W.H. & Moorhouse, A. (1987) Seasonal reproduction, mortality and food of otters (*Lutra lutra* L.) in Shetland. *Symposia of the Zoological Society of London*, **58**, 263–278.

Kruuk, H., Moorhouse, A., Conroy, J.W.H., Durbin, L. & Frears, S. (1989) An estimate of numbers and habitat preference of otters *Lutra lutra* in Shetland, UK. *Biological Conservation*, **49**, 241–254.

Kruuk, H., Wansink, D. & Moorhouse, A. (1990) Feeding patches and diving success of otters (*Lutra lutra* L.) in Shetland. *Oikos*, **57**, 68–72.

Kruuk, H., Conroy, J.W.H. & Moorhouse, A. (1991) Recruitment to a population of otters (*Lutra lutra*) in Shetland, in relation to fish abundance. *Journal of Applied Ecology*, **28**, 95–101.

Kruuk, H., Carss, D.N., Conroy, J.W.H. & Durbin, L. (1993) Otter (*Lutra lutra* L.) numbers and fish productivity in rivers in N.E. Scotland. *Symposia of the Zoological Society of London*, **65**, 171–191.

Kruuk, H., Balharry, E. & Taylor, P.S. (1994) Oxygen consumption of the Eurasian otter *Lutra lutra* in relation to water temperature. *Physiological Zoology*, **67**, 1174–1185.

Kyne, M.J., Smal, C.M. & Fairley, J.S. (1989) The food of otters *Lutra lutra* in the Irish Midlands and a comparison with that of mink *Mustela vison* in the same region. *Proceedings of the Royal Irish Academy (B)*, **89**, 33–46.

Lampe, R. (1979) *Aspects of the Predatory Strategy of the North American Badger*, Taxidea taxus. PhD Thesis, University of Minnesota, Minneapolis.

Libois, R.M. & Rosoux, R. (1989) Écologie de la loutre (*Lutra lutra*) dans le Marais Poitevin. I. Étude de la consommation d'anguilles. *Vie Milieu*, **39**, 191–197.

Loudon, A. & Racey, P. (eds) (1987) Reproductive energetics in mammals. *Symposia of the Zoological Society of London*, **57**, 1–371.

Mason, C.F. & Macdonald, S.M. (1986) *Otters: Ecology and Conservation*. Cambridge University Press, Cambridge.

Mech, L.D. (1966) The wolves of Isle Royale. *Fauna of the National Parks of the U.S. Series*, **7**.

Miller, K.L. (1978) *Energetics of Northern Fur Seal in Relation to Climate and Food Resources of the Bering Sea*. Final report to the US marine mammal commission, contract MM5AC025, National Technical Information Service, Norfolk, VA.

Moors, P.J. (1977) Studies of the metabolism, food consumption and assimilation efficiency of a small carnivore, the weasel (*Mustela nivalis*). *Oecologia*, **27**, 185–202.

Morrison, P.M., Rosemann, M. & Estes, J.A. (1974) Metabolism and regulation in the sea otter. *Physiolological Zoology*, **47**, 218–229.

Nolet, B.A. & Kruuk, H. (1989) Grooming and resting of otters *Lutra lutra* in a marine habitat. *Journal of Zoology, London*, **218**, 433–440.

Nolet, B.A. & Kruuk, H. (1994) Hunting yield and daily food intake of a lactating otter (*Lutra lutra*) in Shetland. *Journal of Zoology, London*, **233**, 326–330.

Nolet, B.A., Wansink, D.E.H. & Kruuk, H. (1993) Diving of otters (*Lutra lutra*) in a marine habitat: use of depths by a single-prey loader. *Journal of Animal Ecology*, **62**, 22–32.

Norman, J.R. (1963) *A History of Fishes*. Ernest Benn Ltd., London.

Oftedal, O.T. & Gittleman, J.L. (1989) Patterns of energy output during reproduction in carnivores. *Carnivore Behavior, Ecology and Evolution* (ed. J.L. Gittleman). pp. 355–378. Cornell University Press, Ithaca.

Ostfelt, R.S. (1991) Measuring diving success in otters. *Oikos*, **60**, 258–260.

Schmidt-Nielsen, K. (1983) *Animal Physiology: Adaptation and Environment*. Cambridge University Press, Cambridge.

Stephens, M.N. (1957) *The Otter Report*. Universities Federation for Animal Welfare, London.

Watson, H.C. (1978) *Coastal Otters in Shetland*. Vincent Wildlife Trust, London.

Watt, J. (1991) *Prey Selection by Coastal Otters* Lutra lutra *L*. PhD Thesis, University of Aberdeen.

Wayre, P. (1979) *The Private Life of the Otter*. Batsford, London.

Webb, J.B. (1976). *Otter Spraint Analysis*. Mammal Society, Reading.

Williams, T.M. (1983) Locomotion in the North American mink, a semi-aquatic mammal. I. Swimming energetics and body drag. *Journal of Experimental Biology*, **103**, 155–168.

Williams, T.M. (1986) Thermoregulation of the North American mink during rest and activity in the aquatic environment. *Physiological Zoology*, **59**, 293–305.

Wise, M.H., Linn, I.J. & Kennedy, C.R. (1981) A comparison of the feeding biology of mink *Mustela vison* and otter *Lutra lutra*. *Journal of Zoology, London*, **195**, 181–213.

CHAPTER 4

Behavioural responses of Arctic skuas *Stercorarius parasiticus* to changes in sandeel availability

R.A. Phillips, R.W. Furness and R.W.G. Caldow

SUMMARY

(1) Analysis of kleptoparasitic interactions of Arctic skuas *Stercorarius parasiticus* foraging within sight of Foula, Shetland, indicated that the skuas were able to switch hosts if particular species (notably Arctic terns *Sterna paradisaea*) were breeding unsuccessfully in a given year.

(2) Adult Arctic skuas spent considerably longer foraging off-territory in 1987, when sandeel (mainly *Ammodytes marinus*) recruitment in Shetland waters was low, than they did in 1979 or from 1992 to 1994.

(3) Both Arctic skua chick growth and fledging success were depressed during the years of low sandeel availability, particularly from 1987 to 1990. However, they were able to breed with moderate success up until at least 1986, in sharp contrast to Arctic terns which failed from 1983 to 1990.

(4) Breeding Arctic skua adults appeared to be in poorer body condition in 1988, the second year of particularly low sandeel recruitment, and there was also strong evidence that many established pairs deferred breeding in that year and in 1990.

(5) Changes in behaviour may therefore act as a buffer allowing Arctic skuas to withstand some degree of reduced prey availability but this appears to involve at least some longer-term costs.

Key-words: adult attendance, body condition, breeding performance, deferred breeding, host availability, *Stercorarius parasiticus*

INTRODUCTION

Seabirds exhibit considerable plasticity both in their behaviour and life-history characteristics (Furness & Monaghan 1987). Generalist seabird predators may switch from previously abundant but subsequently declining target species to others which are more readily obtainable (Montevecchi 1993). Similarly, adults may increase the foraging component of their reproductive effort in order to offset any reduction in their food supply (Cairns 1987, 1992, Montevecchi 1993). This enables them to buffer themselves to an extent from the effects of environmental fluctuations, such as a reduction in prey availability (Cairns 1987, Montevecchi 1993).

Changes in behaviour serve to compensate for short-term variation in fish stocks, and may mask any effects on chick growth rates or fledgling success unless food shortage becomes severe. There is a trade-off between adults' investment in energetically expensive activities such as chick-provisioning and the amount of effort they can expend in maintaining their own body condition (Martin 1987). There is likely to be a limit beyond which any further alteration in behaviour, assuming it represented greater reproductive investment, would have serious repercussions. An adult should abandon a breeding attempt if it is likely to decrease the probability of its survival or jeopardize the resources it can devote to future reproduction (Drent & Daan 1980).

Arctic skuas *Stercorarius parasiticus* L. breeding in Shetland feed almost entirely on sandeels, mainly *Ammodytes marinus* Raitt, stolen from other seabirds, in particular Arctic terns *Sterna paradisaea* Pontoppidan, Atlantic puffins *Fratercula arctica* L., kittiwakes *Rissa tridactyla* L. and common guillemots *Uria aalge* Pontoppidan (Furness 1987). During the mid- to late 1980s and 1990, fisheries' data indicate that sandeel recruitment declined in Shetland waters, with a concomitant decline in the breeding success of a number of surface-feeding sandeel predators, including several of the species on which Arctic skuas depend (Bailey *et al.* 1991). This paper details changes in the reproductive success and behaviour of Arctic skuas between 1975 and 1994, emphasizing the effects of variation in adult attendance and in the proportion of chases directed towards particular kleptoparasitic hosts on the skuas' ability to buffer themselves from the consequences of poor breeding success of their principal victims. The possible longer-term consequences of these changes in behaviour are also discussed with reference to the body weights of adult Arctic skuas and the likelihood of breeding deferral in the subsequent years of low food availability.

METHODS

The study was carried out on Foula, an island to the west of Shetland mainland at 60°08′N 002°05′W. The number of breeding Arctic terns and the total number of fledged Arctic tern chicks were counted in most years from 1974 to 1994. This provides an index, the number of fledglings per pair, to compare annual variation in tern breeding success over the years, plus each of these variables give some indication of the availability of this particular host to Arctic skuas.

Variation in Arctic skua breeding performance was examined in terms of the growth of chicks and the number of chicks fledged per pair. Logistic growth curves were fitted to changes in wing length and weight with age using all measurements from known-age chicks in 1992. Chicks in all years were then aged from their wing length (Hamer *et al.* 1991) using the 1992 curve of wing length versus age. The deviation (or mean deviation if more than one measurement was available) for each chick from its predicted weight (using the 1992 curve of weight versus age) at that age was then calculated, and expressed as a proportion of the expected value. The mean chick growth index was calculated for each year from 1976 to 1994, with the exception of 1990 when no chicks were measured, and used to compare annual

variation in chick body condition. Counts were made of the number of chicks seen fledged at all territories on the island in 1986 and from 1988 to 1994, and from a sample of 28 nests in the north of Foula in 1987.

Arctic skuas obtain most of their food by chasing sandeel-carrying hosts within 1 km of the east coast of Foula (Furness 1978). In most years between 1975 and 1980, and 1986 and 1993, watches were made from a coastal observation site at intervals through June and July (Furness 1978, 1980, Caldow 1988). During each observation period the number of chases of the various host species was recorded. The proportion of chases in each year directed towards Arctic terns was compared (after arcsine transformation) with two indices of that host's availability, the number of breeding adults and the total number of chicks fledged.

Arctic skuas regurgitate freshly-obtained fish to their mate and chicks after return from almost all absences from the territory and time spent away can therefore be used as an index of foraging effort in each year (Furness 1987, Hamer *et al.* 1991). The number of adults present was recorded at each visit to marked territories in 1979, 1987, and 1992 to 1994. This was converted to a mean for each individual territory at four different stages of the season; incubation and 0–9 days, 10–19 days and 20–30 days post-hatching. The mean attendance was then determined for marked territories at each stage in each year. In 1979 territories were not visited beyond twenty days after the chicks had hatched and therefore comparison between then and the other years was limited to the first three of these stages.

Adults were trapped on the nest while incubating in 1987, 1988, 1992, 1993 and 1994. The mean weights of these birds were compared to investigate whether adult body condition during the early part of the season was lower in the former two years when food availability was poor.

The Arctic skua colony was censused in most years by counting apparently occupied territories (Furness 1982). Counts for the years 1974–1982 are from Furness (1983). The remainder are either by Mrs Sheila Gear (unpublished data) or by the authors, with the exception of a 1986 count in Ewins *et al.* (1988). In 1974, 1975, 1985, 1986 and 1989–1991, the population size is the mean of two independent counts. These changes in numbers in successive years were also examined in terms of annual net recruitment to the breeding population, here defined (following the approach of Ollason and Dunnet (1983)) as the difference between the actual population size in a given year and the expected number of surviving birds from the previous year corrected for an estimate of the annual mortality rate. Colour-ringing of Arctic skuas from 1992 to 1994 indicated a mean annual return rate of

86.4%, with no bird absent in 1993 returning in 1994. There are no mortality rate estimates available for Arctic skuas on Foula prior to the 1990s and so a mortality rate of 13.6% (derived from the colour-ringing data) was used in the model. A negative value for the net recruitment index in any year can therefore be interpreted as a consequence of unusually high mortality or that established birds are absent from the colony or potential first-time breeders are refraining from recruiting in a given year.

RESULTS

Breeding data

Arctic tern
Between 1984 and 1990 no Arctic tern chicks fledged from Foula. In addition, less than 0.2 chicks fledged per pair of breeding adults in two earlier years, 1977 and 1978 (Fig. 4.1).

Arctic skua
There was a highly significant difference among years in Arctic skua chick growth (Kruskal–Wallis ANOVA $\chi^2 = 181$, n = 1913, P < 0.0001: Table 4.1). Multiple nonparametric ranges tests (Zar 1984) indicated that chick growth was high in 1983 and to an extent also in 1977, was poor in 1979, 1984 and 1987–1989 (1984 and 1988–1989 in particular) and varied little in most other years. Chicks were not measured in 1990.

Between 1987 and 1990 fewer than 0.25 chicks fledged per pair of breeding adults on Foula. The number of fledglings per pair in 1990 was 0.09. In comparison, from 1991 to 1994 fledgling production was consistent at approximately 0.9 chicks per pair. In 1986, despite this being a poor year for Arctic terns, Arctic skuas fledged 0.63 chicks per pair. These differences correlated with the chick growth index in each year (Fig. 4.2).

Changes in kleptoparasitic hosts

There was considerable variation between years in the proportion of chases directed towards the various species of host (Table 4.2). Arctic terns were important hosts in some years in the late 1970s (1975–1976 and 1979) and in the early 1990s, but were virtually absent as victims from 1986 to 1990. They were chased rela-

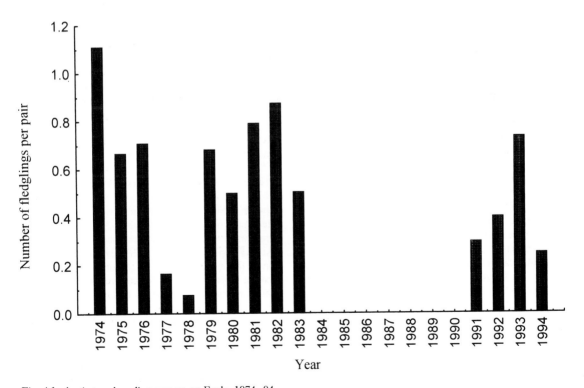

Fig. 4.1. Arctic tern breeding success on Foula, 1974–94.

Table 4.1. Arctic skua chick growth on Foula in 1976–1994. The growth index is the mean difference between the observed and the predicted weight of chicks (derived from the 1992 growth curve) expressed as a proportion of the predicted value. Data are means ± 1SE. Nonparametric ranges tests indicate that chick growth in: (i) 1983 > 1979–1980, 1982, 1984, 1987–1989 and 1993; (ii) 1977 > 1979, 1984 and 1987–1989; (iii) 1976, 1978, 1980–1981, 1985–1986 and 1991–1994 > 1984 and 1988–1989; and (iv) 1979, 1982 and 1987 > 1989.

Year	n	Growth index
1976	134	0.01 ± 0.01
1977	82	0.03 ± 0.01
1978	142	−0.00 ± 0.01
1979	212	−0.03 ± 0.01
1980	120	−0.01 ± 0.01
1981	99	−0.01 ± 0.01
1982	102	−0.02 ± 0.01
1983	73	0.05 ± 0.01
1984	40	−0.09 ± 0.02
1985	48	0.01 ± 0.02
1986	57	−0.00 ± 0.01
1987	58	−0.03 ± 0.01
1988	91	−0.07 ± 0.01
1989	55	−0.14 ± 0.02
1990	0	No data
1991	40	0.03 ± 0.02
1992	202	0.00 ± 0.00
1993	192	−0.01 ± 0.01
1994	166	−0.00 ± 0.01

tively rarely in 1978 and 1980, the former being a very poor year in terms of tern breeding performance (Fig. 4.1). There were highly significant correlations between the proportion of chases directed towards Arctic terns and the indices of their likely availability to Arctic skuas; the total tern breeding population on Foula and the number of tern chicks fledged (Fig. 4.3). In most years when Arctic terns were chased relatively infrequently, puffins tended to be the predominant hosts. Guillemots featured principally in two years, 1989, and 1990. The remainder of the chases were of kittiwakes and razorbills *Alca torda* L., both of which appeared to be relatively unimportant as host species.

Attendance

Table 4.3 shows the data on adult territorial attendance in 1979, 1987 and 1992 to 1994. There were significant differences between these years in attendance at all stages in the season; incubation (Kruskal–Wallis ANOVA $\chi^2 = 17$, n = 588, P < 0.002), 0–9 days post-hatching (Kruskal–Wallis ANOVA $\chi^2 = 38$, n = 453, P < 0.0001), 10–19 days post-hatching (Kruskal–Wallis ANOVA $\chi^2 = 36$, n = 387, P < 0.0001), and 20–30 days post-hatching (Kruskal–Wallis ANOVA $\chi^2 = 44$, n = 285, P < 0.0001). There were no data for this last stage from 1979. Attendance was low in 1987, and to an extent also in 1979 in comparison with the early 1990s. 1979 was the year with the lowest recorded sandeel recruitment between the mid 1970s and 1983 (Bailey *et al.* 1991).

Adult body weight

The weights of the adult Arctic skuas nest-trapped during incubation are shown in Table 4.4. One-way ANOVA indicated that there were no differences among years in adult incubation weights (ANOVA $F_{4148} = 1.6$ ns). However, all six birds trapped in 1988 weighed less than the mean value in all other years (binomial probability distribution P = 0.016).

Net recruitment

The net index shows considerable annual variation between 1975 and 1994 (Table 4.5). Net recruitment was high in several seasons in the latter half of the 1970s, and then constant, but at a much reduced level, through the 1980s until 1987, concurrent with the overall population decline. Net recruitment was negative in 1988 and 1990, but in each of the following years, 1989 and 1991, it was much higher than it had been throughout the 1980s. In 1992, net recruitment was also high in comparison with the very low level in the two subsequent years, 1993 and 1994.

DISCUSSION

Between 1984 and 1990 there was a complete breeding failure on Foula and elsewhere in Shetland of Arctic terns, one of the Arctic skua's major kleptoparasitic hosts (Monaghan *et al.* 1989). Virtual population analysis using fisheries' data indicated a dramatic reduction in sandeel recruitment in Shetland waters during the 1980s, and particularly from 1984 onwards (Bailey *et al.* 1991). Despite this, Arctic skua chick growth was apparently normal in 1985 and 1986, and fledging success was only seriously affected from 1987 to 1990. Given such considerable variation in prey availability, Arctic skua adults must have responded by modifying one or more components of their reproductive or foraging strategies. The considerable temporal gap between

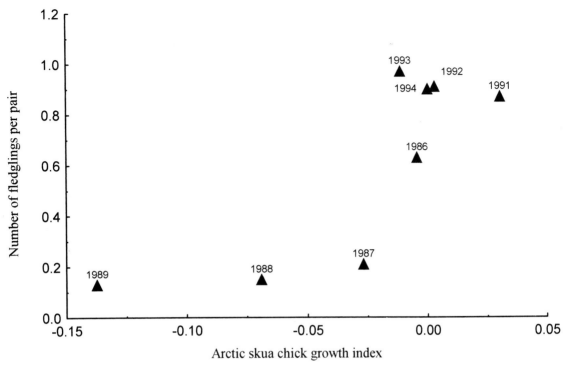

Fig. 4.2. Arctic skua chick growth in relation to fledgling production on Foula (r = 0.80, n = 8, P < 0.02).

Table 4.2. Variation in the species kleptoparasitised by Arctic skuas on Foula: data for 1975–1976 from Furness (1978), for 1978 and 1979 from Furness (1980) and for 1986 and 1987 from Caldow (1988).

Year	Percentage of all chases directed towards:					Total chases observed
	Arctic tern	Puffin	Guillemot	Kittiwake	Razorbill	
1975–6	74	12	0	11	3	117
1978	9	76	14	2	0	198
1979	51	39	5	4	1	741
1980	7	80	13	0	0	314
1986	0	76	14	1	9	1326
1987	1	73	15	0	11	469
1988	0	84	15	1	0	664
1989	1	20	68	5	6	50
1990	0	67	33	0	0	18
1991	35	57	2	5	1	127
1992	26	46	8	20	0	204
1993	42	34	5	18	1	116

the onset of breeding failure for Arctic terns and for Arctic skuas, indicates that, unlike terns, skuas were able to buffer themselves from the effects of declining food availability in these earlier years.

The breeding success of two principal kleptoparasitic hosts of Arctic skuas, Arctic terns and puffins, and less importantly, kittiwakes, was already depressed Shetland-wide because of reduced sandeel availability by 1987

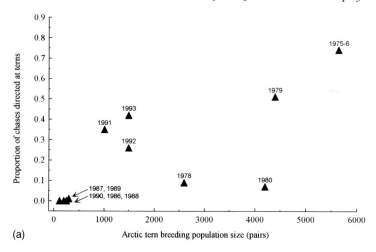

(a)

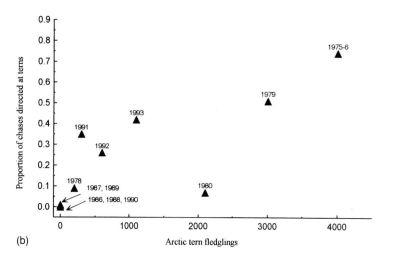

(b)

Fig. 4.3. Variation in the proportion of chases directed towards Arctic terns versus: (a) the size of the tern breeding population in each year (r = 0.73, n = 12, P < 0.01), and (b) the number of tern chicks fledged in each year (r = 0.83, n = 12, P < 0.002).

Table 4.3. Territorial attendance of Arctic skuas on Foula in 1979–1994: data are the mean number of adults present on territory ± 1SE(n).

Year	Stage of season			
	Incubation	0–9 days post-hatching	10–19 days post-hatching	20–30 days post-hatching
1979	1.78 ± 0.02 (175)	1.84 ± 0.03 (115)	1.65 ± 0.05 (83)	No data
1987	1.76 ± 0.02 (28)	1.71 ± 0.03 (25)	1.54 ± 0.06 (24)	1.44 ± 0.06 (23)
1992	1.85 ± 0.02 (122)	1.90 ± 0.02 (104)	1.82 ± 0.03 (87)	1.66 ± 0.04 (85)
1993	1.88 ± 0.01 (136)	1.88 ± 0.02 (111)	1.76 ± 0.03 (102)	1.60 ± 0.03 (100)
1994	1.87 ± 0.01 (127)	1.85 ± 0.02 (98)	1.90 ± 0.02 (91)	1.87 ± 0.03 (77)

Table 4.4. Weights (means ± SE) of incubating Arctic skua adults on Foula in different years.

Year	Number	Weight (g)	Range (g)
1987	9	433.3 ± 13.3	345–475
1988	6	404.3 ± 10.6	360–430
1992	70	442.3 ± 5.0	355–525
1993	49	439.2 ± 6.0	365–530
1994	19	451.1 ± 9.1	375–525

(Martin 1989, Monaghan *et al.* 1989, Harris and Wanless 1990). These former species, as well as guillemots, and to a much lesser extent razorbills, were all chased by Arctic skuas in the mid to late 1970s. During the late 1970s in the years when Arctic terns were breeding successfully, terns were probably chased more frequently than puffins because chases towards them were over twice as likely to be successful (Furness 1978). During the mid-1980s, the majority of chases were directed towards puffins and, particularly in 1989 and also 1990, guillemots. Only from 1991 onwards did Arctic terns and kittiwakes reappear as target species.

Puffin breeding success was consistently good on Foula up until 1985, following which there were five very poor years when few adults were observed carrying sandeels to chicks late in the season (R.W. Furness, unpublished data). Guillemot breeding success on Foula was good in 1988, as in 1981–1983 (Bailey *et al.* 1991) and their breeding success seems to have been generally unaffected by the reduced sandeel recruitment of the late 1980s. The preference of Arctic skuas for chasing puffins rather than guillemots, despite puffins' relatively poor breeding success in the mid 1980s, was again a likely consequence of differential chase success rate. Chases of puffins by dark-phase skuas (the majority of the population) were consistently almost twice as likely to be successful as those of guillemots (Furness 1978, Caldow and Furness 1991). This was despite an overall decline in success rate towards puffins observed since the 1970s (Caldow and Furness 1991).

Arctic terns feed their chicks predominantly on recruiting (0-group) sandeels, and by comparison guillemots specialize on longer and therefore older (mainly 1-group) fish (Furness 1990). While there is a degree of overlap, kittiwakes and puffins tend to feed on sandeels in the size ranges intermediate to those pre-

Table 4.5. Annual variation in the net recruitment of Arctic skuas on Foula between 1975 and 1994 assuming an annual mortality rate for established breeders of 13.6%.

Year	(1) Total number of pairs	(2) Expected population size	Net recruitment index in no. of pairs i.e. (1)–(2)	Net recruitment index as a proportion of actual pop. size
1974	190	–	–	–
1975	247	164	83	0.34
1976	280	213	67	0.24
1977	264	242	22	0.08
1978	252	228	24	0.10
1979	273	218	55	0.20
1980	262	236	26	0.10
1981	245	226	19	0.08
1982	224	212	12	0.05
1983[a]	209	194	15	0.07
1984[a]	194	181	13	0.07
1985	181	168	13	0.07
1986	175	156	19	0.11
1987	169	151	18	0.11
1988	130	146	−16	−0.12
1989	139	112	27	0.19
1990	110	120	−10	−0.09
1991	126	95	31	0.25
1992	159	109	50	0.31
1993	145	137	8	0.06
1994	134	125	9	0.07

[a] Interpolated population size estimates.

ferred by the other two species (Furness 1990). As the availability of 0-group sandeels declined during the 1980s, Arctic skuas therefore modified their foraging strategy to chase the host species specializing on progressively older sandeels. This variation in host selection by Arctic skuas is analogous to the dietary switches apparent in other seabird species when relative availabilities of different prey change (Hislop and Harris 1985, Montevecchi *et al.* 1988, Martin 1989, Hamer *et al.* 1991) and appears to have gone some way towards ameliorating the effects of low sandeel availability for Arctic skuas up until 1987.

In addition to their changes in host selection behaviour, Arctic skuas also have considerable scope to vary the amount of energy they invest feeding themselves and their chicks. In years of below average food abundance (1987 and possibly also 1979) adult Arctic skuas reduced the amount of time they spent on territory. Such increases in foraging effort in poor food years are a common feature in the behavioural repertoire of seabirds experiencing fluctuations in food availability (Gaston and Nettleship 1982, Monaghan *et al.* 1989, Hamer *et al.* 1991, 1993).

However, according to the predictions of life-history theory, any increase in breeding effort ought not to elicit a cost in terms of future reproductive potential. Altering foraging behaviour should have buffered Arctic skuas from the reduction in sandeel availability, but only until a limit beyond which no further increase in effort was feasible without sustaining such future costs. There is however limited evidence (given the small sample size) that adults were lighter in 1988, the first year subsequent to the decline in adult attendance and also chick growth and fledgling success. This may indicate either that sandeels were extremely scarce early in the season in 1988 and adults could not obtain sufficient quantities even to maintain their own body condition, or possibly that birds were suffering repercussions from the energetic costs of more and/or longer foraging trips in the previous year. Similarly, Arctic terns and kittiwakes, but not great skuas, also showed a decline in body condition in years with or following poor food availability (Monaghan *et al.* 1989, Hamer *et al.* 1991, 1993).

The increase in foraging effort in 1987 may also have had considerable bearing on whether adults returned to breed the following year. Net recruitment of breeders to the population was negative in both 1988 and 1990. A net recruitment index of zero in 1988 would have required adult mortality over the 1987–1988 non-breeding season to be 23%, twice that recorded for the Foula population between 1993 and 1994. Assuming adult mortality did not increase to that extent during 1988–1989 and 1990–1991, then established breeders

(and possibly also prospective first-time recruits) refrained from breeding in those years. This hypothesis is backed up by the anomalously high net recruitment indices in each year that followed, strongly suggesting that considerable numbers of these deferring adults returned to the colony in 1989 and 1991. There is doubtless a strong selective pressure for adults that have not been able to recover over a single winter from the effects of the previous poor breeding season not to breed the following year. This corresponds to the low body weights recorded for adults that did attempt to breed in 1988.

High net recruitment in 1992, two years after the previously poor breeding season may indicate that potential first-time breeders refrained from recruiting in 1991 until it was clear from the performances of established pairs that conditions had improved. Similarly, experienced adult kittiwakes breeding in the highly variable environmental conditions in Alaska tend not to attend the colony in years when reproductive performance is poor overall (Murphy *et al.* 1991). Poor net recruitment recorded in 1993 and 1994 is a consequence of the low fledging success in the late 1980s and it would be reasonable to predict an increase in the number of new breeders in 1995 as adults breed on average aged 4.4 years (O'Donald 1983) and some chicks fledged in 1991.

Arctic skuas appear to have considerable scope in their behavioural repertoire to buffer marked fluctuations in prey (or host) availability. However, increasing reproductive investment in a given year may result in birds being in poor body condition or deferring breeding in the following year. There is also limited evidence that potential new recruits may choose not to establish territories until conditions improve.

ACKNOWLEDGEMENTS

Thanks go to the Holbourn family for permission to work on Foula, and to Sheila Gear for allowing us to use her population size estimates. Thanks are also due to the many fieldworkers including, in the more recent years, Morag Campbell, George Porkert and Fiona Stewart who helped find and measure chicks and record chases. Dr Simon Greenstreet, Mark Tasker and two anonymous referees made useful comments on an earlier draft of this manuscript. Part of this research was funded by SOTEAG and NERC. R.A.P. was supported by a NERC Studentship from 1992–1994.

REFERENCES

Bailey, R.S., Furness, R.W., Gauld, J.A. & Kunzlik, P.A. (1991) Recent changes in the population of the sandeel (*Ammodytes marinus* Raitt) at Shetland in relation to estimates of seabird predation. *ICES Marine Science Symposium*, **193**, 209–216.

Cairns, D.K. (1987) Seabirds as indicators of marine food supplies. *Biological Oceanography*, **5**, 261–271.

Cairns, D.K. (1992) Bridging the gap between ornithology and fisheries biology: use of seabird data in stock assessment models. *Condor*, **94**, 811–824.

Caldow, R.W.G. (1988) *Studies on the Morphology, Feeding Behaviour and Breeding Biology of Skuas with Reference to Kleptoparasitism*. PhD Thesis, University of Glasgow.

Caldow, R.W.G. & Furness, R.W. (1991) The relationship between kleptoparasitism and plumage polymorphism in the Arctic skua *Stercorarius parasiticus* (L.). *Functional Ecology*, **5**, 331–339.

Drent, R.H. & Daan, S. (1980) The prudent parent: energetic adjustments in avian breeding. *Ardea*, **68**, 225–252.

Ewins, P.J., Ellis, P.M., Bird, D.B. & Prior, A. (1988) The distribution and status of Arctic and great skuas in Shetland 1985–86. *Scottish Birds*, **15**, 9–20.

Furness, B.L. (1980) *Territoriality and Feeding Behaviour in the Arctic Skua* (Stercorarius parasiticus *L.*). PhD Thesis, University of Aberdeen.

Furness, R.W. (1978) Kleptoparasitism by great skuas (*Catharacta skua* Brünn.) and Arctic skuas (*Stercorarius parasiticus* L.) at a Shetland seabird colony. *Animal Behaviour*, **26**, 1167–1177.

Furness, R.W. (1982) Methods used to census skua colonies. *Seabird*, **6**, 44–47.

Furness, R.W. (1983) *The Birds of Foula*. Brathay Hall Trust.

Furness, R.W. (1987) *The Skuas*. Poyser, England.

Furness, R.W. (1990) A preliminary assessment of the quantities of Shetland sandeels taken by seabirds, seals, predatory fish and the industrial fishery in 1981–83. *Ibis*, **132**, 205–217.

Furness, R.W. & Monaghan, P. (1987) *Seabird Ecology*. Blackie, London.

Gaston, A.J. & Nettleship, D.N. (1982) Factors determining seasonal changes in attendance at colonies of the thick-billed murre *Uria lomvia*. *Auk*, **99**, 468–473.

Hamer, K.C., Furness, R.W. & Caldow, R.W.G. (1991) The effects of changes in food availability on the breeding ecology of great skuas *Catharacta skua* in Shetland. *Journal of Zoology*, **223**, 175–188.

Hamer, K.C., Monaghan, P., Uttley, J.D., Walton, P. & Burns, M.D. (1993) The influence of food supply on the breeding ecology of kittiwakes *Rissa tridactyla* in Shetland. *Ibis*, **135**, 255–263.

Harris, M.P. & Wanless, S. (1990) Breeding success of British kittiwakes *Rissa tridactyla* in 1986–1988: evidence for changing conditions in the northern North Sea. *Journal of Applied Ecology*, **27**, 172–187.

Hislop, J.R.G. & Harris, M.P. (1985) Recent changes in the food of young puffins *Fratercula arctica* on the Isle of May in relation to fish stocks. *Ibis*, **127**, 234–239.

Martin, A.R. (1989) The diet of Atlantic puffin *Fratercula arctica* and northern gannet *Sula bassana* at a Shetland colony during a period of changing prey availability. *Bird Study*, **36**, 170–180.

Martin, T.E. (1987) Food as a limit on breeding birds: a life-history perspective. *Annual Review of Ecological Systematics*, **18**, 453–487.

Monaghan, P., Uttley, J.D., Burns, M.D., Thaine, C. & Blackwood, J. (1989) The relationship between food supply, reproductive effort and breeding success in Arctic terns *Sterna paradisaea*. *Journal of Applied Ecology*, **58**, 261–274.

Montevecchi, W.A. (1993) Birds as indicators of fish stocks. *Birds as Monitors of Environmental Change* (eds R.W. Furness and J.J.D. Greenwood) pp. 217–266. Chapman and Hall, London.

Montevecchi, W.A., Birt, V.L. & Cairns, D.K. (1988) Dietary changes of seabirds associated with local fisheries failures. *Biological Oceanography*, **5**, 153–162.

Murphy, E.C., Springer, A.M. & Roseneau, D.G. (1991) High annual variability in reproductive success of kittiwakes at a colony in western Alaska. *Journal of Animal Ecology*, **60**, 515–534.

O'Donald, P. (1983) *The Arctic Skua*. Cambridge University Press.

Ollason, J.C. & Dunnet, G.M. (1983) Modelling annual changes in numbers of breeding fulmars, *Fulmarus glacialis*, at a colony in Orkney. *Journal of Animal Ecology*, **52**, 185–198.

Zar, J.H. (1984) *Biostatistical Analysis*. Prentice-Hall, New Jersey.

CHAPTER 5

The role of acanthocephalan parasites in the predation of freshwater isopods by fish

A.R. Lyndon

SUMMARY

(1) Alterations in intermediate host phenotype as a result of parasitism are widespread. The alterations caused by cystacanths of *Acanthocephalus lucii* and *Acanthocephalus anguillae* in *Asellus aquaticus* are investigated in relation to fish predation.

(2) Both species of parasite are distributed randomly in *Asellus* populations.

(3) The distributions of both species in fish definitive hosts deviated significantly from a Poisson distribution ($P < 0.01$ in all cases).

(4) The non-random distribution of parasites in fish hosts was at least partially explained by the effects of parasites on the manoeuvrability, behaviour and appearance of the intermediate host.

(5) The two parasite species had different effects on infected *Asellus* which were related to the feeding methods of their respective fish hosts.

(6) Parasitism appears to play an important role in the predation of infected crustaceans, but is unlikely to have noticeable effects on crustacean population dynamics overall at the levels of infection observed in nature.

Key-words: *Acanthocephalus* spp., *Asellus aquaticus*, behavioural alteration, colouration, transmission enhancement

INTRODUCTION

Many parasites with indirect life-cycles rely on the predation of intermediate hosts, containing parasite larval stages, in order to complete their development into reproducing adults. The potential benefits to such a parasite resulting from any increased susceptibility of infected intermediate hosts to predators has long been recognized (Holmes & Bethel 1972) and many studies have investigated the effects of parasites, especially acanthocephalans, on host behaviour (Moore 1984 and references therein, Gotelli & Moore 1992). However, in many cases there has not been a systematic analysis of the effects of parasites on a particular host, far less an

assessment of how any alterations in the host might augment parasite transmission in the wild. A notable exception is the work of Bethel & Holmes (1973, 1977), which showed the behavioural changes observed in *Gammarus lacustris* harbouring two different acanthocephalans to be related to the different feeding methods of appropriate duck hosts. Such subtleties of parasite–host systems have rarely been addressed, primarily because most work has involved analysis of only single intermediate–definitive host pairs, and the presence of any alteration(s) in infected intermediate hosts has been taken to be advantageous to the parasite. That such a level of analysis is inadequate becomes apparent when one considers that parasite-induced changes in a host may render that host susceptible to predation by unsuitable definitive hosts as well as those in which the parasite can achieve maturity (Brassard *et al.* 1982).

Given that there will be a strong selection for parasites which are successfully transmitted to a suitable definitive host and, conversely, a strong selection against those which are ingested by unsuitable hosts, it seems reasonable to expect the occurrence of parasite specific changes in intermediate host phenotype which lead to increased susceptibility to predation by particular species of definitive host(s) (Dobson 1988). Such a situation appears to pertain for the host–parasite associations examined by Bethel & Holmes (1973, 1977), but no other examples are available, although there is a considerable need for more field studies (Dobson 1988). To investigate further the suggestion that parasites specifically alter their intermediate hosts so as to enhance transmission to a limited range of definitive hosts, rather than utilizing intermediate hosts predisposed to certain phenotypic alterations, it is necessary to study other systems where two (or more) parasites with different definitive host specificities have the same intermediate host. Such a system exists with *Acanthocephalus anguillae* and *Acanthocephalus lucii*, which both use *Asellus aquaticus* as their intermediate host, but which have broadly differing definitive host preferences (Brown *et al.* 1986). This paper reports findings regarding the effect of these parasites on *Asellus aquaticus* in relation to their occurrence in populations of different fish hosts, namely chub *Leuciscus cephalus*, perch *Perca fluviatilis*, sticklebacks *Gasterosteus aculeatus* and eels *Anguilla anguilla*.

MATERIALS AND METHODS

Chub, perch and eels were obtained by electro-fishing from the River Lee (Herts., England; National Grid Reference (NGR) TL2112) and the Hogsmill Brook (Surrey, England; NGR TQ0069). Both these rivers are tributaries of the River Thames. Sticklebacks were collected using a sweep net from the River Tame (Staffs., England; NGR SK1813), a tributary of the River Trent. The fish were either frozen or preserved in 10% formalin until examination. The alimentary canal was dissected from the fish, opened longitudinally and all acanthocephalans found identified using the key of Brown *et al.* (1986). *Asellus aquaticus* were obtained by kick sampling from the Rivers Tame, Meden (Notts., England; NGR SK6873), Trent (Staffs., England; NGR SK1615) and Bristol Avon (Wilts., England; NGR ST7964) and by sweep netting from the Bridgewater–Taunton Canal (Somerset, England; NGR ST3232), the Kennet and Avon Canal (Avon,

England; NGR ST7860) and the Fazeley Canal (Staffs., England; NGR SK1999). Samples used for analysis of frequency distributions were preserved in formal acetic acid at the waterside, whereas those used for behavioural studies were placed in large, plastic bottles of local water containing plastic bags of ice to reduce deoxygenation due to warming. Preserved *Asellus* were dissected under water using mounted needles on a binocular microscope at $\times 120$–$\times 310$ magnification. Any larval parasites recovered were identified by their colouration (cystacanths of *Acanthocephalus anguillae* turned bright orange when preserved inside *Asellus*, whereas *Acanthocephalus lucii* cystacanths did not) and proboscis morphology in squash preparations, while fresh cystacanths were identified as before after manual eversion of the proboscis using mounted needles.

Effects of the parasites on infected *Asellus* were assessed in terms of the colouration of the hosts and their behaviours. The preference of *Asellus* for light or dark (photophilia; Allely *et al.* 1992) was examined by placing isopods individually in a half blacked-out tank with an angle-poise lamp (40 W bulb, 21 cm above water surface) situated directly over its centre and measuring the time spent in the light half over a total period of 30 min subsequent to 15 min acclimation. The response of *Asellus* to disturbance was measured in a plastic tank in which a glass rod was agitated at one end, their actions being recorded over a 3 min period as (1) no response, (2) movement away, (3) movement towards, and (4) movement directly through the focus of the disturbance. The last category included animals which approached the disturbance sufficiently closely that they lost their grip on the tank. In addition, the ability of *Asellus* to right themselves was compared between infected and uninfected (both non-reproductive individuals and ovigerous females with marsupia) individuals by up-turning them in a petri dish of aerated fresh water, using a fine paint-brush, and recording the time (up to a maximum of 5 min) taken to regain an upright position. Results from animals using the sides of the dish to assist in this were excluded from the analysis and the animals retested on a subsequent day.

RESULTS

The distributions of both *Acanthocephalus anguillae* and *Acanthocephalus lucii* in *Asellus aquaticus* are indistinguishable from Poisson distributions calculated using their mean abundance (for definitions of all terms used, see Margolis *et al.* 1982) as an estimate of μ (Zar 1984) (see Table 5.1). In contrast, the distributions of *Acanthocephalus anguillae* in chub, eels and perch deviate highly significantly from the predicted Poisson distri-

Table 5.1. Observed frequency distribution (O) and expected Poisson distribution (E) for *Acanthocephalus* spp. infecting *Asellus aquaticus* at selected sites of sympatric occurrence.

Site	Species	μ^a	Sample size	Probability (%) of infection						χ^2	P
				Zero		Single		Multi			
				O	E	O	E	O	E		
R. Meden	*A. anguillae*	0.05	154	95	95	5.2	4.9	0	0.13	0.15	>0.99
	A. lucii	0.01		99	99	1.3	1.3	0	0.02	0.02	0.99
R. Tame	*A. anguillae*	0.15	201	87	86	11	13	2	1	1.29	>0.50
	A. lucii	0.02		99	99	1.5	1.5	0	0.01	0.01	>0.99

[a] Mean abundance.

Table 5.2. Observed frequency distribution (O) and expected Poisson distribution (E) for *Acanthocephalus* spp. infecting fish; μ = 16.9 for chub, 12.19 for eels and 5.13 for perch.

Host	Sample size	Number of fish infected								χ^2	P
		0–9		10–19		20–29		30+			
		O	E	O	E	O	E	O	E		
Chub	23	14	4	4	18	1	1	4	0	>35.9	<0.001
Eel	31	20	1	2	22	2	8	7	0	>383	<0.001
Perch	15	12	14.45	3	0.55	0	0	0	0	11.3	<0.01

bution (Table 5.2), suggesting some degree of selection for parasitized intermediate hosts.

The colouration of parasitized *Asellus* was quite marked, and differed between the two parasite species. The colouration of *Asellus* infected with cystacanths of *Acanthocephalus lucii* has previously been described by Brattey (1983) and was similar to that observed here, i.e. melanization of the respiratory opercula, whilst *Asellus* harbouring cystacanths of *Acanthocephalus anguillae* exhibited melanization of the whole body, as more recently noted by Dezfuli *et al.* (1994).

With regard to behavioural differences between uninfected *Asellus* and those infected with either of the two parasites, the results of the various experiments undertaken are shown in Tables 5.3 and 5.4 and Figure 5.1. The photophilic responses of the *Asellus* indicate that, overall, each group spent more time in the dark half of the tank (i.e. they were photophobic) (Table 5.3; chi-squared test, $P < 0.001$ in all cases). However, there were large differences between groups in the degree of their photophobia. *Asellus* infected with *Acanthocephalus lucii* were statistically indistinguishable from uninfected individuals, whereas *Acanthocephalus anguillae*-infected *Asellus* spent on average more than twice as long in the light-zone as uninfected animals

Table 5.3. Mean times spent by uninfected and infected *Asellus* in the light half of a half blacked-out arena. (The duration of each trial was 30 min (= 1800 s); results were compared using Wilcoxon's Rank Sum test.)

Parasite	n	Mean time in light (s) (±SEM)	P
Uninfected	20	207 ± 78	–
A. anguillae	12	471 ± 148	0.15
A. lucii	13	223 ± 101	1

Table 5.4. Mean times taken by uninfected, uninfected ovigerous female, and infected *Asellus* to regain an upright position after being overturned.

Group	n	Mean time to right (s) (± SEM)
Uninfected	20	17.65 ± 5.87
Ovigerous	16	40.81 ± 18.76
A. anguillae	20	89.50 ± 25.97
A. lucii	20	56.80 ± 22.05

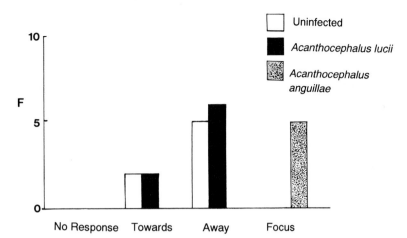

Fig. 5.1. Qualitative responses of uninfected and *Acanthocephalus* spp. infected *Asellus aquaticus* to disturbance.

(Table 5.3), although this was not statistically significant (Rank Sum test, P = 0.15). This lack of statistical difference was due to the occurrence of some zero values in the *Acanthocephalus anguillae* infected *Asellus*. That the difference is of biological significance is indicated by the fact that only 2 out of 20 uninfected *Asellus* spent more than 400 s in the light, whereas 6 out of 12 *Acanthocephalus anguillae*-infected *Asellus* spent at least this time in the light-zone.

The responses of *Asellus* to disturbance showed a similar pattern to that seen for photophilia; that is, animals infected with *Acanthocephalus lucii* showed very similar responses to those which were uninfected, while those infected with *Acanthocephalus anguillae* exhibited a markedly different behaviour pattern (Fig. 5.1). In short, infection with *Acanthocephalus anguillae* seems to cause an attraction to disturbance not observed in the other two groups.

In contrast to these results, the effect of parasitism on the ability of infected animals to right themselves was similar for both parasites (Table 5.4). *Asellus* infected with either species took three to five times longer, on average, to right themselves compared with uninfected individuals. That this is a non-specific, physical effect of parasitism is suggested by the fact that uninfected, ovigerous female *Asellus* (i.e. with a ventral brood-pouch) also took longer to right themselves than male or non-ovigerous female animals (uninfected group) (Table 5.4). There were no significant differences between the righting times of ovigerous and infected *Asellus* (Kruskal–Wallis test, P > 0.33) but a significant difference was apparent between these groups and uninfected, non-ovigerous *Asellus* (P < 0.05).

Differences in behaviour and appearance between

Asellus infected with the two parasites, such as those described above, could be related to variations between fish species in their suitability for these parasites. That such variations occur is supported by both field and laboratory findings. Analysis of sticklebacks and *Asellus* from the R. Tame (Table 5.5) shows that *Acanthocephalus lucii* is the only acanthocephalan present in the former, despite its having a much lower prevalence in the intermediate host at this site than *Acanthocephalus anguillae*. Furthermore, *Acanthocephalus anguillae* achieves high infrapopulations in barbel in the field (A.R. Lyndon, personal observation), but *Acanthocephalus lucii* has not been recorded from this fish (Kennedy 1974, A.R. Lyndon, personal observation). Laboratory infections revealed that *Acanthocephalus anguillae* established readily in chub (97%) but not in perch (0%), whereas for *Acanthocephalus lucii* the opposite pattern was observed (0% in chub; 92.5% in perch).

Table 5.5. Occurrence of *Acanthocephalus anguillae* and *Acanthocephalus lucii* in *Asellus* and sticklebacks at a site (R. Tame) where the parasites are sympatric.

	Host	
	Asellus aquaticus	3-Spine stickleback
A. anguillae (%)	9.41[a]	0[b]
Range	5.85–12.93	–
A. lucii (%)	0.66[a]	25[b]
Range	0–1.49	14–36

[a] Mean value over two years.
[b] Values from two samples over six months.

DISCUSSION

The results show that for the *Acanthocephalus–Asellus* system studied, two parasite species have different effects on a single species of intermediate host. The apparent absence of behavioural alterations in *Acanthocephalus lucii*-infected *Asellus* is unusual for an acanthocephalan–arthropod system (Moore 1984), although not unique (see Allely *et al.* 1992). As pointed out by Poulin (1994), the paucity of data relating to acanthocephalans which do not cause behavioural changes in their intermediate hosts may stem from a reluctance to publish such studies. In the present case, the consistent occurrence of different alterations in the same host species brought about by two closely related parasites gives strong support to the idea that such alterations are specifically induced by the parasite, rather than being random pathological effects. If this is the case, we might expect such changes to enhance parasite transmission to suitable definitive hosts. However, it is necessary to sound a note of caution, since non-specific parasite effects (change in righting ability) were also evident (Table 5.4), meaning that the presence of a single alteration in a host is not necessarily indicative of a parasite-specific ('adaptive') modification. It is therefore important to study a wide range of host characteristics (as studied here) in context if the presence of subtle and potentially interacting effects is to be detected.

Having established the presence of parasite-specific modifications, the question arises as to whether these are related in any way to the feeding ecology of the various fish hosts. Table 5.6 gives a synopsis of the known feeding methods of the main species in question. From this it is clear that there is a distinction to be made between chub and barbel on the one hand and perch and sticklebacks on the other, both in terms of preferred flow regimes and feeding techniques. It is not suggested, however, that this scheme is hard and fast, but rather that the majority of individuals within a given host species will feed for most of the time using the

method indicated for that species (Table 5.6). Eels are not included in Table 5.6 because of doubts surrounding their feeding behaviour. Until recently, they were considered to be benthic feeders (e.g. Mann & Blackburn 1991), but evidence from their parasite fauna suggests that they may also forage in the water column (Kennedy *et al.* 1992), leading to the possibility that their feeding behaviour combines aspects of both the chub/barbel and the perch/stickleback groups.

Perch and sticklebacks are primarily visual predators and, as such, the most important factors influencing their prey selection are likely to be colour, movement, shape and size of the prey (FitzGerald & Wootton 1993). The effect of *Acanthocephalus lucii* on infected *Asellus* is to increase both the colour contrast of the prey and the conspicuousness of a constantly moving part of the organism (the respiratory opercula responsible for ventilation of the gills). It does not seem unreasonable to infer from this that the changes induced by *Acanthocephalus lucii* in *Asellus* are likely to make parasitized individuals more obvious to predators most suitable for the continued development of the parasite. In contrast, the main effects of *Acanthocephalus anguillae* on *Asellus* are of a behavioural nature, for although a marked colour change is evident, overall dark colouration is within the normal repertoire of pigmentation for un-infected *Asellus* in certain situations (e.g. Irish loughs) (A.R. Lyndon, personal observation), whereas melanized opercula never occur in the absence of *Acanthocephalus lucii* (Brattey 1983, A.R. Lyndon, personal observation). The attraction of *Acanthocephalus anguillae*-infected *Asellus* to disturbance (Fig. 5.1) will tend to increase their susceptibility to predation by barbel and to a lesser extent chub, whilst in combination with their increased photophilia this may lead to an enhanced occurrence in the drift compared to uninfected individuals which would lay them open to predation by station holding chub. Interestingly, *Pomphorhynchus laevis*, which is also an acanthocephalan parasite of chub, causes a significant increase in the drift of infected, compared to

Table 5.6. Preferred flow-regime and major feeding methods of the principal fish species acting as hosts for *Acanthocephalus anguillae* and *Acanthocephalus lucii* (see text for details).

Species	Preferred flow regime	Main feeding method(s)
Perch	Still to slow	Visual selection of single food items (from substrate)
Stickleback	Still to medium	Visual selection of single food items
Chub	Medium to fast	Station-holding, food taken from stream; 'disturbance' feeding in margins
Barbel	Medium to fast	Benthic 'vacuuming'

uninfected, *Gammarus* (McCahon *et al.* 1991). Clearly, it would be of great interest to obtain data on the drift of *Acanthocephalus anguillae* infected *Asellus*.

Eels probably exhibit feeding behaviour incorporating both visual selection of individual benthic prey and predation of invertebrates in the water column (see above), so that they are likely to encounter and ingest both parasite species. The implication is that eels should ingest both parasites roughly in proportion to their occurrence in the environment, assuming that all feeding behaviours are equally represented. It would be interesting to investigate the extent to which individual deviation from the most common mode of feeding might affect exposure of fish hosts to parasites, and consequently the dispersion of parasites within host populations.

Figure 5.2 shows a schematic representation of the effects of *Acanthocephalus anguillae* and *Acanthocephalus lucii* on the effective availability of infected *Asellus* to the different categories of host. The top bar in the diagram represents the absolute occurrence (observed

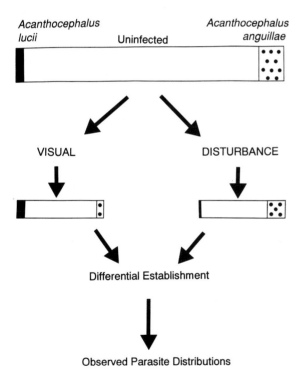

Fig. 5.2. Schematic diagram representing the absolute availability (observed prevalence) and the effective availability of infected *Asellus* to two categories of predator as a consequence of the latters' different feeding techniques. See text for full explanation.

prevalence) of the two parasites in the *Asellus* population, while the two bars below this represent the effective prevalence as perceived by the respective predators as a result of parasite-induced changes in the *Asellus*. What this conveys is not only that uninfected *Asellus* are less likely to be detected by a predator than parasitized ones, but also that visual predators are much more likely to detect *Acanthocephalus lucii*-infected individuals, with *Acanthocephalus anguillae*-infected *Asellus* essentially forming part of the 'uninfected' population as far as they are concerned. The converse situation applies to disturbance predators. Thus, ingestion of the 'wrong' parasite is reduced to random occurrence, whilst predation of the 'right' parasite is inflated. It is indicated that there may be further screens to successful establishment even if the 'wrong' parasite is ingested, so that the observed distribution of parasite species in different host species is likely to reflect a combination of these processes. However, for the parasite, which effectively faces death (at least in a reproductive sense) even if it can establish in an unsuitable host, it is clearly the predation of the intermediate host which is the critical step and the only one over which it can exert any degree of control.

Finally, it is worth considering the potential effects of these parasites on the population dynamics of the *Asellus*. Do they have a deleterious effect on the production of the affected *Asellus* populations? Is there any evidence of a coevolution of host and parasite in terms of host pathology and/or killing of the parasite? Although previous work has suggested that *Acanthocephalus lucii* prevents reproduction in female *Asellus aquaticus* (Brattey 1983), three occurrences of cystacanth-infected females with eggs (11–35) in the marsupium were observed during the present study, suggesting that parasitism by this species does not necessarily preclude reproduction, while Dezfuli *et al.* (1994) have observed ovigerous *Asellus aquaticus* infected with *Acanthocephalus anguillae*. In the field, isopods of both sexes infected with either parasite were commonly found in precopulatory amplexus (A.R. Lyndon, personal observation) in spite of previous claims to the contrary (Dezfuli *et al.* 1994). Taking these observations in combination with the very low parasite abundances seen in most localities (Table 5.1) leads to the conclusion that any impact of these parasites upon their intermediate host populations is likely to be very small, if not negligible, and that there will be little selection pressure for resistance in the *Asellus* (Dobson 1988). This interpretation is supported by the fact that no evidence of pathological change (e.g. melanization of the parasite; tissue proliferation or inflammation) was seen in either the hosts or the parasites, whilst all of the infective stages recovered from live *Asellus* were themselves alive.

ACKNOWLEDGEMENTS

This work was supported by a Natural Environment Research Council grant (GR3/8049). Thanks are due to Phil and Jan Shears, Lisa Trevethick, Leo Aguirre-Macedo and Victor Vidal-Martinez for invaluable assistance in obtaining infected *Asellus*. The fish used in this study were kindly supplied by Mark Pilcher and Janet Moore of the National Rivers Authority (Thames East Region).

REFERENCES

Allely, Z., Moore, J. & Gotelli, N.J. (1992) *Moniliformis moniliformis* infection has no effect on some behaviours of the cockroach *Diploptera punctata. Journal of Parasitology*, **78**, 524–526.

Bethel, W.M. & Holmes, J.C. (1973) Altered evasive behaviour and responses to light in amphipods harboring acanthocephalan cystacanths. *Journal of Parasitology*, **59**, 945–956.

Bethel, W.M. & Holmes, J.C. (1977) Increased vulnerability of amphipods to predation owing to altered behaviour induced by larval acanthocephalans. *Canadian Journal of Zoology*, **55**, 110–115.

Brassard, P., Rau, M.E. & Curtis, M.A. (1982) Parasite-induced susceptibility to predation in diplostomiasis. *Parasitology*, **85**, 495–501.

Brattey, J. (1983) The effects of larval *Acanthocephalus lucii* on the pigmentation, reproduction and susceptibility to predation of the isopod *Asellus aquaticus. Journal of Parasitology*, **69**, 1172–1173.

Brown, A.F., Chubb, J.C. & Veltkamp, C.J. (1986) A key to the species of Acanthocephala parasitic in British freshwater fishes. *Journal of Fish Biology*, **28**, 327–334.

Dezfuli, B.S., Rosetti, E., Fano, E.A. & Rossi, R. (1994) Occurrence of larval *Acanthocephalus anguillae* (Acanthocephala) in the *Asellus aquaticus* (Crustacea, Isopoda) from the River Brenta. *Bollettino di Zoologia*, **61**, 77–81.

Dobson, A.P. (1988) The population biology of parasite-induced changes in host behaviour. *Quarterly Review of Biology*, **63**, 139–165.

FitzGerald, G.J. & Wootton, R.J. (1993) The behavioural ecology of sticklebacks. *Behaviour of Teleost Fishes* (ed. T.J. Pitcher), pp. 537–572. Chapman and Hall, London.

Gotelli, N.J. & Moore, J. (1992) Altered host behavior in a cockroach – acanthocephalan association. *Animal Behaviour*, **43**, 949–959.

Holmes, J.C. & Bethel, W.M. (1972) Modification of intermediate host behaviour by parasites. *Zoological Journal of the Linnaean Society*, **51** (Suppl. 1), 123–149.

Kennedy, C.R. (1974) A checklist of British and Irish fish parasites with notes on their distribution. *Journal of Fish Biology*, **6**, 613–644.

Kennedy, C.R., Nie, P., Kaspers, J. & Paulisse, J. (1992) Are eels (*Anguilla anguilla* L.) planktonic feeders? Evidence from parasitic communities. *Journal of Fish Biology*, **41**, 567–580.

McCahon, C.P., Maund, S.J. & Poulton, M.J. (1991) The effect of the acanthocephalan parasite (*Pomphorhynchus laevis*) on the drift of its intermediate host (*Gammarus pulex*). *Freshwater Biology*, **25**, 507–513.

Mann, R.H.K. & Blackburn, J.H. (1991) The biology of the eel *Anguilla anguilla* L. in an English chalk stream and interactions with juvenile trout *Salmo trutta* L. and salmon *Salmo salar* L. *Hydrobiologia*, **218**, 65–76.

Margolis, L., Esch, G.W., Holmes, J.C., Kuris, A.M. & Schad, G.A. (1982) The use of ecological terms in parasitology (report of an ad hoc committee of the American Society of Parasitologists). *Journal of Parasitology*, **68**, 131–133.

Moore, J. (1984) Altered behavioural responses in intermediate hosts: an acanthocephalan parasite strategy. *American Naturalist*, **123**, 572–577.

Poulin, R. (1994) Meta-analysis of parasite-induced behavioural changes. *Animal Behaviour*, **48**, 137–146.

Zar, J.H. (1984) *Biostatistical Analysis*. Prentice-Hall International, London.

CHAPTER 6

Dippers *Cinclus cinclus* as predators in upland streams

S.J. Ormerod

SUMMARY

(1) The role of birds as predators in streams is poorly known but the dippers *Cinclus* spp., common along hillstreams on four continents, are particularly suited to investigation.

(2) Dipper diet consists of aquatic insects and small fish. Prey use, switching and selection is influenced by: (i) prey availability, for example due to variation in water quality; (ii) prey quality, with large and small prey selected under different circumstances, and mineral content possibly important; (iii) time of year, particularly due to chick rearing; (iv) ontogenetic development, immediately after fledging; (v) prey behaviour, through which escape responses vary.

(3) Bottom-up effects of prey abundance on dipper distribution are well known. Likely top-down effects by dippers as consumers have been estimated from energetic and dietary data. There are indications of substantial effects on some prey which have been confirmed only in the American dipper *C. mexicanus* Swainson during short-term exclosures. Effects on interactions between prey, or interactions between dippers and other predators, are possibilities that cannot be excluded.

(4) Marked omnivory in the dipper, together with complex prey refuges in rivers and the absence of obviously strong interactions between prey species, may prevent dippers having 'keystone' effects. Contrasting possibilities for the ecological role of dippers thus require resolution by experimental investigation.

Key-words: birds, *Cinclus cinclus*, diet, predation, rivers

INTRODUCTION

The majority of studies of predator–prey interactions in freshwaters have involved fish, amphibians or predatory invertebrates (Hildrew 1992). By comparison, the role of birds has been neglected, which is surprising in view of their ubiquity, species richness, and abundance on both lakes and streams (see Ormerod & Tyler 1993a). As a result, important reviews and theoretical developments about predation in freshwaters have occurred in the absence of potentially important data, and potentially important organisms (e.g. Kerfoot & Sih 1987, Hildrew 1992, Sih & Wooster 1994, Dodson *et al*. 1994).

Recently, attention has been drawn to the possible role of common and abundant river passerines, the dippers *Cinclus* spp., as predators of invertebrates and fish in upland streams (Ormerod & Tyler 1991). Five species from this genus occur along hillstreams on four continents, and are particularly suited to investigations of predator–prey interactions. Throughout the year, prey abundances are easily measured (e.g. Ormerod *et al.* 1985), diet has been easily assessed in chicks and adults (Ormerod & Tyler 1991), behaviour has been easily observed in dippers and prey (Yoerg 1994; Jenkins & Ormerod, in press), and questions about predator–prey interactions are conducive to experimental investigation (Harvey & Marti 1993; Jenkins & Ormerod, in press). In addition, the complete dependence of dippers on production from the stream ecosystem means they are sensitive to the degradation of rivers, for example through pollution (Ormerod *et al.* 1991), a feature which makes them valuable indicators of environmental change (Ormerod & Tyler 1993a).

In this paper, I review some of the work carried out on predatory interactions between dippers and stream organisms with emphasis on studies on *Cinclus cinclus* L. in Wales. I outline their diet, factors influencing prey selection, behaviour of both dippers and their prey, and illustrate the potential importance of dippers as consumers. I allude also to some unknown features of the role of dippers as ecological agents in stream communities, which require further research.

DIET, PREY SELECTION AND BEHAVIOURAL STUDIES

Activity and foraging methods

Feeding is the single most frequent activity in dippers, taking on average 45–55% of the active day (Bryant & Tatner 1988, O'Halloran *et al.* 1991). On acidic streams, where food is scarcer, this time is significantly greater than on circumneutral streams (55% versus 38%) (O'Halloran *et al.* 1991). It increases also during winter, when foraging takes up over 60% of the active day, but falls to 30–40% during the moult in July and August; under these circumstances, much of the daily energy requirement must be gleaned during this shorter time. Circadian patterns are apparent from changes in body mass (Ormerod & Tyler 1990) and activity data (O'Halloran *et al.* 1991), which both indicate a peak in feeding during early daylight.

Foraging activity is divided into diving, swimming, wading at the river margins turning stones and leaves (wade prying), and gleaning from stone surfaces. Activities such as fly-catching and gleaning from plant surfaces also occur, but account for fewer than 1.5% of foraging bouts (Spitznagel 1985). The major foraging techniques are diving (around 56% of foraging bouts) and wade prying (37% of bouts), with diving most common during the period November–March (O'Halloran *et al.* 1991). In part, this reflects the strong effect of increasing river discharge, which increases the need for diving. At the highest discharges however,

diving begins to decline, probably because rivers in spate become too turbid and turbulent for effective prey capture (O'Halloran *et al.* 1991). Dippers under these circumstances switch to feeding in smaller streams, concentrating also on epibenthic insects such as blackfly larvae and *Baetis* mayflies which are common in the diet during the winter (see Ormerod & Tyler 1986).

Diet

The relatively restricted array of prey available in rivers, the ability to identify prey remains in faeces or pellets, and the ability to reconstruct individual prey weights (e.g. Ormerod & Tyler 1986), have combined to allow detailed dietary studies in dippers. The diet consists overwhelmingly of the aquatic stages of mayflies, stoneflies, caddis, blackfly larvae (Simuliidae) and fish from the Salmonidae or, more often, the Cottidae (Ormerod & Tyler 1986, Ormerod *et al.* 1987, Ormerod & Tyler 1991). Crustaceans and molluscs sometimes occur, but generally provide less than 10% of the prey eaten (Ormerod & Tyler 1986). Among the less frequent prey, cottid fish make a disproportionately large contribution to the biomass in the diet in Wales, particularly in winter, because of their relatively large mean dry weight of around 220 mg (Table 6.1). This is true also of caddis larvae (usually 6–60 mg dry weight), and this group dominates the diet by weight for much of the year (e.g. Table 6.1). By contrast, smaller baetid mayflies (about 0.5–2 mg dry weight) and blackfly larvae (0.5–1 mg dry weight) each contribute around 30–35% of the items eaten in late summer and in winter (Ormerod & Tyler 1986). Both these groups are epibenthic, forming dense patches in the case of blackfly, which make for frequent encounters and easy capture by moulting adults whose flight and swimming ability is impaired. Recently fledged juveniles who have yet to develop their foraging skills also use these prey (see below and Yoerg 1994).

Prey selection

The selection of prey types and exact make up of the diet is affected by factors including prey availability, time of year, prey quality, ontogenetic development, and prey behaviour. Among these, the effect of prey availability is most easily demonstrated and occurs, for example, because of marked differences in the benthic communities of acidic and circumneutral streams. At acidic sites, prey such as fish, mayfly nymphs, molluscs and crustaceans are extremely scarce, and never occur in dipper diet (see Table 6.1). In fact, during the

Table 6.1. The estimated percentage contributions (by weight) to the diets of dippers on acidic and circumneutral streams in Wales at different times of the year (from Ormerod & Tyler 1991). Breeding is taken as April–June inclusive, moult as July–September inclusive, and winter as October–March. No data were available from acidic streams during moult. Data from the breeding season have been combined for nestlings >5 days old (ON) as opposed to those <5 days old (YN).

| | Circumneutral streams | | | | | | Acid streams | | | | |
| | Breeding | | | | Moult | Winter | Breeding | | | | Winter |
	YN	Adults	ON	Adults			YN	Adults	ON	Adults	
Molluscs	0	0	0	0	0	6.1	0	0	0	0	0
Crustacea	0	0	0	0	0	3.2	0	0	0	0	0
Mayflies	35.3	19.6	3.8	17.1	7.3	3.2	6.1	0	0.2	1.3	0
Stoneflies	18.8	4.9	0.8	1.7	0.7	1.3	20.2	33.5	4.2	36.6	32.1
Caddis	45.7	75.4	80.9	51.4	69.4	19.3	73.6	64.6	95.5	61.3	60.1
True-flies	0.2	0.1	0.1	0.8	3.4	3.1	0.1	1.9	0.1	0.8	7.7
Others	0	0	8.5	1.6	0.1	0.3	0	0	0	0	0.1
Fish	0	0	5.9	27.4	19.1	63.8	0	0	0	0	0

breeding season, there is a marked switch in the commonest prey type from stoneflies along acid streams to mayfly nymphs on circumneutral streams, even though stoneflies remain abundant (Fig. 6.1). At all but acidic sites, where they form the only numerous prey, dippers strongly select against several stonefly families (see also Ormerod & Tyler 1986; Smith & Ormerod 1986). This is despite the absence of any marked escape behaviour which would help them avoid predation (see below and Jenkins & Ormerod in press). In this case, other intrinsic qualities of the prey or low encounter rates may be involved caused by, for example, the interstitial distribution of some stoneflies in the substratum (Hildrew *et al.* 1980, Ernst & Stewart 1986).

Some of the biggest changes in diet and selection occur seasonally, and relate to chick rearing. Here, the selection of both large and small prey are important as the diet changes with chick age. Young nestlings are fed predominantly with *Baetis* mayflies (<1–2 mg individual dry weight), whereas increasingly older nestlings are fed with limnephilid and hydropsychid caddis (6–60 mg individual dry weight) (Fig. 6.2). Adults take increasingly smaller prey for themselves over this period (Ormerod *et al.* 1987). While such strong positive selection for caddis larvae might reflect their low chitin content (Spitznagel 1985), probably the single feature of importance is that their large size allows efficient food delivery to the nest: prey size is an increasingly strong correlate

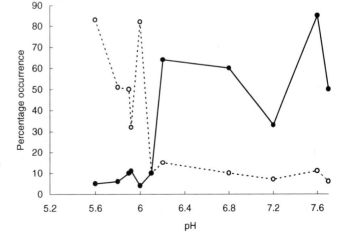

Fig. 6.1. The percentage contribution made to the combined diet of adult and nestling dippers by stonefly nymphs (○) and mayfly nymphs (●) in relation to stream pH in upland Wales; each nest site is represented by one datum point (after Ormerod & Tyler 1991).

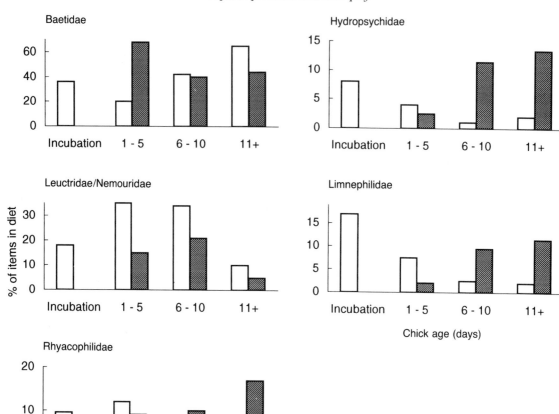

Fig. 6.2. Changes in the contribution made by selected families of insect to the diet of adult (open) and nestling (shaded) dippers with the age of nestlings in south-western Norway (after Ormerod *et al.* 1987). Rhyacophilidae, Hydropsychidae and Limnephilidae are all caddis larvae with large individual weight.

with prey selection for growing nestlings (Fig. 6.3) and small prey probably could not be harvested or carried to the nest quickly enough to satisfy the brood's energy demand. By day 11, for example, load size reaches around 75 mg dry weight (Tyler & Ormerod 1994), equivalent to around 4–9 caddis or 50–150 mayflies. This loading in a typical brood of five young is sustained at around 30 visits per hour throughout daylight (Tyler & Ormerod 1994), so that even using large caddis, each adult has to find, catch, handle and carry 60–135 prey per hour (see Ormerod *et al.* 1987). In this context, it is possible that brood provisioning in the dipper is subject to a particular energetic constraint because of the linear shape of the riverine habitat. A dipper territory of 5000 m² on a river 8 m wide will be 625 m long, so that a nest at its centre could involve round trips for foraging

equal to this distance. This contrasts with a bird in a territory of similar area, but circular shape, which will never be more than about 40 m from a centrally placed nest.

Outside the breeding season, during the moult and in the winter, the diet of dippers varies between sites and regions (Tyler & Ormerod 1994). At these times, they feed more opportunistically (Ormerod & Tyler 1986). However, one interesting feature is that prey such as fish, *Gammarus* shrimps and molluscs, all rich in calcium, occur in the diet predominantly in the late winter and pre-breeding period (see Table 6.1). There is evidence that, in the absence of these sources on acid streams, calcium may be in short supply for dippers during the formation of eggs (Ormerod *et al.* 1991, Ormerod & Tyler 1993b). However, the exact way that

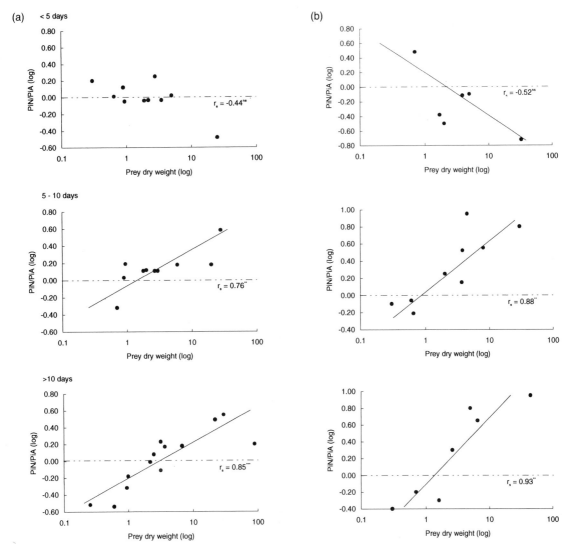

Fig. 6.3. Prey selection by breeding dippers during nest provisioning at different nestling ages: (a) in nests along circumneutral streams in Wales (Ormerod & Tyler 1991); and (b) in nests on streams in south-western Norway (Ormerod *et al.* 1987). The figure shows a selection index (PiN/PiA) for each prey type plotted against prey weight when nestlings are <5 days old, 5–10 days old, and >10 days old (PiN and PiA are the percentages of prey type i in the diets of nestlings and adults respectively). Thus larger prey are selected by adults for their nestlings once they are greater than about 5 days old.

dietary and endogenous calcium sources are involved in egg production in dippers, for example through fluxes in the formation of medullary bone (Krementz & Arkney 1995), are still far from clear. Selection for calcareous materials requires further investigation, and will be particularly interesting where favoured prey, such as cottid fishes, are naturally scarce – as in Scotland and Ireland.

Behavioural studies on predator and prey

Some of the most recent studies of interactions between dippers and prey have focused directly on the behaviour of predator and prey. Both appear to have some influence on prey selection.

In mid Wales, Yoerg (1994) assessed the development of foraging skills in recently fledged dippers from over 230 hours' observation and 40 000 prey captures in 22 individually marked birds. As adult provisioning declined between days 5 and 9 after fledging, juveniles increased the time they gave to foraging from 4% to 30%, and eventually to 50% by day 21. Although there were distinct improvements in juvenile foraging skills over this time period (Fig. 6.4, Table 6.2), juveniles fed predominantly on the smallest prey ('crumbs'), mostly blackfly larvae and small mayflies (74–94% of prey compared with 57% in adults); larger items provided only 0.5–3% of the diet (compared with 7–15% in adults). There were strong contrasts with adults in almost all the parameters recorded, juveniles for example foraging at lower water speeds, shallower depths, and being more likely to drop prey or stumble. Interestingly, the specialization by juveniles on blackfly occurred despite much greater profitability (as energy

gained per unit handling time) when they ate larger prey. Yoerg (1994) ascribed this feature to the high risks of fumbling larger prey, and to the skill and energetic expense involved in catching them. Such high dependence by recently fledged juveniles on certain abundant and easily caught prey types might have hitherto unexamined repercussions where such prey are scarce, for example on acidic streams.

In addition to active selection by dippers as predators, the use of different foods is liable in part to reflect escape behaviour or risk-avoidance strategies by the prey (e.g. Jeffries & Lawton 1984, Sih 1987, Scrimgeour & Culp 1994). The behavioural response of stream macroinvertebrates to real and simulated predatory activity by dippers has thus been investigated by a mix of field and laboratory method (Jenkins & Ormerod, in press). In the field, experimental enclosures were used

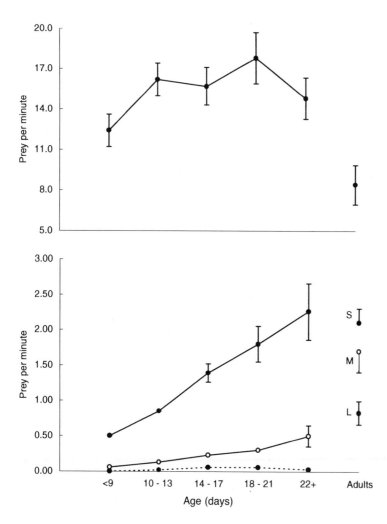

Fig. 6.4. Capture rates of prey of different sizes by developing juvenile dippers of increasing age (in days) and adults. Upper panel: 'crumbs' (the smallest prey); lower panel L, large prey; M, medium prey; S, small prey (after Yoerg 1994).

Table 6.2. A summary of the developmental changes in dippers during 22 days after fledging, and differences between independent young and adults (after Yoerg 1994).

	Prey size	Developmental change	Independent young relative to adults
Proportion in diet of:	'Crumbs'	Decreases	Higher
	Small prey	Increases	Same
	Large prey	Increases	Lower
Capture rate of:	'Crumbs'	Increases	Higher
	Small prey	Increases	Same
	Large prey	Increases	Lower
Handling time for:	Small prey	Decreases	Same
	Medium prey	No change	Longer
	Large prey	No change	Same
Proportion dropped of:	Small and medium prey	Decreases	None by adults
Stumbling rate:		Decreases	Higher

to assess whether dippers influenced invertebrate drift. This behaviour involves animals being swept by the river current for short distances, and is sometimes used to escape other riverine predators (see Jenkins & Ormerod, in press). Although drift varied between streams and periods of the field experiment, dippers had no additional significant effects (Fig. 6.5). In a laboratory stream, prey were observed directly during encounters with a model dipper which simulated foraging activities such as flight, swimming, bill contact with the prey, and stone turning. In these experiments, invertebrate families varied markedly in their response to the model dipper. Blackfly larvae, limnephilid and hydropsychid caddis were mostly immobile, and lacked effective escape behaviour of any type, consistent with frequent use of these prey by dippers in the wild (Fig. 6.6). Some families drifted (baetid mayflies, gammarid shrimps) or moved away (heptageniid and ephemerellid mayflies, perlid stoneflies) from the model, at least indicating some potential escape mechanisms. However, most individual prey deferred response until the point of physical contact from the model bird's bill (Fig. 6.6), which might be too late to avoid capture. This pattern explained the lack of gross drift responses in the field experiment, and contrasts strongly with normal behavioural responses shown by aquatic invertebrates to other riverine predators. A range of hypotheses has been advanced to explain this contrast, including for example the possibility that invertebrates do not perceive dippers as important predators, or that premature escape through drift carries a higher risk of predation by fish

(see Jenkins & Ormerod, in press). These hypotheses require further examination.

DIPPERS AS CONSUMERS: AN INFLUENCE ON BENTHIC COMMUNITIES?

Bottom-up effects by the distribution and abundance of prey on dippers are now well known. Along acidic streams, the reduced abundance of mayfly nymphs, caddis larvae and fish is accompanied by markedly reduced dipper density, and impaired breeding performance by comparison with circumneutral streams (Ormerod *et al.* 1991, Vickery 1991, 1992). Dipper territories also increase in length between about 0.3 and 2.0 km as prey become scarcer. Such a situation raises the possibility that prey densities could, under such circumstances, be limiting. Is it possible that top-down effects might also occur from dippers as consumers of prey?

Averaged throughout the year, a dipper in a temperate region such as Wales needs around 173–213 kJ of energy daily for basic activities (Bryant & Tatner 1988, O'Halloran *et al.* 1991). These values are adjusted to take account of average climatic daylength and climatic conditions (O'Halloran *et al.* 1991). Moult and feather renewal take about 1000 kJ, so that the annual energy requirements of a pair are around 153 000–158 000 kJ. The adults need to harvest a further 10 000 to 20 000 kJ to feed one or two broods of young. These total requirements, allowing also for assimilation efficiency of 70%,

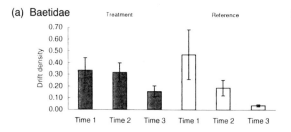

(a) Baetidae

(b) Leuctridae

(c) Heptageniidae

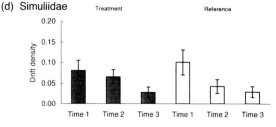

(d) Simuliidae

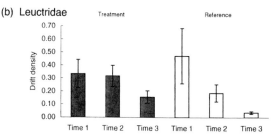

(e) All Taxa

Fig. 6.5. Changes in the density (n per m^3 of filtered volume) of aquatic invertebrates drifting before (Time 1), during (Time 2) and after (Time 3) the introduction of a live dipper to a treatment enclosure (shaded bars), showing also contemporaneous values in a reference enclosure (unshaded). The values are means ($\pm$ SE) from eight replicate streams. Dippers had no statistically significant effect on drift (after Jenkins & Ormerod in press).

are equivalent to around 10.5–11 kg dry weight of food per year. Detailed knowledge of how such energetic needs are met by different prey (Table 6.1) allows some calculation of likely cropping rates along Welsh streams (Ormerod & Tyler 1991). For example, a pair of dippers is liable to take up to 6.6 kg dry weight of caddis larvae annually, 3.7 kg of fish (mostly bullheads *Cottus gobio* L.), 3.2 kg of stonefly nymphs, and up to 1 kg of mayflies (Table 6.3). On territories of typical size, these values represent requirements up to 1.1, 0.8, 0.3 and 0.2 g·m^{-2}, respectively, which are potentially substantial relative to typical production (Table 6.3). For example, the annual production of bullheads along Welsh streams ranged from 0.08 to 0.46 g dry mass·m^{-2}, less than the total which dippers might crop (Ormerod & Tyler 1991). Production by hydropsychid caddis ranged from 0.32–1.18 g dry mass·m^{-2}, of which Ormerod & Tyler (1991) estimated that dippers could consume 11–24% even at the highest production values. However, calculations of this type are subject to well known uncertainties: they allow no real assessment of impacts on prey abundance or standing crop by dippers either alone, or more realistically alongside other predators (e.g. see Soluk 1993). So far, only

Harvey & Marti (1993) have attempted to verify these predicted losses experimentally, by short-term (18 day) exclusion of American dippers *Cinclus mexicanus* at the reach scale. For the large caddis *Dicosmoecus*, the effects of dipper predation were indeed substantial. These features raise the possibility that dippers might influence interactions between prey species, or that they might interact either additively or competitively with other riverine predators such as fish (e.g. see discussion by Harvey & Marti 1993).

Does this mean that dippers could have a keystone influence on benthic communities similar, for example, to that of sea otters? (See Estes & Palmisano 1974, Estes, this volume.) A clear outline of the role of any vertebrate predator in rivers, either through keystone effects on the competitive interaction between other organisms, or through cascading effects on the abundance of animals at lower trophic levels (Hairston *et al.* 1960, Carpenter *et al.* 1985) is elusive. In lakes (e.g. Bronmark *et al.* 1992), and sometimes in streams with simple communities (e.g. Power 1990, 1992), such cascading effects have been demonstrated, but more often they have not. Even if such effects were intense, there is strong evidence that the experimental tech-

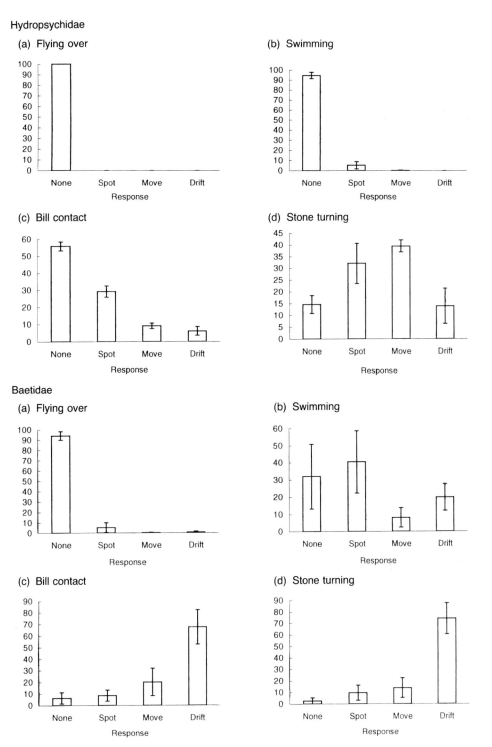

Fig. 6.6. Examples of the behavioural response of two invertebrate families to a model dipper simulating various foraging activities in a laboratory stream. The histograms show the mean numbers of animals ($\pm$ SD), from five sets of trials of n = 100, showing either no response, body movement on the spot, movement away, or swimming and drifting when encountering dippers simulating flight, swimming over the prey, bill contact, and turning stones in the substratum (after Jenkins & Ormerod in press).

Table 6.3. Estimated total annual exploitation of prey (g dry mass) by an average pair of dippers along acidic and circumneutral Welsh streams allowing for the production of one and two broods. The values in parentheses are the calculated consumption per m^2 of each prey type, assuming territory areas of $11\,250\,m^2$ and $4680\,m^2$ on acidic and circumneutral streams respectively (after Ormerod & Tyler 1991).

	Acidic	Circumneutral	Circumneutral (two broods)
Mollusca	0	287 (0.061)	287 (0.061)
Crustacea	0	149 (0.032)	149 (0.032)
Stoneflies	3 277 (0.291)	248 (0.053)	265 (0.052)
Mayflies	18 (0.001)	975 (0.208)	1 042 (0.222)
Caddis	6 606 (0.587)	4 868 (1.040)	5 203 (1.112)
True-flies	566 (0.050)	210 (0.045)	214 (0.046)
Fish	0	3 512 (0.751)	3 673 (0.785)
Other prey	0	69 (0.014)	149 (0.032)
Total	10 467 (0.930)	10 319 (2.205)	10 980 (2.346)

Table 6.4. A demonstration of omnivory in the dipper along Welsh streams. Each of the families listed from the different trophic groups occur commonly in the diet of dippers on acidic or neutral Welsh streams. The values in parentheses are geometric mean prey weights (mg) during the breeding season (largely after Ormerod & Tyler 1991).

Vertebrate predator	Dipper
Vertebrate predators	*Salmonidae* (fry of ca. 220), *Cottidae* (ca. 220)
Invertebrate predators	*Perlidae/Perlodidae* (3.0), *Rhyacophilidae* (6.3), *Polycentropodidae* (3.0), *Cordulegasteridae* (> 100)
Herbivore/detritivores (occasional predators)	*Limnephilidae* (26.0), *Hydropsychidae* (6.3)
Herbivores/detritivores	*Leuctridae/Nemouridae* (0.9), *Heptageniidae* (2.6), *Baetidae* (0.7), *Ephemerellidae* (0.4), *Simuliidae* (0.5), *Glossosomatidae* (3.6)

niques employed to look for them, often exclosures or enclosures, might be confounded by design difficulties: the patchiness of prey in rivers demands levels of replication seldom attempted, while drifting or crawling prey often replace those consumed during experiments (see Hildrew 1992, Sih & Wooster 1994). In addition, the ecological effects of predation in rivers are sometimes rendered unimportant in the longer term by disturbances such as large spates (Hildrew 1992). One other important factor is that vertebrates often feed at more than one trophic level simultaneously (i.e. they are omnivores). The latter features are known in terrestrial systems (e.g. Spiller & Schoener 1994), and in rivers (Hildrew 1992), so that obvious control on single trophic levels might not occur. Dippers feed omnivorously in this way, using prey across a wide range of body sizes (Table 6.4). Their river habitats are also characterized often by complex habitat structures in which prey refuges will be common. None of their prey interact strongly or obviously with primary producers in a way equivalent to urchins in the sea otter–urchin–kelp system described in this volume by Estes.

Apparently, therefore, there are contrasting reasons to surmise that dippers might or might not have an ecological role as keystone consumers. On the one hand, impacts on individual populations seem likely from estimates and experiments on production, but the full ecological effects are not clearly predictable from our available information. Only an experimental approach is liable to resolve more fully the predatory role of dippers in streams.

ACKNOWLEDGEMENTS

My thanks are due to those who have contributed to the primary data reviewed here. The editors and two referees commented on the manuscript.

REFERENCES

Bronmark, C., Klosiewski, S.P. & Stein, R.A. (1992) Indirect effects of predation in a freshwater benthic food chain. *Ecology*, **73**, 1662–1674.

Bryant, D.M. & Tatner, P. (1988) Energetics of the annual cycle of dippers (*Cinclus cinclus*). *Ibis*, **130**, 17–38.

Carpenter, S.R., Kitchell, J.F. & Hodgson, J.R. (1985) Cascading trophic interactions and lake productivity. *BioScience*, **35**, 634–639.

Dodson, S.I., Crowl, T.A., Peckarsky, B.L., Kats, L.B., Covich, A.P. & Culp, J.M. (1994) Non-visual communication in freshwater benthos: an overview. *Journal of the North American Benthological Society*, **13**, 268–282.

Ernst, M.R. & Stewart, K.W. (1986) Microdistribution of eight stonefly species (Plecoptera) in relation to organic matter in an Ozark foothill stream. *Aquatic Insects*, **8**, 237–254.

Estes, J.A. & Palmisano, J.F. (1974) Sea otters: their role in structuring nearshore communities. *Science*, **185**, 1058–1060.

Hairston, N., Smith, F.E. & Slobodkin, L. (1960) Community structure, population control and competition. *American Naturalist*, **94**, 421–425.

Harvey, B.C. & Marti, C.D. (1993) The impact of dipper *Cinclus mexicanus* predation on stream benthos. *Oikos*, **68**, 431–436.

Hildrew, A.G. (1992) Food webs and species interactions. *The Rivers Handbook Volume 1* (eds P. Calow & G.E. Petts), pp. 309–330, Blackwell Scientific, Oxford.

Hildrew, A.G., Townsend, C.R. & Henderson, J. (1980) Interactions between larval size, microdistribution and substrate in the stoneflies of an iron-rich stream. *Oikos*, **35**, 387–396.

Jeffries, M.J. & Lawton, J.H. (1984) Enemy-free space and the structure of ecological communities. *Biological Journal of the Linnean Society*, **23**, 269–286.

Jenkins, R.K.B. & Ormerod, S.J. (in press) The influence of a river bird, the dipper *Cinclus cinclus* on the behaviour and drift of its invertebrate prey. *Freshwater Biology*.

Kerfoot, W.C. & Sih, A. (1987) *Predation: Direct and Indirect Effects on Aquatic Communities*. University Press of New England, Hanover, New Hampshire.

Krementz, D.G. & Arkney, C.D. (1995) Changes in total body calcium and diet of breeding house sparrows. *Journal of Avian Biology*, **26**, 162–167.

O'Halloran, J., Gribbin, S.D., Tyler, S.J. & Ormerod, S.J. (1991) The ecology of dippers *Cinclus cinclus* in relation to stream acidity in upland Wales: time activity patterns and energy use. *Oecologia*, **85**, 271–280.

Ormerod, S.J., Boilstone, M.A. & Tyler, S.J. (1985) Factors influencing the abundance of breeding dippers *Cinclus cinclus* in the catchment of the River Wye, mid-Wales. *Ibis*, **127**, 332–340.

Ormerod, S.J., Efteland, S. & Gabrielsen, L. (1987) The diet of breeding dippers *C. cinclus cinclus* and their nestlings in south western Norway. *Holarctic Ecology*, **10**, 201–205.

Ormerod, S.J., O'Halloran, J., Gribbin, S.D. & Tyler, S.J. (1991) The ecology of dippers *Cinclus cinclus* in relation to stream acidity in upland Wales: breeding performance, calcium physiology and nestling growth. *Journal of Applied Ecology*, **28**, 419–433.

Ormerod, S.J. & Tyler, S.J. (1986) The diet of dippers *Cinclus cinclus* wintering in the catchment of the River Wye, Wales. *Bird Study*, **33**, 36–45.

Ormerod, S.J. & Tyler, S.J. (1990) Assessments of body condition in dippers *Cinclus cinclus*: potential pitfalls in the derivation and use of conditions indices. *Ringing and Migration*, **11**, 31–41.

Ormerod, S.J. & Tyler, S.J. (1991) Predatory exploitation by a river bird, the dipper *Cinclus cinclus* along acidic and circumneutral streams in upland Wales. *Freshwater Biology*, **25**, 105–116.

Ormerod, S.J. & Tyler, S.J. (1993a) Birds as indicators of changes in water quality. *Birds as Indicators of Environmental Change* (eds R.W. Furness & J.J.D. Greenwood), pp. 179–216. Chapman and Hall, London.

Ormerod, S.J. & Tyler, S.J. (1993b) The adaptive significance of brood size and time of breeding in the dipper *Cinclus cinclus* as seen from post-fledging survival. *Journal of Zoology*, **231**, 371–381.

Power, M.E. (1990) Effects of fish in river food webs. *Science*, **250**, 811–814.

Power, M.E. (1992) Habitat heterogeneity and the functional significance of fish in river food webs. *Ecology*, **73**, 1675–1688.

Scrimgeour, G.J. & Culp, J.M. (1994) Foraging and evading predators: the effects of predator species on a behavioural trade-off by a lotic mayfly. *Oikos*, **69**, 71–79.

Sih, A. (1987) Predators & prey lifestyles: an evolutionary and ecological overview. *Predation: Direct and Indirect Impacts on Aquatic Communities* (eds W.C. Kerfoot & S. Sih), pp. 201–225. University Press of New England. Hanover, New Hampshire.

Sih, A. & Wooster, D.E. (1994) Prey behaviour, prey dispersal and predator impacts on stream prey. *Ecology*, **75**, 1199–1207.

Smith, R.P. & Ormerod, S.J. (1986) The diet of moulting dippers in the catchment of the Welsh River Wye. *Bird Study*, **33**, 138–139.

Soluk, D.A. (1993) Multiple predator effects: predicting combined functional response of stream fish and invertebrate predators. *Ecology*, **74**, 219–225.

Spiller, D.A. & Schoener, T.W. (1994) Effects of top and intermediate predators in a terrestrial food web. *Ecology*, **75**, 182–196.

Spitznagel, A. (1985) Seasonal variation in food supply and food choice of the dipper *Cinclus cinclus*. *Okologie der Vogel*, **7**, 239–327.

Tyler, S.J. & Ormerod, S.J. (1994) *The Dippers*. Poyser, London.

Vickery, J.A. (1991) Breeding densities of dippers *Cinclus cinclus*, grey wagtails *Motacilla cinerea* and common sandpipers *Actitis hypoleucos* in relation to the acidity of streams in south west Scotland. *Ibis*, **133**, 178–185.

Vickery, J.A. (1992) The reproductive success of the dippers *Cinclus cinclus* in relation to the acidity of streams in south western Scotland. *Freshwater Biology*, **23**, 195–206.

Yoerg, S.I. (1994) Development of foraging behaviour in the Eurasian dippers *Cinclus cinclus* from fledging until dispersal. *Animal Behaviour*, **47**, 577–588.

CHAPTER 7

Between-year variations in the diet and behaviour of harbour seals *Phoca vitulina* in the Moray Firth; causes and consequences

P.M. Thompson, D.J. Tollit, S.P.R. Greenstreet, A. Mackay and H.M. Corpe

SUMMARY

(1) The relationship between changes in the winter diet, distribution and abundance of Moray Firth harbour seals and the abundance and distribution of their clupeid prey during the period 1987 and 1994 is assessed.

(2) Between-year differences in the seals' diet, haulout distribution and foraging locations were closely linked to between-year variation in the abundance and distribution of clupeid fish, and this had an effect on seal body condition.

(3) The consequences of these between-year variations in food availability are discussed in terms of their impact on the seals' population dynamics and in relation to assessments of the ecological impacts of these seal populations.

Key-words: clupeids, fisheries interaction, *Phoca vitulina*, spatial variation, temporal variation

INTRODUCTION

The abundance of marine pelagic fish stocks is known to vary widely between years. Consequently, if we are to develop models which can predict the impact of predators on these stocks, it is essential that we understand how predator populations respond to temporal changes in prey abundance.

In the case of central place foragers such as lactating otariid seals and nesting seabirds, between-year changes in food availability are known to affect behaviour and population dynamics, both of which will influence the population's energy requirements. Several species have been shown to make longer feeding trips in years of low prey abundance (e.g. Watanuki *et al.* 1993, Boyd *et al.* 1994, Uttley *et al.* 1994). Long absences from breeding sites and insufficient provisioning of young under these conditions have also resulted in high chick or pup mortality, and have even lead to the loss of complete year-classes (e.g. Trillmich & Limberger 1985, Hamer *et al.* 1993). For some tropical pinnipeds, extreme reductions in the level of up-welling of nutrient rich

water (i.e. El Niño situations) can also lead to high adult mortality, particularly amongst larger males, leading to skewed sex ratios (Trillmich & Ono 1991, Trillmich 1993). In most of these cases, however, predators depend upon one or two key prey species and their maximum foraging ranges are constrained by their need to return regularly to breeding sites to feed young or to mate. In contrast, many phocid seals have a more catholic diet and can fast throughout much of the breeding season, resulting in separation of their foraging and reproductive activity (Costa 1991). Consequently, behavioural responses of phocids to changes in food availability may be more flexible, resulting in less severe effects on their population dynamics. Nevertheless, a switch to less favoured prey items, or a change in activity levels, may still have energetic consequences which result in more subtle impacts at the population level. If so, such between-year changes in food availability must be taken into account when considering the role of phocids as predators in marine ecosystems.

Harbour seals *Phoca vitulina* are coastal phocids, whose diet has been shown to vary geographically

(Härkönen 1987, Pierce *et al.* 1990), seasonally (Brown & Mate 1983, Härkönen 1987, Pierce *et al.* 1991) and between years (Tollit & Thompson in Review). In this paper, we use data from a long-term study in north-east Scotland to explore the causes and consequences of between-year variations in the diet of this species. Initial studies in this area highlighted marked seasonal changes in diet composition (Pierce *et al.* 1991), and seals exhibited behavioural and dietary changes in response to a winter increase in the inshore abundance of the clupeids herring *Clupea harengus* and sprat *Sprattus sprattus* (Thompson *et al.* 1991). Subsequently, it has been shown that the importance of these clupeid prey differed between years (Tollit & Thompson in Review). In this paper, we determine the relationship between changes in winter diet and the distribution and abundance of the seals and their prey. We then explore the consequences of the observed changes in diet and behaviour, first in terms of their impact on the seals'

population dynamics and, secondly, in relation to assessments of the ecological impacts of seal populations.

METHODS

The study was carried out between 1987 and 1994 in the Moray Firth, north-east Scotland (57°40′N, 004°00′W). Harbour seals remain in this area throughout the year, hauling out to breed and to rest on intertidal sites within the Beauly, Cromarty, Dornoch and Inverness Firths (Fig. 7.1).

Estimates of diet composition were based on the analyses of prey hard parts from faecal samples collected at haul-out sites in the Beauly and Inverness Firths. Samples were collected throughout the year in 1988 and 1989, but winter sampling subsequently focused on the period November to February. Prey were identified to species using the characteristics of fish otoliths and

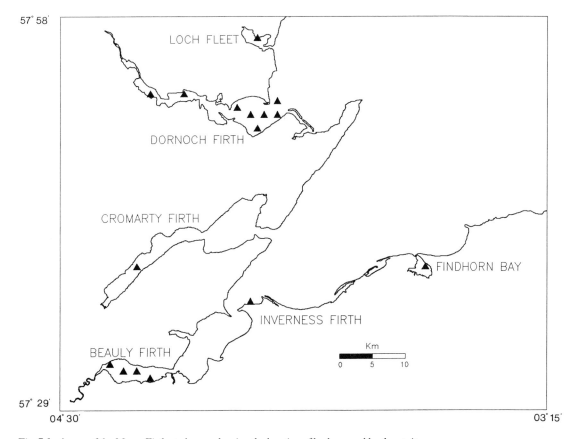

Fig. 7.1. A map of the Moray Firth study area showing the location of harbour seal haul-out sites.

cephalopod beaks. Measurements of the size of otoliths and beaks also provided estimates of prey weight. Full details of the analytical techniques used are presented in Pierce *et al.* (1991) and Tollit & Thompson (in Review). All estimates of diet composition presented in this paper are expressed as a percentage of total prey weight.

Key winter foraging areas were identified in two years of the study (1988/89 and 1989/90) by obtaining daily locations of VHF-tagged seals. Individual foraging ranges were estimated by carrying out harmonic mean home-range analysis on all locations which were more than 2 km from haul-out sites. The relative importance of each 1 km square within the study area was then determined by counting how many individual foraging ranges overlapped that square (P.M. Thompson, unpublished data).

Previous work on the relationship between haul-out and foraging distribution had suggested that counts made during January provided a good indicator of between-year differences in foraging distribution (P.M. Thompson, unpublished data). Trends in the seals' winter haul-out distribution were therefore assessed by counting seals at all haul-out sites in the study area on two to five occasions during January of each year. These data were then used to provide mean January counts for each of the firths and were combined to give a mean total for the whole Moray Firth. The relative importance of each firth in any particular year was then expressed as a proportion of the mean count for the whole Moray Firth.

Data on the condition of harbour seals from the Moray Firth were collected from all seals captured and released during the study. Information on capture and handling techniques is available in Thompson *et al.* (1992). The analyses in this paper were carried out only on animals caught in the spring or early summer (March–May), prior to pupping and the moult. Standard nose–tail length was recorded and seals were weighed using a suspended scale. A condition index was subsequently calculated following the technique used by Kruuk *et al.* (1987). Treating males and females separately, weight was first regressed against length. The condition index was then calculated by dividing observed weight by the predicted weight.

Quantitative data on the distribution and abundance of pelagic fish were available from Scottish Office Agriculture and Fisheries Department research cruises made in January 1992 and January 1994. Survey transects were steamed in the Moray Firth in a north–south direction, approximately 5 km apart, within an area south of latitude 58°15′ N and west of longitude 03°00′ W. In the inner firths, transects necessarily followed deep water channels. The biomass of pelagic

fish along each transect was determined using acoustic integration methods (MacLennan & Simmonds 1991) and pelagic trawl samples were obtained in areas where fish were abundant to determine species and size composition. The data were worked up for 5′ latitude by 5′ longitude rectangles and combined for all rectangles to provide an estimate of the total pelagic fish biomass in the study area.

Echo-sounding and trawl surveys were also made in the inner Moray Firth in January or February of 1988, 1989, 1990 and 1991. These earlier surveys provided no quantitative data on abundance, but descriptions of the number, size and density of shoals were used to provide an index of clupeid abundance in different years. Although these data provide only an extremely crude index with which to compare between-year differences in prey availability, they were carried out independently from analyses of the seals' behavioural and dietary data with which they were later compared.

RESULTS

Seal distribution and diet

Harbour seals in the Moray Firth exhibited two distinct foraging patterns in winter. In three winters (1987/88, 1988/89 and 1993/94) a high percentage (>60%) of the diet consisted of clupeids (Fig. 7.2), and over 40% of the population used haul-out sites in the Beauly and Cromarty Firths (Fig. 7.3). In the remaining winters (1989/90, 1990/91, 1991/92 and 1992/93), less than 30% of the population used the Beauly and Cromarty Firth sites whilst numbers using a haul-out site at the mouth of the Inverness Firth increased (Fig. 7.3).

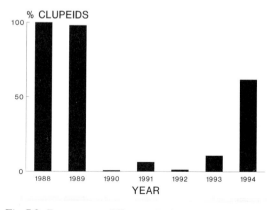

Fig. 7.2. Between-year differences in the percentage (by weight) of clupeids in the winter diet of harbour seals in the Beauly and Inverness Firths.

Inverness Firth

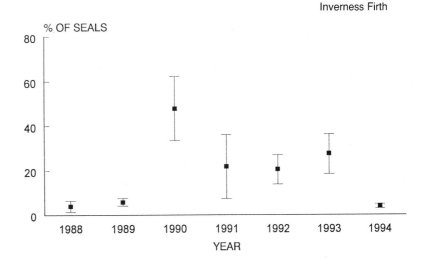

Cromarty and Beauly Firths

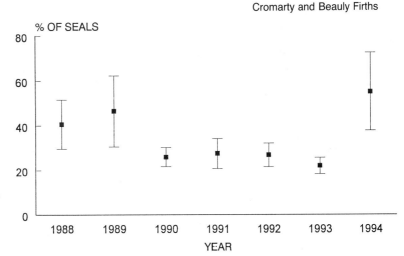

Fig. 7.3. Between-year differences in the terrestrial distribution of Moray Firth harbour seals. Data for each site or group of sites are presented as the mean percentage (±1 SE) of the Moray Firth total for January of each year.

Clupeids were virtually absent from the diet (<11%), with sandeels and gadids being the most important prey in these years. The importance of the Beauly and Cromarty Firth haul-out sites and the percentage of clupeids in the seals' winter diet was positively correlated (Spearman Rank Correlation, 5 df, r_s = 0.679, p < 0.05).

Because of between-year changes in the seals' use of haul-out sites, it is possible that between-year differences in diet could be confused with local geographical variation. Data from the Beauly and Inverness Firths were therefore examined separately. In years

when the overall importance of clupeids was low, the clupeid fraction of the diet declined at both sites but, at all times, clupeids were less important at the Inverness site (Fig. 7.4). The importance of alternative prey also differed between sites, with gadids predominating in the Beauly Firth and sandeels in the Inverness Firth (Fig. 7.4).

Data on foraging distribution were available in one winter when clupeid prey predominated (1988/89) and one year when they were virtually absent (1989/90). Seals foraged much further out to sea during the winter in which they were not feeding on clupeids (Fig. 7.5),

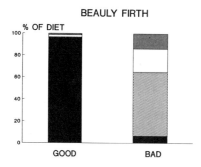

Fig. 7.4. Local geographical variation in the winter diet of Moray Firth harbour seals in 'good' years when clupeids predominated in the overall diet (1987/88, 1988/89 & 1993/94) and in 'bad' years when clupeids were virtually absent (1989/90, 1990/91, 1991/92 & 1992/93).

and the increased use of the Inverness Firth haul-out site in 1989/90 reflected the fact that this site was closer to their foraging areas.

Relationships with prey distribution and abundance

Fish surveys conducted in January 1992 and January 1994 suggest that both the abundance and distribution of clupeids differed between these two winters (Table 7.1). Estimates of sprat abundance in 1994 were two orders of magnitude higher than those for 1992. In 1994, sprat were also found much further into the inner firths in comparison to 1992 (Fig. 7.6). Differences in the abundance and distribution of herring were less marked, with an estimated biomass of 397 tonnes in 1992 and 1202 tonnes in 1994 (Fig. 7.6).

In these two years, high clupeid abundance was reflected by an increase in the importance of clupeids in the diet of Moray Firth seals. Our index of clupeid abundance in these and other years in the study ranged between 2 and 14 (Table 7.1). Sample sizes remain small, but this index of abundance was significantly higher in the three winters when clupeids formed a high proportion of the seals' diet (Mann–Whitney $U_{3,3}$ p < 0.05). In subsequent analyses and discussion, 1987/88, 1988/89 and 1993/94 are regarded as 'good' clupeid years and 1989/90–1992/93 as 'bad' clupeid years. The index of clupeid abundance was also significantly correlated with the proportion of seals which were found in the Beauly and Cromarty Firths in each of these years (Spearman Rank Correlation, 4 df, r_s = 0.77, p < 0.05).

Variations in seal body condition

82 females and 64 males were captured and measured during March, April or May of 1988–1994. There was a good relationship between weight and length for both females (r^2 = 0.92, n = 82) and males (r^2 = 0.94, n = 64), and comparison of fitted and measured weights for each individual yielded condition indices which ranged from 0.65 to 1.6. After combining the data for both sexes, a significant difference in the condition index of seals caught in the spring following 'good' and 'bad' clupeid years was apparent (Mann–Whitney $U_{68,78}$, p < 0.01) (Fig. 7.7).

DISCUSSION

Causes

These data illustrate that between-year differences in the winter diet of the Moray Firth harbour seal population are related to the distribution of seals in both their resting and foraging areas. Furthermore, these data suggest that the observed changes in seal diet and behaviour result from variations in the abundance of overwintering clupeids in the inner Moray Firth. This study therefore lends further support to suggestions that these clupeid stocks provide an important prey resource for wintering seals and seabirds in the Moray Firth (Aspinall & Dennis 1988, Thompson *et al.* 1991). In the 1960s and 1970s, the inner Moray Firth also supported a commercial fishery on sprat and herring, with annual landings exceeding 60 000 tonnes in some years (Hopkins 1986). This fishery was closed in the 1980s in order to protect juvenile herring stocks. Nevertheless, landings from this period do illustrate the extreme between-year variability in the size of this

(a)
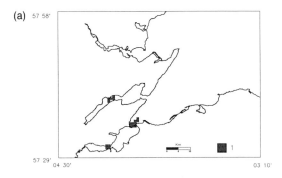

Fig. 7.5. The foraging distribution of radio-tagged seals (a) in winter 1988/89 and (b) 1989/90.

(b)
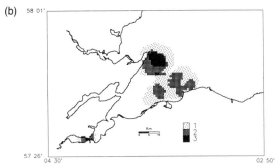

Table 7.1. Estimates of clupeid biomass and indices of clupeid abundance based on fishery surveys in the Moray Firth (1987–1994).

Year	Clupeid index	Clupeid biomass (tonnes)
1987/88	6	No data
1988/89	14	No data
1989/90	4	No data
1990/91	3	No data
1991/92	2	1275
1992/93	No data	No data
1993/94	12	8240

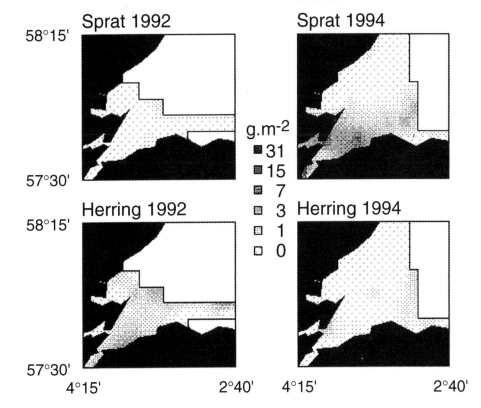

Fig. 7.6. The estimated distribution of pelagic schooling fish in the Moray Firth in January 1992 and in January 1994.

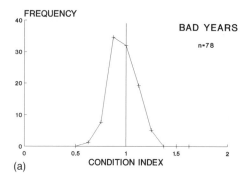

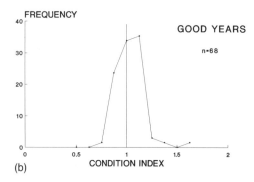

Fig. 7.7. Frequency distributions of seal condition indices (a) after winters when clupeids were virtually absent from the diet and (b) after winters in which clupeids were predominant in the diet.

resource (Fig. 7.8), albeit over a much higher range of values than recorded in our quantitative surveys in 1992 and 1994 (Table 7.1). The overwintering herring in the inner Moray Firth appear to originate from the west coast of Scotland, whilst the sprat probably spawn in deep water to the west of Orkney (Hopkins 1986). The size of these wintering stocks may therefore depend on variations in the currents which transport larvae around the coast, potentially affecting either recruitment or the advection of larvae into the Moray Firth. Although we cannot yet determine the extent to which such environmental factors might affect variations in clupeid abundance, this does serve to illustrate the potential involvement of large-scale oceanographic and climatic cycles in this particular predator–prey system.

Consequences

That seals travel further to feed, and switch to alternative prey, when clupeids are absent from inshore areas is perhaps of no great surprise. Of more interest, is the question of whether or not this change in foraging behaviour is of any consequence to individual seals. Our preliminary results on variations in body condition (Fig. 7.7) suggest that there were indeed energetic costs to individuals in those years when clupeids were less abundant. Such a difference may have been related to differences in the calorific value of clupeids and alternative prey (Hislop *et al.* 1991), or it may have resulted from the increased foraging costs associated with the use of more distant feeding areas (Fig. 7.5).

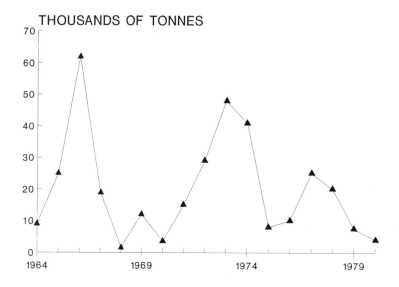

Fig. 7.8. Between-year variation in the landings of sprat in the Moray Firth (1964–1980). Data are from Hopkins (1986).

In common with other long-lived vertebrates, first-year survival is relatively low for harbour seals (Reijnders 1978, Boulva & McLaren 1979, Bigg 1981). The impact of variation in food availability may be greatest in young seals as they learn to forage over their first winter. In addition, harbour seals show marked reductions in body condition during the summer breeding and moult periods (Drescher 1979, Pitcher 1986). Animals in better condition in the spring following a good clupeid winter may have higher chances of survival over the relatively stressful summer period.

The extent to which breeding females allocate resources to reproduction may also be affected by adult body condition in spring. Female harbour seals appear to forage during the latter third of lactation (Boness *et al.* 1994) and the amount of feeding is related to female body size (Thompson *et al.* 1994). Variation in condition during the spring may therefore affect average birth weights or lead to variation in maternal investment during lactation, both of which may result in between-year variations in pre- and post-weaning survival. Resulting between-year variations in survival rates, and differences in year-class strength are similar in nature, but milder in impact, to those seen in tropical otariids (Trillmich & Ono 1991). Nevertheless, such variations, together with factors such as epizootics (Thompson & Hall 1993), suggest that age-structure of temperate populations of phocids may vary in time.

Implications for assessments of seal–fisheries interactions

Estimates of the impact of pinniped populations upon fish stocks require information on the seal population's energy requirements, the diet composition, and the abundance and structure of the prey stocks (Lavigne *et al.* 1982, Harwood & Croxall 1988). This study demonstrates that all three of these parameters can vary considerably between years, highlighting the need to take inter-annual variability into account when assessing the potential impact of seals on fish stocks.

Firstly, between-year variation in body condition and activity may cause population energy requirements to vary. Variation in body condition may result in changes in the age-structure of the population, but previous studies suggest that differences in population age-structure have relatively little impact on population energy requirements (Hiby & Harwood 1985, Härkönen & Heide-Jørgensen 1991). In contrast, current models have not explored the effect of changes in activity costs on either individual or population energy requirements. Indeed, activity costs have usually only been estimated crudely due to a lack of empirical data on the activity

budgets of most species (e.g. Lavigne *et al.* 1985, Härkönen & Heide-Jørgensen 1991, Olesiuk 1993). The increased foraging range shown by seals during years of low clupeid abundance suggests that the relationship between prey availability and the seals' energy requirements should be explored further. On the other hand, while individual activity costs may have increased in bad clupeid years, differences in the condition of seals suggest that growth costs may have decreased in bad years. Although these data (Fig. 7.7) refer to variation in weight, related data indicate that the body lengths of yearling seals were also significantly greater after a good clupeid year (H.M. Corpe, unpublished data). Whether this is a result of variations in prey quality or in the redirection of energy from activity to growth remains uncertain. Nevertheless, the implications of such variations in growth rate for overall population energy requirements clearly require further exploration. The potential for more subtle effects, for example through the influence which growth rates may have on the average age of sexual maturity (Laws 1956), should also be considered.

Secondly, our data on variation in diet composition illustrate that, whatever the energy requirements of a population, its predatory impact on a particular prey species may differ markedly from year to year (Figs 7.2 and 7.4). The relationship between winter diet and clupeid abundance further illustrates that the seals' impact upon a particular prey species may result from changes in the abundance either of that species or of other potential prey species. Consequently, accurate predictions concerning the impact of seals on particular prey stocks will require accurate predictions of the relative abundance of all potential prey.

ACKNOWLEDGEMENTS

We would like to thank the many colleagues who have assisted on this project over the years. The work has been supported by a series of research contracts to Professor Paul Racey and Dr Paul Thompson from the Scottish Office Agriculture and Fisheries Department. We thank two anonymous referees for their comments.

REFERENCES

Aspinall, S.J. & Dennis, R.H. (1988) Goosanders and red-breasted mergansers in the Moray Firth. *Scottish Birds*, **15**, 65–70.

Bigg, M.A. (1981) Harbour seal, *Phoca vitulina* and *P. largha*. *Handbook of Marine Mammals* (eds S.H. Ridgway & R.J. Harrison), pp. 1–28. Academic Press, New York.

Boness, D.J., Bowen, W.D. & Oftedal, O.T. (1994) Evidence of a maternal foraging cycle resembling that of otariid seals in a small phocid, the harbor seal. *Behavioural Ecology and Sociobiology*, **34**, 95–104.

Boulva, J. & McLaren, I.A. (1979) Biology of the harbor seal, *Phoca vitulina*, in Eastern Canada. *Bulletin of the Fisheries Research Board of Canada*, **200**, 1–24.

Boyd, I.L., Arnould, J.P.Y., Barton, T. & Croxall, J.P. (1994) Foraging behaviour of Antarctic fur seals during periods of contrasting prey abundance. *Journal of Animal Ecology*, **63**, 703–713.

Brown, R.F. & Mate, B.R. (1983) Abundance movements and feeding habits of harbor seals (*Phoca vitulina*) at Netarts and Tillamook Bays, Oregon. *Fishery Bulletin*, **81**, 291–301.

Costa, D.P. (1991) Reproductive and foraging energetics of pinnipeds: implications for life history patterns. *Behaviour of Pinnipeds* (ed. D. Renouf), pp. 300–344. Chapman and Hall, London.

Drescher, H.E. (1979) Biology, ecology and conservation of harbour seals in the tidelines of Schleswig-Holstein. *Beitrage zur Wildbiologie*, **1**, 1–73.

Hamer, K.C., Monaghan, P., Uttley, J.D., Walton, P. & Burns, M.D. (1993) The influence of food supply on the breeding ecology of kittiwakes *Rissa tridactyla* in Shetland. *Ibis*, **135**, 255–263.

Härkönen, T. (1987) Seasonal and regional variations in the feeding habits of the harbour seal, *Phoca vitulina*, in the Skagerrak and the Kattegat. *Journal of Zoology*, **213**, 535–543.

Härkönen, T. & Heide-Jørgensen, M.-P. (1991) The harbour seal *Phoca vitulina* as a predator in the Skagerrak. *Ophelia*, **34**, 191–207.

Harwood, J. & Croxall, J.P. (1988) The assessment of competition between seals and commercial fisheries in the North Sea and the Antarctic. *Marine Mammal Science*, **4**, 13–33.

Hiby, A.R. & Harwood, J. (1985) The effects of variations in population parameters on the energy requirements of a hypothetical grey seal population. *Interactions Between Marine Mammals and Fisheries* (eds J.R. Beddington, R.J.H. Beverton & D.M. Lavigne), pp. 337–343. G. Allen and Unwin, London.

Hislop, J.R.G., Harris, M.P. & Smith, J.G.M. (1991) Variation in the calorific value and total energy content of the lesser sandeel (*Ammodytes marinus*) and other fish preyed on by seabirds. *Journal of Zoology*, **224**, 501–517.

Hopkins, P.J. (1986) Exploited fish and shellfish populations in the Moray Firth. *Proceedings of the Royal Society of Edinburgh*, **91B**, 57–72.

Kruuk, H., Conroy, J.H.W. & Moorhouse, A. (1987) Seasonal reproduction, mortality and food of otters (*Lutra lutra* L.) in Shetland. *Symposium of the Zoological Society of London*, **58**, 263–278.

Lavigne, D.M., Barchard, W., Innes, S. & Oritsland, N.A. (1982) Pinniped bioenergetics. *Mammals in the Seas*, pp. 191–235. FAO, Rome.

Lavigne, D.M., Innes, S., Stewart, R.E.A. & Worthy, G.A.J. (1985) An annual energy budget for north-west Atlantic harp seals. *Marine Mammals and Fisheries* (eds J.R. Beddington, R.J.H. Beverton & D.M. Lavigne), pp. 319–336. G. Allen and Unwin, London.

Laws, R.M. (1956) Growth and sexual maturity in aquatic mammals. *Nature*, **179**, 193–194.

MacLennan, D.N. & Simmonds, E.J. (1991) *Fisheries Acoustics*. Chapman and Hall, London.

Olesiuk, P.F. (1993) Annual prey consumption by harbor seals (*Phoca vitulina*) in the Strait of Georgia, British Columbia. *Fishery Bulletin*, **91**, 491–515.

Pierce, G.J., Boyle, P.R. & Thompson, P.M. (1990) Diet selection in seals. *Trophic Relationships in the Marine Environment* (eds M. Barnes & R.N. Gibson), pp. 222–238. Aberdeen University Press, Aberdeen.

Pierce, G.J., Thompson, P.M., Miller, A., Diack, J.S.W., Miller, D. & Boyle, P.R. (1991) Seasonal variation in the diet of common seals (*Phoca vitulina*) in the Moray Firth area of Scotland. *Journal of Zoology*, **223**, 641–652.

Pitcher, K.W. (1986) Variation in blubber thickness of harbor seals in southern Alaska. *Journal of Wildlife Management*, **50**, 463–466.

Reijnders, P.J.H. (1978) Recruitment in the harbour seal (*Phoca vitulina*) population in the Dutch Waddensea. *Netherlands Journal of Sea Research*, **12**, 164–179.

Thompson, P.M. & Hall, A.J. (1993) Seals and epizootics – what factors might affect the severity of mass mortalities? *Mammal Review*, **23**, 149–154.

Thompson, P.M., Pierce, G.J., Hislop, J.R.G., Miller, D. & Diack, J.S.W. (1991) Winter foraging by common seals (*Phoca vitulina*) in relation to food availability in the inner Moray Firth, N.E. Scotland. *Journal of Animal Ecology*, **60**, 283–294.

Thompson, P.M., Cornwell, H.J.C., Ross, H.M. & Miller, D. (1992) Serologic study of phocine distemper in a population of harbor seals in Scotland. *Journal of Wildlife Diseases*, **28**, 21–27.

Thompson, P.M., Miller, D., Cooper, R. & Hammond, P.S. (1994) Changes in the distribution and activity of female harbour seals during the breeding season: implications for their lactation strategy and mating patterns. *Journal of Animal Ecology*, **63**, 24–30.

Tollit, D.J. & Thompson, P.M. (in Review) Seasonal and between year variation in the diet of harbour seals *Phoca vitulina* in the Moray Firth, Scotland. *Canadian Journal of Zoology*.

Trillmich, F. (1993) Influence of rare ecological events on pinniped social structure and population dynamics. *Symposium of the Zoological Society of London*, **66**, 95–114.

Trillmich, F. & Limberger, D. (1985) Drastic effects of El Niño on Galapagos pinnipeds. *Oecologia*, **67**, 19–22.

Trillmich, F. & Ono, K.A. (1991) *Pinnipeds and El Niño*. Springer-Verlag, Berlin.

Uttley, J.D., Walton, P., Monaghan, P. & Austin, G. (1994) The effects of food abundance on breeding performance and adult time budgets of guillemots. *Uria aalge. Ibis*, **136**, 205–213.

Watanuki, Y., Kato, A., Mori, Y. & Naito, Y. (1993) Diving performance of Adelie penguins in relation to food availability in fast sea-ice areas: comparison between years. *Journal of Animal Ecology*, **62**, 634–646.

CHAPTER 8

The daily food requirements of fish-eating birds: getting the sums right

M.J. Feltham and J.M. Davies

SUMMARY

(1) Estimates of damage to fisheries caused by fish-eating birds are very sensitive to assumptions about their daily food requirements (DFR).

(2) We reviewed 18 studies that reported on the DFR of cormorants *Phalacrocorax carbo* and found that estimates varied between 100 and 963 $g \cdot day^{-1}$, resulting in up to a ten-fold difference in damage estimates derived from these data.

(3) The racial identity of birds and biases in the methods used to estimate food intake both affected estimates of DFR, with estimates derived from pellets producing values up to 37% less than other methods.

(4) Measuring the daily energy expenditure of birds in the field using doubly-labelled water (DLW) provides a more accurate way of estimating intake. Data from one such study on goosanders *Mergus merganser* are presented and an allometric equation is subsequently used to estimate the DFR of cormorants, for which no DLW data are currently available.

(5) This method provided estimates of cormorant DFR ranging from 692 to 881 $g \cdot day^{-1}$, depending on age and racial identity.

Key-words: cormorant, daily food requirements, doubly-labelled water, goosander, *Mergus merganser*, *Phalacrocorax carbo*

INTRODUCTION

Cormorants *Phalacrocorax carbo* and goosanders *Mergus merganser* eat almost entirely fish and their presence at British inland fisheries has brought them into conflict with anglers and fisheries managers (Davies & Feltham 1994, Marquiss & Carss 1994). Although both species are fully protected under the 1981 Wildlife and Countryside Act, the Ministry of Agriculture, Fisheries and Food (MAFF), the Scottish Office Agriculture and Fisheries Department (SOAFD) and the Welsh Office Agriculture Department (WOAD) can, under section 16 of the Act, issue licences to shoot these birds '*for the purposes of preventing serious damage . . . to fisheries*'. The extent to which cormorants may inflict such losses is currently being studied on the river Ribble, north-west Lancashire, but estimates have been found to be highly sensitive to assumptions about the daily food requirements (DFR) of birds (Davies & Feltham 1994). In this paper, we briefly review 18 published studies that have estimated the DFR of cormorants and present a predictive equation based on recent doubly-labelled water (DLW) studies to enable fisheries' managers and anglers to produce their own estimates of DFR with a view to improving future estimates of damage to their fisheries.

Data from the published studies we reviewed were estimated in several different ways: (i) from the analysis of regurgitated pellets; (ii) from the analysis of regurgitated fish and stomach contents; (iii) from measurements of the food requirements of captive adults or captive chicks; and (iv) from a variety of energy demand calculations. These studies highlighted two problems: (i) DFR estimates were extremely variable, between $100\,g\cdot bird^{-1}\cdot day^{-1}$ (Reicholf 1990) and $963\,g\cdot bird^{-1}\cdot day^{-1}$ (derived from data in van Dobben 1952); and (ii) they were likely to be too low. On the basis of these estimates, if birds on the Ribble met their daily food requirements exclusively on the river (assuming a standing crop of $250\,kg\cdot ha^{-1}$) (M. Diamond, personal communication), variation in DFR estimates alone would mean that cormorants could, in theory, be taking anywhere between 4.2% and 41.1% of the standing crop from this river each year – a tenfold difference!

VARIATION IN ESTIMATES OF DFR

There were two reasons for the observed variation in published DFR estimates: (i) the racial identity of the birds on which the data were collected; and (ii) the method used to estimate DFR. Estimates of DFR for cormorants varied between 5.5% and 41.0% of their body mass. Adults of the north-western European race *P. c. carbo* Linnaeus may, however, be up to 37.5% heavier than the southern and central European race, *P. c. sinensis* Blumenbach (Cramp & Simmons 1977). Mean ($\pm$SE) intake estimates for *P. c. carbo* were, therefore, not surprisingly 26% greater than for *P. c. sinensis* ($498 \pm 86\,g$ (n = 5 studies) and $395 \pm 40\,g$ (n = 17 studies), respectively) (references in Table 8.1).

The different methods used to estimate DFR also produced variable results (means shown in Table 8.1). In particular, estimates derived from pellets were on average 32–37% lower than estimates obtained using other methods, lending some support to previous claims about the biases inherent in data derived from such material (Duffy & Laurensen 1983, Johnstone *et al.* 1990, Zijlstra, in press).

CURRENT ESTIMATES OF DFR MAY BE TOO LOW

The use of stomach contents and/or regurgitates to estimate DFR is likely to cause underestimates for one or more of the following reasons: (i) stomachs are rarely 'full'; (ii) regurgitates are often incomplete; (iii) the original weight of partially digested food is not always estimated; (iv) the extent to which single regurgitates or 'full' stomachs represent the entire DFRs of adult birds is unclear (Duffy & Jackson 1986, Marquiss & Carss 1994, Feltham & Davies 1995). DFR estimates from

Table 8.1. The mean daily food requirements (DFR) of cormorants *Phalacrocorax crabo*: data from 20 estimates from 18 published studies (sources shown below).

Method used to estimate DFR	Range of mean intake values (g)	Overall mean intake value (g)	Number of studies
Captive/chick rearing	319–601	460	n = 3
Pellets	227–373	300	n = 8
Stomach samples/ regurgitates	420–533	477	n = 6
Energy demand calculations	366–516	441	n = 3

Sources: *P. c. sinensis* – van Dobben 1952 (three estimates); Gremillet & Plos 1994; Heinroth 1927 (two estimates cited in van Dobben 1952); Keller in press; Madsen & Sparck 1950; Marteijn & Dirksen 1991; Martyniak *et al.* in press; Mellin *et al.* in press; Noordhuis *et al.* in press; Platteeuw 1991; Reicholf 1990; Sato *et al.* 1988 (two estimates); Voslamber 1988. (Two studies did not specify method of estimation and so are excluded from the above table.)
P. c. carbo – Hartley 1948; Kennedy & Greer 1988; McIntosh 1978; Mills 1965; Rae 1969.

captive birds or chicks are likely to underestimate those of wild adult birds because they rarely experience the diet, environmental conditions, and particularly the energy budgets, of wild adult birds.

Energy studies of cormorants have either (i) predicted basal metabolic rate (BMR) from allometric equations relating it to body mass and assumed daily energy expenditure (DEE) to be $3 \times$ BMR (e.g. Platteeuw 1991) or, (ii) applied theoretical energetic costs to time-budget data (e.g. Sato *et al.* 1988). These methods may also substantially underestimate DFR for several reasons. Firstly, empirical data on the BMR (and costs of flight, diving, etc.) of birds >1000 g are relatively few, so that the exponents of equations usually used for such estimates (e.g. Lasiewski & Dawson 1967, Aschoff & Pohl 1970, Kendeigh *et al.* 1977, Walsberg 1983) are predominately based on small birds for which the majority of data exist. Empirical measurements of BMR in both cormorants and goosanders are, however, rather greater than predictions from these equations (Sato *et al.* 1988, Feltham 1995). Secondly, the use of the $3 \times$ BMR figure to derive DEE (and hence DFR) is itself an assumption about *possible* food intake derived from a limited number of feeding studies (e.g. Ebbinge *et al.* 1975). Thirdly, the energetically expensive flight component in very mobile birds such as cormorants is often under-represented in time-budgets (see e.g. Martin & Bateson (1993) focal animal sampling).

USING DOUBLY-LABELLED WATER (DLW) TO IMPROVE ESTIMATES OF DFR

The use of stable isotopes to measure CO_2 production in free-ranging birds overcomes many of the problems mentioned previously and provides by far the most accurate way of estimating DEE, and hence daily food requirements (DFR). The doubly-labelled water (DLW) technique (Lifson *et al.* 1955, Lifson & McClintock 1966) is now becoming an increasingly common tool in studying energy expenditure of free-living animals. There have been at least 34 validation studies, in species ranging in size from scorpions to humans, and the method, applied to vertebrates, typically provides estimates of energy expenditure accurate to within 5% (Nagy 1987, Speakman & Racey 1988). Whilst no study has yet used DLW to estimate the daily food requirement of cormorants, recent work on goosanders *Mergus merganser* highlights its advantages over previous methods (see Feltham (1995) for full methodology).

Using this method the mean ($\pm$ SE) DEE of nine captive goosanders released on two Scottish rivers was $1939 \pm 184 \, \text{kJ} \cdot \text{day}^{-1}$ and estimated DFRs of males and females were 522 g and 480 g, respectively (Table 8.2). This was considerably greater, by a factor of 1.5, than previous estimates which assumed DEE to be $3 \times$ BMR (Carter & Evans 1986). Annual predation of smolts by goosanders during this study was therefore estimated to be between 8000–15 000 fish or 3–16% of annual production (Feltham 1995).

HOW MUCH DO CORMORANTS EAT EACH DAY?

As the DLW method has none of the biases associated with predicting DFR from prey remains (e.g. regurgitates and pellets) and provides empirical data on the DEE of free-living birds as opposed to estimating it from captives or assumptions about BMR, a study applying it to cormorants would be invaluable. As no such study has been published we derived a regression of $\log_{10}$ daily food requirement (g) on $\log_{10}$ body mass (g) from eight other DLW studies on cold water piscivores that use flapping flight (from Birt-Friesen *et al.* 1989) and the goosander study mentioned above as an aid to help fisheries' managers and anglers to get the sums right. Estimates for adult and juvenile cormorants were then predicted from body mass using data in Cramp & Simmons (1977) for *P. c. sinensis* and data from birds shot under licence in Britain (n = 74) for *P. c. carbo* (source MAFF) (Table 8.3). Estimates assume a nitrogen corrected metabolizable energy coef-

Table 8.2. Estimated DFR of male and female goosanders *Mergus merganser* from doubly-labelled water measurements of field metabolic rate (FMR). Calculations for males and females were based on the mean body masses of 37 wild goosanders shot during the study multiplied by the predicted FMR in $\text{mlCO}_2 \cdot \text{g}^{-1} \cdot \text{h}^{-1}$ from nine captives released on two Scottish rivers. SEs for estimates of intake are not shown as they reflect only variation in bird mass (data from Feltham 1995).

	Males (n = 21)	Females (n = 16)
Mean ($\pm$ SE) body mass (g)	1620 ± 24	1215 ± 43
DEE $(\text{kJ} \cdot \text{day}^{-1})^a$	2037	1867
Energy intake $(\text{kJ} \cdot \text{day}^{-1})^b$	2444	2243
DFR $(\text{g} \cdot \text{day}^{-1})^c$	522	480
DFR (% body mass)	32	40
Smolt intake $(\text{g})^d$	123.5	113.3
Parr intake $(\text{g})^e$	218.5	200.5
Estimated number of salmon smolts eaten per day^f	11	10
Estimated number of salmon parr eaten per day^g	52	48

^{a} Assumes male and female FMR are 2.087 & $2.553 \, \text{mlCO}_2 \cdot \text{g}^{-1} \cdot \text{h}^{-1}$ respectively, based on FMR = 3.950 − 0.00115 mass (g) from nine captive goosanders (Feltham 1995).
^{b} DEE multiplied by 1.2. Assumes assimilation efficiency for fish diet of 83%.
^{c} Assumes calorific values of $4.18 \, \text{kJ} \cdot \text{g}^{-1}$ wet mass for smolts, $5.02 \, \text{kJ} \cdot \text{g}^{-1}$ wet mass for parr and $4.6 \, \text{kJ} \cdot \text{g}^{-1}$ wet mass for other species.
^{d} Smolt intake (g) = [FMR $\times$ bird mass (g) $\times$ 0.0251 $(\text{kJ} \cdot \text{mlCO}_2^{-1}) \times 24$ (h) $\times 1.2 \times 0.66 \times 0.32]/4.18$ $(\text{kJ} \cdot \text{g}^{-1})$.
^{e} Parr intake (g) = [FMR $\times$ bird mass (g) $\times$ 0.0251 $(\text{kJ} \cdot \text{mlCO}_2^{-1}) \times 24$ (h) $\times 1.2 \times 0.66 \times 0.68]/5.02$ $(\text{kJ} \cdot \text{g}^{-1})$. (Equations d and e assume a proportion of salmon in the diet of 66%, of which 32% consists of smolts and 68% of parr; from Feltham 1995.)
^{f} Smolt intake divided by the mean mass of smolts eaten by goosanders = 11.3 g.
^{g} Parr intake divided by the mean mass of parr eaten by goosanders = 4.2 g.

ficient of 77.65%, the mean of empirical measurements on double crested cormorants *P. auritus* (Brugger 1993) and a calorific value for prey items of $5.42 \, \text{kJ} \cdot \text{g}^{-1}$ wet mass (Cooper 1978, Gardiner & Geddes 1980, Brugger 1993). The allometric equation so derived was:

$$\text{DFR} = 2.7644 \, \text{Body Mass}^{0.7229}$$

(SE $\log_{10}a$ = 0.0538, SE $\log_{10}b$ = 0.0120) (r^2 = 95.3, $P < 0.0001$). Using this equation DFR estimates based on DLW studies were almost twice as great as previous

Table 8.3. The mean daily food requirements (DFR) of cormorants estimated from a predictive equation based on nine doubly-labelled water studies relating DEE to body mass. Species incuded in the $log_{10}-log_{10}$ regression of DFR on body mass were *Pelacanoides georgicus* (109 g), *P. urinatrix* (137 g), *Sula bassanus* (3210 g), *Rissa tridactyla* (386 g), *Uria aalge* (940 g), *U. lomvia* (834 g), *Cepphus grylle* (420 g), *Aethia pusilla* (83 g), *Mergus merganser* (1393 g). Data as quoted in Birt-Freisen *et al.* (1989) and Feltham (1995). Sample size refers to the number of birds on which the assumed mean body mass is based. Body mass data for *P. c. sinensis* are from Cramp & Simmons (1977) and for *P. c. carbo* from British birds shot under licence (source MAFF).

Race	Age	Mean DFR (g)	Mean body mass assumed (g)	Sample size (n)
P. c. carbo	Adult	881	2901	32
P. c. carbo	Juv/imm	826	2657	42
P. c. sinensis	Adult	739	2275	122
P. c. sinensis	Juv/imm	692	2079	10

estimates (Table 8.1) and DFRs of *P. c. carbo* adults were, on average, 20% greater than those of *P. c. sinensis* adults (Table 8.3).

Clearly, predicting DFR from the above equation is sensitive to the value chosen for body mass. This choice is not, moreover, a trivial one, especially as there are clear differences between races, sex, ages of birds and even season. Until empirical DLW measurements are available on cormorants, fisheries' managers and anglers should wherever possible, therefore, base estimates of damage to their fisheries on DFR figures derived from the above equation using the body masses of birds sampled locally. Perhaps then we may be a little closer to getting the sums right.

REFERENCES

Aschoff, J. & Pohl, H. (1970) Der Ruheumsatz con Vögeln als Funktion der Tageszeit und der Köopergrösse. *Journal für Ornithologie, Lieptzig*, **111**(1), 38–47.

Birt-Friesen, V.L., Montevecchi, W.A., Cairns, D.K. & Macko, S.A. (1989) Activity-specific metabolic rates of free-living northern gannets and other seabirds. *Ecology*, **70**, 357–367.

Brugger, K.E. (1993) Digestibility of three fish species by double-crested cormorants. *Condor*, **95**, 25–32.

Carter, S.P. & Evans, P.R. (1986) Goosander and merganser studies in Scotland, 1986. *Final Report on contract HP3-03-208/9*. Nature Conservancy Council, Peterborough.

Cooper, J. (1978) Energetic requirements for growth and maintenance of the Cape gannet (Aves: Sulidae). *Zoologica Africana*, **13**, 305–317.

Cramp, S. & Simmons, K.E.L. (eds) (1977) *Handbook of the Birds of Europe, the Middle East and North Africa: The Birds of the Western Palaearctic 1*, pp. 200–207. Oxford University Press, Oxford.

Davies, J.M. & Feltham, M.J. (1994) *The impact of cormorants* (Phalacrocorax carbo carbo) *on angling catches on the river Ribble, Lancashire*. Interim Report to the National Rivers Authority.

Duffy, D.C. & Jackson, S. (1986) Diet studies of seabirds: a review of methods. *Colonial Waterbirds*, **9**, 1–17.

Duffy, D.C. & Laurenson, L.J.B. (1983) Pellets of Cape cormorants as indicators of diet. *Condor*, **85**, 305–307.

Ebbinge, B., Canters, K. & Drent, R.H. (1975) Foraging routines and estimated daily food intake of barnacle geese wintering in the Northern Netherlands. *Wildfowl*, **26**, 5–19.

Feltham, M.J. (1995) Consumption of Atlantic salmon *Salmo salar* L. smolts and parr by goosanders *Mergus merganser* L: estimates from doubly-labelled water measurements of captive birds released on two Scottish rivers. *Journal of Fish Biology*, **46**, 273–281.

Feltham, M.J. & Davies, J.M. (1995) How much do cormorants and goosanders eat? *Proceedings of the Institute of Fisheries Management Annual Study Course*, **25**, 143–166.

Gardiner, R.W. & Geddes, P. (1980) The influence of body composition on the survival of juvenile salmon. *Hydrobiologia*, **69**, 67–72.

Gremillet, D.J.H. & Plos, A.L. (1994) The use of stomach temperature records for the calculation of daily food intake in cormorants. *Journal of Experimental Biology*, **189**, 105–115.

Hartley, P.H.T. (1948) The assessment of the food of birds. *Ibis*, **90**, 361–381.

Johnstone, J., Wanless, S. & Harris, M.P. (1990) The usefulness of pellets for assessing the diet of adult shags (*Phalocracorax aristotelis*). *Bird Study*, **37**, 5–11.

Keller, T. (in press) The food of wintering cormorants in Bavaria. *Proceedings of the 1993 Workshop on Cormorants* Phalacrocorax carbo. 13–17 April 1993, Gdansk, Poland.

Kendeigh, S.C., Dol'nik, V.R. & Gavrilov, V.M. (1977) Avian energetics. *Granivorous Birds in Ecosystems* (eds S.C. Kendeigh & J. Pinowski), pp. 127–424. International Biological Programme 12, Cambridge University Press, Cambridge.

Kennedy, G.J.A. & Greer, J.E. (1988) Predation by cormorants, *Phalacrocorax carbo* L., on the salmonid populations of an Irish river. *Aquaculture and Fisheries Management*, **19**, 159–170.

Lasiewski, R.C. & Dawson, W.R. (1967) A re-examination of the relation between standard metabolic rate and body weight in birds. *Condor*, **69**, 13–23.

Lifson, N., Gordon, G.B. & McClintock, R. (1955) Measurement of total carbon dioxide production by means of D_2O^{18}. *Journal of Applied Physiology*, **7**, 704–710.

Lifson, N. & McClintock, R. (1966) Theory of use of the turnover rates of body water for measuring energy and materials balance. *Journal of Theoretical Biology*, **12**, 46–74.

McIntosh, R. (1978) Distribution and diet of the cormorant on the lower reaches of the river Tweed. *Fisheries Management*, **9**, 107–113.

Madsen, F.J. & Sparck, R. (1950) On the feeding habits of the southern cormorant (*Phalacrocorax carbo sinensis* Shaw) in Denmark. *Danish Review of Game Biology*, **1(3)**, 45–75.

Marquiss, M. & Carss, D.N. (1994) Avian piscivores: basis for policy. *National Rivers Authority Research and Development Report 461/8/N&Y*. 104 pp. National Rivers Authority, Newcastle.

Marteijn, E.C.L. & Dirksen, S. (1991) Cormorants *Phalacrocorax carbo sinensis* feeding in shallow eutrophic lakes in The Netherlands in the nonbreeding period: prey choice and fish consumption. *Proceedings of the 1989 Workshop on Cormorants* Phalacrocorax carbo (eds M.R. Van Eerden & M. Zijlstra). pp. 135–155. Rijkswaterstaat Directorate Flevoland, Lelystad.

Martin, P. & Bateson, P. (1993) *Measuring Behaviour: an Introductory Guide*, 2nd edn. Cambridge University Press, Cambridge.

Martyniak, S., Mellin, M., Stachowiack, P. & Wittke, A. (in press) Food composition of cormorants in two colonies in NE Poland from pellet analysis. *Proceedings of the 1993 Workshop on Cormorants* Phalacrocorax carbo. 13–17 April 1993, Gdansk, Poland.

Mellin, M., Ibron-Mirowska, I. & Martyniack, A. (in press) Food composition of cormorants shot in two fish-farms in NE Poland. *Proceedings of the 1993 Workshop on Cormorants* Phalacrocorax carbo. 13–17 April 1993, Gdansk, Poland.

Mills, D.H. (1965) The distribution and food of the cormorant *Phalacrocorax carbo* in Scottish inland waters. *Freshwater & Salmon Fisheries Research*, **35**. HMSO, Edinburgh.

Nagy, K.A. (1987) Field metabolic rate and food requirement scaling in mammals and birds. *Ecological Monographs*, **57**, 111–128.

Noordhuis, R., Marteijn, E.C.L. & Noordhuis, R. (in press) Cormorants along the Dutch river and freshwater lakes in the non-breeding period: distribution and feeding habits.

Proceedings of the 1993 Workshop on Cormorants Phalacrocorax carbo. 13–17 April 1993, Gdansk, Poland.

Platteeuw, M. (1991) Time and energy constraints of fishing behaviour in breeding cormorants *Phalacrocorax carbo sinensis*. *Proceedings of the 1989 Workshop on Cormorants* Phalacrocorax carbo (eds M.R. Van Eerden & M. Zijlstra), pp. 192–203. Rijkswaterstaat Directorate Flevoland, Lelystad.

Rae, B.B. (1969) The food of cormorants and shags in Scottish estuaries and coastal waters. *DAFS Marine Research*, **1**. HMSO, Edinburgh.

Reicholf, J. (1990) Verzehren überwinternde Kormorane (*Phalacrocorax carbo*) abnorm hohe Fischmengen? *Mitteilungen Zoologische Gesellschaft Braunau*, **5(9/12)**, 165–174.

Sato, K., Hwang-Bo, J., & Okumura, J. (1988) Food consumption and basal metabolic rate in common cormorants *Phalacrocorax carbo*. *Laboratory of Animal Physiology, Nagoya University*, **8**, 58–62.

Speakman, J.R. & Racey, P.A. (1988) Validation of the doubly labelled water technique in small insectivorous bats by comparison with indirect calorimetry. *Physiological Zoology*, **61(6)**, 514–526.

van Dobben, W.H. (1952) The food of the cormorant in The Netherlands. *Ardea*, **40**, 1–63.

Voslamber, B. (1988) Visplaatskeuze, fourageerwijze en voedselkeuze van aalschovers *Phalacrocorax carbo sinensis* in het Ijsselmeerebied in 1982. *Flevobericht 286*. Rijksdienst voor de Ijsselmeerpolders, Lelystad.

Walsberg, G.E. (1983) Avian ecological energetics. *Avian Biology 7* (eds D.S. Farner, J.R. King & K.C. Parkes), pp. 161–220. Academic Press, New York.

Zijlstra, M. (in press) Pellet production and the usage of fish remains in determining the cormorants diet. *Proceedings of the 1993 Workshop on Cormorants* Phalacrocorax carbo. 13–17 April 1993, Gdansk, Poland.

CHAPTER 9

Trophic interactions of squid *Loligo forbesi* in Scottish waters

G.J. Pierce and M.B. Santos

SUMMARY

(1) Squid are an increasingly important fishery resource in the north-east Atlantic but their role as predators and prey of other marine organisms is poorly known. The main squid species of commercial importance in the northern north-east Atlantic is the veined squid *Loligo forbesi*. The present paper reviews trophic interactions involving this species in Scottish waters.

(2) Fishery and population data are used to develop a simple model of the squid population. Stock size estimates, combined with available data on diet and energetic requirements are used to estimate consumption of the main prey species for two annual cohorts (1990/91, 1991/92).

(3) *Loligo forbesi* is primarily piscivorous, although the diet also includes crustaceans and there is some cannibalism. In 1990/91, depending on assumptions about natural mortality, the Scottish population of *Loligo forbesi* may have eaten between 9000 and 16 000 tonnes of food, including 1700–3000 tonnes of sandeels (Ammodytidae), a similar amount of *Trisopterus* spp. and 1000–1700 tonnes of whiting *Merlangius merlangus*.

(4) *Loligo forbesi* is eaten by various predatory fish, seabirds and marine mammals in the north-east Atlantic. Harbour seals *Phoca vitulina* and grey seals *Halichoerus grypus* eat *Loligo forbesi* in small amounts. *Loligo forbesi* is more important in the diets of harbour porpoise *Phocoena phocoena* and these three mammal species alone may remove more squid from Scottish waters every year than the commercial fishery.

(5) The importance of *Loligo forbesi* in fish and seabird diets is not well documented, although many studies indicate that other squid species are eaten. Many toothed cetaceans are specialist feeders on cephalopods and good quantitative data on the diets of more of these potential predators is needed to obtain realistic estimates of natural mortality in *Loligo forbesi*.

Key-words: food consumption, *Loligo forbesi*, marine mammals, north-east Atlantic, population size

INTRODUCTION

The veined squid *Loligo forbesi* is widely distributed in the north-east Atlantic, primarily in shelf and shelf edge waters and is the subject of increasingly important fisheries, both directed and by-catch. Recent studies in Scottish waters have examined distribution and abundance, population ecology, diet and the nature of the commercial fishery (Pierce *et al.* 1994a,b,c) but no previous attempts have been made to quantify the trophic interactions. The diet of *Loligo forbesi* has been documented in Scottish, Irish and Spanish waters (Ngoile 1987, Collins *et al.* 1993, Rocha *et al.* 1993, Pierce *et al.* 1994b) and the species is primarily piscivorous. However, there are no estimates of population size and the potential impact on fish stocks remains unknown. Data on the importance of squid as prey are patchy. There are many relevant studies on diets of fish, seabirds and marine mammals in the north-east Atlantic, but few attempts to quantify the amounts of cephalopods eaten, although Furness (1994) estimates consumption of squids by seabird populations.

The present paper uses existing data to quantify some of the trophic interactions involving *Loligo forbesi* in the north-east Atlantic, focusing on the fished population in Scottish waters.

SOURCES OF DATA

Consumption by squid

Squid energy requirements
Based on data for *Illex illecebrosus* and *Loligo* spp. compiled by O'Dor & Wells (1987), daily food intake (I, $g \cdot day^{-1}$) may be expressed in relation to body weight (W, g) as:

$$I = 0.0683 + 0.0474 \times W \qquad (1)$$

Using this regression, daily food intake was estimated for all individuals in monthly population samples (July 1990–July 1992) (Pierce *et al.* 1994a), and the mean daily food requirement per individual calculated for each month.

Squid population size
A simple difference equation model is used here, including terms for recruitment, fishing mortality and natural mortality. Immigration and emigration are assumed to be negligible. Data on the monthly landings in Scotland (SOAFD, unpublished data) and monthly size composition (Pierce *et al.* 1994a) were used to estimate the number of *Loligo* landed monthly between July 1990 and July 1992. This assumes that all squid caught in Scottish waters are *Loligo forbesi* and are landed in Scotland, and the size composition in monthly samples was representative of the population. Recruitment in *Loligo forbesi* occurs during much of the year, although with varying seasonal peaks (Lum-Kong *et al.* 1992, Boyle & Ngoile 1993; Pierce *et al.* 1994a). The proportion of recruits (animals of <150 mm mantle length) in monthly samples is assumed to be representative of the rate of recruitment to the population:

$$N_t = (N_{t+1} + C_t)/(1 + r_t - m) \qquad (2)$$

where N_t is the population size at the start of month t, r_t is the recruitment in month t expressed as a proportion of N_t, C_t is the commercial catch in month t and m is the monthly natural mortality, expressed as a (fixed) proportion of N_t.

Pauly (1985) gives values in the range 0.53–2.13 for instantaneous (annual) natural mortality (M) in squids. This corresponds to values for m as defined above between 0.043 and 0.163. Most adults disappear from the fished population in July, while in 1991 recruitment reached a peak in August (Pierce *et al.* 1994a). Setting the population level at the end of July in each year to zero, population sizes for the preceding months were estimated using Equation (2) for different levels of monthly natural mortality (m = 0, 0.04, 0.16). Input catch parameters and population results for m = 0.16 are illustrated in Figure 9.1.

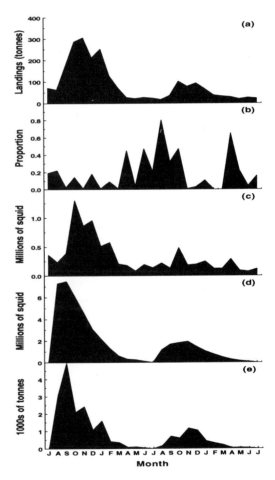

Fig. 9.1. Population model for *Loligo forbesi* in Scottish waters July 1990–July 1992 using m = 0.16 for illustration purposes: (a) monthly landings (tonnes) by the commercial fishery in Scotland; (b) monthly recruitment to the fished population based on the proportion of squid with mantle length <150 mm in monthly samples; (c) estimated numbers (millions) of squid landed monthly; (d) estimated monthly population size (millions of squid); (e) estimated monthly population food consumption (thousands of tonnes).

Squid diet

Data on the diet of *Loligo forbesi* in Scottish waters were taken from Pierce *et al.* (1994b). Variation in diet with body size and season is ignored. Assuming that meals of different prey taxa are normally of equivalent weight, the proportion by weight in the diet is equivalent to the modified frequency of occurrence: fish (75.7%), crustacean (17.4%), cephalopod (6.9%). Within taxa, relative weights eaten are assumed to be proportional to the number of individuals identified from bones, otoliths (fish) and beaks (cephalopods). Fish identified only as *Gadidae* were assigned *pro rata* to those gadid species which were identified, while beaks of decapod cephalopods were similarly assigned.

Population food intake

Monthly food intake by the squid population (I, tonnes) for prey type i is given by:

$$I = N_t \times P_i \times F_t \times T_t \qquad (3)$$

where N_t is the squid population size at the start of month t, T_t is the number of days in month t, P_i is the proportion by weight of prey-type i in the diet, F_t is the average weight of food eaten daily per individual. Monthly food intake was summed to give annual food consumption (Table 9.1).

Consumption of squid by marine mammals and other predators

Population consumption is calculated for marine mammals on which we have some dietary data: harbour seals *Phoca vitulina*, grey seals *Halichoerus grypus* and harbour porpoise *Phocoena phocoena* (Table 9.2). For other potential predators on squid, data in the literature are summarized (Table 9.3). Ignoring within-year variation and effects of population structure, annual consumption of squid by a population of marine mammals (I, tonnes) is given by:

$$I = \frac{365 \times N \times E}{10^6 \times D_s} \times \frac{P_s \times D_s}{P_o \times D_o} \qquad (4)$$

where N is the estimated population size, E is the average individual daily energy intake (kcal), D_s is the calorific density of squid (kcal $\cdot$ g^{-1}), D_o is the average calorific density of other prey in the diet (kcal $\cdot$ g^{-1}), P_s is the proportion (by weight) of squid in the diet and P_o is the proportion (by weight) of other prey.

Not all data are precisely known. Estimated energetic intake for the harbour porpoise (Table 9.2) assumes a normal daily energy expenditure of approximately three times the basal metabolic rate predicted, using the Kleiber relationship, from body weight (after Worthy *et al.* 1987, Murie 1987). Taking average body weight to be 45 kg (at the lower end of the range for mean adult body size) (Leatherwood *et al.* 1983), this is equivalent to approximately 3600 kcal $\cdot$ day^{-1}. At 90% assimilation efficiency (a realistic figure for seals) (Prime & Hammond 1987) this would require a daily intake of 4000 kcal $\cdot$ day^{-1}.

Croxall & Prince (1982) give a figure of 3.5 kJ $\cdot$ g^{-1} for average calorific density of cephalopods, i.e. 836 kcal $\cdot$ kg^{-1}, noting that values for fish are usually higher. However, although calorific densities often quoted for fish such as sandeels (1367), sprat ($\leqslant$1587) and *Tri-*

Table 9.1. Amounts of food eaten by *Loligo forbesi* 1990/91 and 1991/92, based on estimated importance of main prey species in the diet and estimates of squid population size using various values for natural mortality (m).

Prey species	Proportion of diet	Weights (tonnes) eaten 1990/91 for m =			Weights (tonnes) eaten 1991/92 for m =		
		0	0.04	0.16	0	0.04	0.16
Trisopterus	0.166	1559	1754	2678	467	524	800
Haddock	0.060	563	634	968	169	190	289
Whiting	0.104	976	1099	1678	292	329	501
Cod	0.009	85	95	145	25	28	43
Sandeel	0.183	1718	1933	2952	515	579	881
Clupeidae	0.088	826	930	1419	247	278	423
Mackerel/scad	0.022	207	232	355	62	70	106
Other fish	0.126	1183	1331	2032	354	398	606
Crustacea	0.174	1634	1838	2806	489	550	838
Loligo forbesi	0.049	460	518	790	138	155	236
Other ceph.	0.020	188	211	323	56	63	96
Total	1.000	9389	10563	16130	2812	3161	4817

Table 9.2. Estimated annual cephalopod consumption for seals and porpoises.

	Harbour seal		Grey seal		Harbour porpoise
	UK 1991	Scotland 1991	UK 1991	Scotland 1991	North Sea 1994
Population size	24 640 (1)	23 089 (1)	93 500 (1)	85 400 (1)	352 500 (5)
Daily energy intake (kcal)	4 680 (2)		5 530 (4)		4 000 (6), (7)
Daily food intake (kg)	5.6 (2)		6.6 (4)		4.8 (6), (7)
%Diet = Cephalopod	1.7 (3)		9.4 (3)		3.8 (3), (8)
%Diet = *Loligo*	0.06 (3)		0.25 (3)		1.98 (3), (8)
Annual population Cephalopod consumption (tonnes)	874	819	21 211	19 373	23 680
Annual population *Loligo* consumption (tonnes)	32	30	552	504	12 229

Sources: (1) Hiby *et al.* (1993); (2) Härkönen & Heide-Jørgensen (1991); (3) G.J. Pierce *et al.* unpublished data; (4) Fedak & Hiby (1985); (5) Hammond *et al.* (1995); (6) Worthy *et al.* (1987); (7) Leatherwood *et al.* (1983); (8) Santos *et al.* (1994).

sopterus spp. (1058–1146) are higher than values for most Gadidae and flatfish (700–800) (Murray & Burt 1969, Prime & Hammond 1987), Hislop *et al.* (1991a) show that there are very large seasonal and size-related changes in calorific density. Thus small sprat may on average be no more energy rich than cod or whiting. We make the simplifying assumption that fish and cephalopods are on average of equal calorific value.

Diet was quantified by analysis of harbour seal faeces (N = 658), grey seal faeces (N = 495) and porpoise stomachs (N = 108). Weights of fish and cephalopods eaten were calculated from measurements on otoliths and beaks (Pierce *et al.* 1991a,b, unpublished data, Santos *et al.* 1994).

RESULTS AND DISCUSSION

Predation by *Loligo forbesi*

Maximum population size for *Loligo forbesi* in Scottish waters during the study period is estimated to have ranged between 4 million (for m = 0) and 7.5 million (for m = 0.16). Depending on m, the population would have eaten between 9000 and 16 000 tonnes of food in 1990/91 and 3000 to 5000 tonnes in 1991/92 (Table 9.1). In 1990/91, the food eaten included 1700–3000 tonnes of sandeels (Ammodytidae), 1600–2700 tonnes of *Trisopterus* spp. and 1000–1700 tonnes of whiting *Merlangius merlangus*. This excludes any estimate of consumption by pre-recruit and paralarval squid. Empirical estimates of natural mortality, on a seasonal or monthly basis, are needed to provide meaningful confidence limits for the size of the squid population. Scottish squid landings vary widely from year to year, possibly reflecting cyclic variations in population size

(Pierce *et al.* 1994c). From a low of 88 tonnes in 1979, annual landings rose to approximately 1900 tonnes in 1989. Landings in 1990, 1991 and 1992 were relatively large (approximately 1400, 900 and 1100 tonnes, respectively) and estimated consumption by squid in these years would therefore be towards the upper end of the natural range of values. To put the amount of food eaten by squid in Scottish waters in perspective, predators in the North Sea are estimated to have eaten approximately 2.2 million tonnes of fish and 6.1 million tonnes of other food during 1991 (Anon. 1994), and these figures may even be underestimates (Greenstreet, in press).

Predation by seals and propoises on *Loligo forbesi*

Seals ate very little *Loligo forbesi* (Table 9.2); in fact only single specimens were found in faeces of both species. Most of the cephalopods eaten by grey seals were octopus *Eledone cirrhosa*. Porpoises probably have a more significant impact on *Loligo forbesi*, although population estimates are for the North Sea rather than for Scotland (Table 9.2). Mortality of *Loligo forbesi* due to marine mammals appears to be of at least a similar order of magnitude to fishing mortality (see Pierce *et al.* 1994c) and probably rather higher, implying that the higher values for natural mortality in the preceding squid model are more realistic.

Predation by other predators on cephalopods

As the selective list of predators in Table 9.3 makes clear, there is a long way to go before natural mortality of *Loligo forbesi* can be fully quantified. Large-scale

Table 9.3. Marine predators of cephalopods in the north-east Atlantic.

Area	Predator	% of diet = cephalopods[a]	Species eaten[b]; other comments	Sources
North Sea	Cod *Gadus morhua*	0–11	Not stated but includes A Cephalopods mainly eaten in 1st quarter of year	Daan (1981), Anon. (1994), J.R.G. Hislop, personal communication
	Haddock *Melanogrammus aeglefinus*	0–11		
	Whiting *Merlangius merlangus*	0–14		
	Saithe *Pollachius virens*	0–30		
	Mackerel *Scomber scombrus*	0–80		
	Grey gurnard *Eutrigla gurnardus*	0–11		
	Ray *Raja radiata*	0–1		
NE Atlantic	Swordfish *Xiphias gladius*	(4)	H, T, ocs	Guerra *et al.* (1993)
Scotland	Puffin *Fratercula artica*	(1)	A	Harris & Hislop (1978)
Skomer (Wales)	Guillemot *Uria aalge*	(1)	Squid	Hatchwell (1990)
Rum	Manx shearwater *Puffinus puffinus*	40	Omm, ocs	Furness (1994)
Foula, St Kilda	Northern fulmar *Fulmarus galcialis*	1	Omm	Furness (1994)
	Leach's stormpetrel *Oceanodroma leucorhoa*	(1)	Omm	
	British stormpetrel *Hydrobates pelagicus*	(1)	Omm	
Foula	Great skua *Catharcta skua*	0.1	L	Furness (1994)
	Various other seabirds	Absent		
UK waters	Harbour seal *Phoca vitulina*	(1)–(2)	E, L, S	Steven (1934), Sergeant (1951), Rae (1968), Pierce *et al.* (1991b), Thompson *et al.* (1993)
	Grey seal *Halichoerus grypus*	(1)	E, L, R, T	Rae (1968), SMRU (1984), Pierce *et al.* (1991a)
UK and other NE Atlantic waters, Mediterranean	Fin whale *Balaenoptera physalus*	(1)		Rae (1965), Rae (1973), Clarke & MacLeod (1974), Clarke & Kristensen (1980), Leatherwood *et al.* (1983), Clarke & Pascoe (1985), Clarke (1986), Martin & Clarke (1986), Corbet & Harris (1991), Bello (1992), Würtz *et al.* (1992), Clarke *et al.* (1993), Haug *et al.* (1993), González *et al.* (1994a), Santos *et al.* (1994)
	Sei whale *Balaenoptera borealis*	(2)		
	Minke whale *Balaenoptera acutorostrata*	(1)–(2)		
	Sperm whale *Physeter macrocephalus*	(4)	B, G, H, T, ocs	
	Pygmy sperm whale *Kogia breviceps*	(4)	E, H, L, Sl, ocs	
	White whale *Delpinapterus leucas*	(4)		
	Narwhal *Monodon monocerus*	(2)		
	Northern bottlenose whale *Hyperoodon ampullatus*	(4)	G, H, L, S, T, ocs	
	Cuvier's beaked whale *Ziphius cavirostris*	(3)	B, H, ocs	
	Sowerby's beaked whale *Mesoplodon bidens*	(3)	Omm?	
	True's beaked whale *Mesoplodon mirus*	(4)		
	Harbour porpoise *Phocoena phocoena*	(1)	L	
	Common dolphin *Delphinus delphis*	(2)	A, L, Sa	
	Striped dolphin *Stenella coeruleoalba*	(2)	A, G, H, I, L, Sl, T	
	Bottle-nosed dolphin *Tursiops truncatus*	(1)	L, S	
	Atlantic white-sided dolphin *Lagenorhynchus acutus*	(2)	L	
	White-beaked dolphin *Lagenorhynchus albirostris*	(2)	L	
	Melon-headed whale *Peponocephala electra*	(2)–(3)		
	False killer whale *Pseudorca crassidens*	(1)		
	Killer whale *Orcinus orca*	(1)	I, L, S, T	
	Long-finned pilot whale *Globicephala melaena*	(3)	B, E, G, H, L, O, Sl, S, T	
	Risso's dolphin *Grampus griseus*	(4)	E, G, H, L, R, S, St, T, Te	

[a] Where no data on %weight were available, importance of cephalopods in the diet is expressed as: (1) some, (2) considerable, (3) large proportion, (4) main food (after Clarke, 1986).
[b] Key to cephalopods: A, *Alloteuthis*; B, *Brachioteuthis*; E, *Eledone*; G, *Gonatus*; H, *Histioteuthis*; L, *Loligo* (*L. forbesi* + *L. vulgaris*); I, *Illex*; O, *Ommastrephes*; Omm, Ommastrephidae; ocs, oceanic squids; R, *Rossia*; S, *Sepia*; Sl, *Sepiola*; St, *Sepietta*; T, *Todarodes*; Te, *Todaropsis*.

studies on fish diets in the North Sea (Daan 1981, Anon. 1994) showed that cephalopods are seasonally important prey of some commercial fish. The squid *Alloteuthis subulata* was the principal cephalopod prey at least in the case of whiting (Hislop *et al.* 1991b). For younger fish, any *Loligo forbesi* eaten would probably be pre-recruits. Although squid are among the main prey of the swordfish *Xiphias gladius*, it takes oceanic species rather than the (neritic) loliginids (Guerra *et al.* 1993). Furness (1994) estimated that seabirds may remove 100 000 tonnes of squid annually from the north-east Atlantic, but most squid eaten are probably ommastrephids rather than *Loligo*. Also, the smaller loliginid *Alloteuthis subulata* is a more likely prey for most seabirds than adult *Loligo forbesi*, which may reach weights of more than 1 kg (Pierce *et al.* 1994a).

Many cetaceans of north-east Atlantic waters eat *Loligo* spp. (Clarke, 1986, Corbet & Harris 1991) and some, e.g. Risso's dolphin and sperm whales, are specialist feeders on cephalopods. Some of the other cephalopods eaten are also fished commercially in Europe, e.g. the cuttlefish *Sepia officinalis* and the ommastrephid squids *Todaropsis eblanae* and *Illex coindetii* (Pierce *et al.* 1994c, González *et al.* 1994a,b).

Quantitative information on consumption by a wider array of predators is needed to provide realistic estimates of natural mortality in *Loligo forbesi*.

ACKNOWLEDGEMENTS

We wish to acknowledge funding from the Commission of the European Communities, under Contracts No. AIR 1 CT92 0573 (GJP) and ERB 4001 GT 93 3630 (MBS). Nick Bailey and two anonymous referees provided useful comments on the manuscript.

REFERENCES

Anon. (1994) *Report of the Multispecies Assessment Working Group, Copenhagen, 23 November–2 December 1993.* ICES CM 1994/Assess:9.

Bello, G. (1992) Stomach content of a specimen of *Stenella coerulealba* (Cetacea: Delphinidae) from the Ionian Sea. *Atti della Società Italiana di Scienze Naturali e del Museo Civico di Storia Naturale di Milano*, 133, 41–48.

Boyle, P.R. and Ngoile, M.A.K. (1993) Assessment of maturity state and seasonality of reproduction of *Loligo forbesi* (Cephalopoda: Loliginidae) from Scottish waters. *Recent Advances in Cephalopod Fisheries Biology* (eds T. Okutani, R. O'Dor, & T. Kubodera), pp. 37–48. Tokai University Press, Tokyo.

Clarke, M.R. (1986) Cephalopods in the diet of odontocetes. *Research on Dolphins* (eds M.M. Bryden & R. Harrison). pp.

281–321. Clarendon Press, Oxford.

Clarke, M.R. & Kristensen, T.K. (1980) Cephalopod beaks from the stomachs of two northern bottlenosed whales (*Hyperoodon ampullatus*). *Journal of the Marine Biological Association of the United Kingdom*, 60, 151–156.

Clarke, M.R. & MacLeod, N. (1974) Cephalopod remains from a sperm whale caught off Vigo, Spain. *Journal of the Marine Biological Association of the United Kingdom*, 54, 959–968.

Clarke, M.R., Martins, H.R. & Pascoe, P. (1993) The diet of sperm whales (*Physeter macrocephalus* Linnaeus 1758) off the Azores. *Philosophical Transactions of the Royal Society of London B*, 339, 67–82.

Clarke, M.R. & Pascoe, P.L. (1985) The stomach contents of a Risso's dolphin (*Grampus griseus*) stranded at Thurlestone, south Devon. *Journal of the Marine Biological Association of the United Kingdom*, 65, 663–665.

Collins, M.A., Lordan, C., De Grave, S., Burnell, G.M. & Rodhouse, P.G. (1993) *Aspects of the Diet of Loligo forbesi Steenstrup in Irish Waters.* ICES CM 1993/K:44.

Corbet, G.B. & Harris, S. (eds) (1991) *The Handbook of British Mammals.* 3rd edn. Blackwell Scientific Publications, Oxford.

Croxall, J.P. & Prince, P.A. (1982) Calorific content of squid (Mollusca: Cephalopoda). *British Antarctic Survey Bulletin*, 55, 27–31.

Daan, N. (1981) Data Base Report of the Stomach Sampling Project 1981. *Cooperative Research Report 164.* 144 pp. International Council for the Exploration of the Sea, Copenhagen.

Fedak, M.A. & Hiby, A.R. (1985) Population energy requirements of seals. *The Impact of Grey Seals on North Sea Resources* (eds P.S. Hammond & J. Harwood), pp. 50–58. Report to the Commission of the European Communities on Contract ENV 665 UK, SMRU, NERC, Cambridge.

Furness, R.W. (1994) An estimate of the quantity of squid consumed by seabirds in the eastern North Atlantic and adjoining seas. *Fisheries Research*, 21, 165–177.

González, A.F., López, A., Guerra, A. & Barreiro, A. (1994a) Diets of marine mammals stranded on Northwestern Spanish Atlantic coast with special reference to Cephalopoda. *Fisheries Research*, 21, 179–191.

González, A.F., Rasero, M. & Guerra, A. (1994b) Preliminary study of *Illex coindetii* and *Todaropsis eblanae* (Cephalopoda: Ommastrephidae) in Northern Spanish Atlantic waters. *Fisheries Research*, 21, 115–126.

Greenstreet, S.P.R. (in press) Estimation of the daily consumption of food by fish in the North Sea in each quarter of the year. *Scottish Fisheries Research Report*, 55.

Guerra, A., Simon, F. & González, A.F. (1993) Cephalopods in the diet of the swordfish, *Xiphias gladius*, from the Northeastern Atlantic Ocean. *Recent Advances in Cephalopod Fisheries Biology* (eds T. Okutani, R. O'Dor & T. Kubodera), pp. 159–164. Tokai University Press, Tokyo.

Hammand, P., Benke, H., Berggren, P., Collet, A., Heide-Jørgensen, M.-P., Heimlich-Boran, S., Leopold, M. & Øien, N. (1995) *The distribution and abundance of harbours porpoises and other small cetaceans in the North Sea and adjacent waters.* ICES CM 1995/N:10.

Härkönen, T. & Heide-Jørgensen, M.-P. (1991) Harbour seals as predators in the Skaggerak. *Ophelia*, **34**, 191–207.

Harris, M.P. & Hislop, J.R.G. (1978) The food of young puffins *Fratercula arctica*. *Journal of Zoology, London*, **185**, 213–236.

Hatchwell, B.J. (1990) The feeding ecology of young guillemots *Uria aalge* on Skomer island, Wales. *Ibis*, **133**, 153–161.

Haug, T., Gjøsæter, H., Lindstrøm, U. & Nilssen, K.T. (1993) *Studies of Minke Whale* Balaenoptera acutorostrata *Ecology in the Northeast Atlantic: Preliminary Results from Studies of Diet and Food Availability during Summer 1992*. ICES CM 1993/N:7.

Hiby, L., Duck, C. & Thompson, D. (1993) Seal stocks in Great Britain: surveys conducted in 1991. *NERC News*, (January 1993), 30–31.

Hislop, J.R.G., Harris, M.P. & Smith, J.G.M. (1991a) Variation in the calorific value and total energy content of the lesser sandeel (*Ammodytes marinus*) and other fish preyed on by seabirds. *Journal of Zoology, London*, **224**, 501–517.

Hislop, J.R.G., Robb, A.P., Bell, M.A. & Armstrong, D.W. (1991b) The diet and food consumption of whiting (*Merlangius merlangus*) in the North Sea. *ICES Journal of Marine Science*, **48**, 139–156.

Leatherwood, S., Reeves, R.R. & Foster, L. (1983) *The Sierra Club Handbook of Whales and Dolphins*. Sierra Club Books, San Fransisco.

Lum-Kong, A., Pierce, G.J. & Yau, C. (1992) Timing of spawning and recruitment in *Loligo forbesi* (Cephalopoda: Loliginidae) in Scottish waters. *Journal of the Marine Biological Association of the United Kingdom*, **72**, 301–311.

Martin, A.R. & Clarke, M.R. (1986) The diet of sperm whales (*Physeter macrocephalus*) captured between Iceland and Greenland. *Journal of the Marine Biological Association of the United Kingdom*, **66**, 779–790.

Murie, D.J. (1987) Experimental approaches to stomach content analyses of piscivorous marine mammals. *Approaches to Marine Mammal Energetics* (eds A.C. Huntley, D.P. Costa, G.A.J. Worthy & M.A. Castellini), pp. 147–163. Society for Marine Mammalogy, Lawrence, Kansas.

Murray, J. & Burt, J.R. (1969) *The Composition of Fish*. Torry Advisory Note No. 38, Ministry of Agriculture, Fisheries and Food, Torry Research Station, Aberdeen.

Ngoile, M.A.K. (1987) *Fishery biology of the squid* Loligo forbesi *Steenstrup (Cephalopoda: Loliginidae) in Scottish waters*. PhD Thesis, University of Aberdeen.

O'Dor, R.K. & Wells, M.J. (1987) Energy and nutrient flow. *Cephalopod Life Cycles Vol. II. Comparative Reviews* (ed. P.R. Boyle), pp. 109–133. Academic Press, London.

Pauly, D. (1985) Population dynamics of short-lived species with emphasis on squids. *NAFO Scientific Council Studies*, **9**, 143–154.

Pierce, G.J., Miller, A., Thompson, P.M. & Hislop, J.R.G. (1991a) Prey remains in grey seal (*Halichoerus grypus*) faeces from the Moray Firth, north-east Scotland. *Journal of Zoology, London*, **224**, 337–341.

Pierce, G.J., Thompson, P.M., Miller, A., Diack, J.S.W., Miller, D. & Boyle, P.R. (1991b) Seasonal variation in the diet of common seals (*Phoca vitulina*) in the Moray Firth

area of Scotland. *Journal of Zoology, London*, **223**, 641–652.

Pierce, G.J., Boyle, P.R., Hastie, L.C. & Key, L. (1994a) The life history of *Loligo forbesi* in Scottish waters. *Fisheries Research*, **21**, 17–41.

Pierce, G.J., Boyle, P.R., Hastie, L.C. & Santos, B. (1994b) Diets of squid *Loligo forbesi* and *Loligo vulgaris* in the northeast Atlantic. *Fisheries Research*, **21**, 149–163.

Pierce, G.J., Boyle, P.R., Hastie, L.C. & Shanks, A. (1994c) Distribution and abundance of the fished population of *Loligo forbesi* in UK waters: analysis of fishery data. *Fisheries Research*, **21**, 193–216.

Prime, J.H. & Hammond, P.S. (1987) Quantitative assessment of grey seal diet from fecal analysis. *Approaches to Marine Mammal Energetics* (eds A.C. Huntley, D.P. Costa, G.A.J. Worthy & M.A. Castellini), pp. 165–182. Society for Marine Mammalogy, Lawrence, Kansas.

Rae, B.B. (1965) The food of the common porpoise (*Phocoena phocoena*). *Journal of Zoology, London*, **146**, 114–122.

Rae, B.B. (1968) The food of seals in Scottish waters. *Marine Research*, **2**. 23 pp. Department of Agriculture and Fisheries for Scotland.

Rae, B.B. (1973) Additional notes on the food of the common porpoise (*Phocoena phocoena*). *Journal of Zoology, London*, **169**, 127–131.

Rocha, F., Castro, B.G., Gil, M.S. & Guerra, A. (1993) *The Diet of* Loligo vulgaris *and* Loligo forbesi *(Cephalopoda: Loliginidae) in the Galician Waters (NW Spain)*. ICES CM 1993/K:15.

Santos, M.B., Pierce, G.J., Ross, H.M., Reid, R.J., Wilson, B. (1994) *Diets of Small Cetaceans from the Scottish Coast*. ICES CM 1994/N:11.

Sergeant, D.E. (1951) The status of the common seal (*Phoca vitulina* L.) on the East Anglian coast. *Journal of the Marine Biological Association of the United Kingdom*, **29**, 707–717.

SMRU (1984) *Interactions between Grey Seals and UK Fisheries. A Report on Research conducted for the Department of Agriculture and Fisheries Scotland by the Natural Environment Research Council's Sea Mammal Research Unit 1980 to 1983*. SMRU, NERC, Cambridge.

Steven, G.A. (1934) A short investigation into the habits, abundance, and species of seals on the north Cornwall coast. *Journal of the Marine Biological Association of the United Kingdom*, **19**, 489–501.

Thompson, P.M., Wood, D., Tollit, D., Corpe, H. & Racey, P.A. (1993) The behaviour and ecology of common and grey seals in the Moray Firth. *Report to the Scottish Office Agriculture and Fisheries Department*. Department of Zoology, University of Aberdeen.

Worthy, G.A.J., Innes, S., Braune, B.M. & Stewart, R.E.A. (1987) Rapid acclimation of cetaceans to an open-system respirometer. *Approaches to Marine Mammal Energetics* (eds A.C. Huntley, D.P. Costa, G.A.J. Worthy & M.A. Castellini), pp. 115–126. Society for Marine Mammalogy, Lawrence, Kansas.

Würtz, M., Poggi, R. & Clarke, M.R. (1992) Cephalopods from the stomachs of a Risso's dolphin (*Grampus griseus*) from the Mediterranean. *Journal of the Marine Biological Association of the United Kingdom*, **72**, 861–867.

CHAPTER 10

The influence of large, mobile predators in aquatic food webs: examples from sea otters and kelp forests

J.A. Estes

SUMMARY

(1) Although large, mobile predators occur in many aquatic communities, their ecological functions are poorly understood in all but a few of these. This paper is about one of these few exceptions – the effects of sea otters in kelp forests.

(2) By preying on herbivorous invertebrates, sea otters exert strong forces downwards through the food web, in turn releasing kelps and other macroalgae from limitation by herbivory. Our research has focused on three questions: (1) How general are these interactions? (2) What is their breadth of influence in coastal food webs? (3) What are their evolutionary consequences?

(3) Quasi-random surveys of sublittoral reef habitats across the North Pacific rim demonstrate that the expected impacts of sea otter predation on sea urchins and kelps is highly predictable from at least the western Aleutian Islands to British Columbia.

(4) Broader ranging impacts on coastal food webs are known or suspected in several cases. Vertical food web effects, arising through enhanced primary production brought about by sea otter predation, include increased growth of suspension feeding invertebrates, increased fish production and, possibly, population enhancement of such high trophic-level species as bald eagles and sea otters themselves.

(5) Lateral food web effects include competitive interactions among kelp species, population reductions in benthic feeding sea ducks (a likely competitor with sea otters) and alterations to diet and foraging behaviour of glaucous-winged gulls.

(6) Herbivore limitation by sea otters and their recent ancestors appears to have selected for marine plants that are strong competitors while being poorly defended against herbivores.

(7) Social and economic consequences of these interactions are discussed, as are their implications to other aquatic predators.

Key-words: *Enhydra lutris*, evolution, food webs, herbivory, predation

INTRODUCTION

Predators may affect biological communities in two general ways: (1) as the recipients of upward flowing energy and materials, or (2) as the initiators of downward acting forces that limit or enhance populations lower in the food web. There is little evidence that top-level predators exert important community-level effects by recycling materials or dissipating energy. However, downward acting forces by top-level predators are known from many studies. Some of the best examples are from aquatic communities (Strong 1992), and, because of their experimental tractability, involve sedentary or weakly mobile organisms.

The extreme mobility of marine birds and mammals has hindered the study of their roles as predators in marine communities. The only convincing examples for marine birds involve coastal species (e.g. Hockey & Branch 1984, Marsh 1987) that feed on sessile or weakly motile prey. Top-down effects by marine mammals are known or suspected in several cases, for example baleen whales *Balaenoptera* spp. in the Southern Ocean (Laws 1977), lake-bound harbour seals *Phoca vitulina* in eastern Canada (Power & Gregoire 1978), benthic feeding walruses *Odobenus rosmarus* and grey whales *Eschrichtius robustus* in the Bering Sea (Oliver & Slattery 1985, Oliver *et al.* 1985) and sea otters *Enhydra lutris* in the North Pacific Ocean (Estes & Palmisano 1974, Duggins 1980).

This paper summarizes the effects of sea otter predation in kelp forest communities. It has long been known that sea otters regulate herbivorous invertebrate populations, in turn allowing macroalgal assemblages to flourish. Our subsequent research has built on this general theme to consider three main questions:

(1) How commonly and under what conditions do these effects on invertebrates and algae occur (i.e. how general are they)?
(2) Are other species affected, and if so, in what ways (i.e. how broad are they)?
(3) What characters of the affected species have been shaped by the predatory forces of otters and their recent ancestors (i.e. what are their evolutionary consequences)?

GENERALITY

Although top-down effects from predation have been demonstrated repeatedly, there is little information on how these effects vary in space and time. This lack of information has raised a question over whether the top-down effect of sea otters on herbivores and plants is rare or common (Foster & Schiel 1988, Foster 1990). We addressed this issue by conducting random surveys of benthic habitats in areas where sea otters were present or absent. Such comparisons have been completed or are underway in the Aleutian Islands, south-east Alaska, British Columbia, and California, thus including much of the sea otter's range in the eastern North Pacific Ocean. In each region, and with no *a priori* knowledge of the habitats, 20 to 40 sites were randomly selected in areas with or without sea otters. A random sample of 10 to 20 $0.25\,\mathrm{m}^{-2}$ plots was then selected at each site, from which the abundance of macroinvertebrates (mostly sea urchins *Strongylocentrotus* spp.) and macroalgae (mostly kelps) were measured. These measurements were used to determine the extent to which sites with and without sea otters differed. Sampling and analytical procedures are described in detail by Estes & Duggins (1995). Only the data for Alaska (Estes & Duggins 1995) and Canada (Watson 1993) have been analysed and will be presented here.

Patterns of variation in community structure between areas with and without sea otters were found to be broadly similar for the Aleutian Islands, south-east Alaska, and the west coast of Vancouver Island (Fig. 10.1). At locations lacking sea otters, herbivore abundance was high but variable and kelp abundance

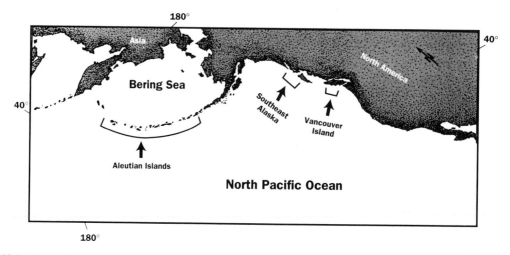

Fig. 10.1. Chart of the North Pacific region in which the study locations are shown.

was uniformly low. In contrast, sites with sea otters were characterized by uniformly low herbivore abundance and high but variable plant abundance (Fig. 10.2). Regional differences probably relate to differing patterns of echinoid recruitment (Estes & Duggins 1995) and length of time the sites had been occupied by sea otters.

BREADTH OF ECOLOGICAL EFFECTS

Evidence for the widespread occurrence of strong top-down linkages among sea otters, herbivorous invertebrates, and macroalgae is unequivocal. Kelp forests, however, cor .n many other species and functional assemblages, some of which participate in strong food web interactions of their own. The influence of sea otter predation on these other species, assemblages, and interactions are imagined to be of two general kinds (Fig. 10.3). One is trophic enrichment, i.e. the enhancement of energy and materials flowing upward from autotrophs through higher trophic forms. Because kelps and other macroalgae, with their high growth rates, significantly contribute to the primary production of coastal ecosystems (Mann 1982), this enrichment effect may be of considerable importance. This possibility was evaluated by contrasting the production, fate, and influence of kelp-derived organic carbon between islands with and without sea otters in the western and central Aleutian archipelago (Duggins *et al.* 1989). Two features of the autotrophs were utilized in this investigation. One is that benthic macroalgae and microalgae (i.e. phytoplankton and diatoms) are the only significant sources of primary production in this region; the other

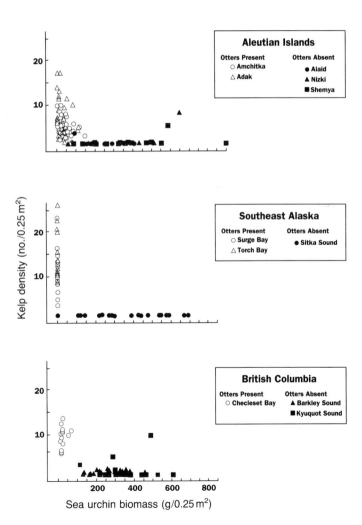

Fig. 10.2. The abundance of kelps and sea urchins at sites with and without sea otters in the Aleutian Islands, south-east Alaska, and Vancouver Island, British Columbia. Data from Alaska are from Estes and Duggins (1995); those from British Columbia are from Watson (1993).

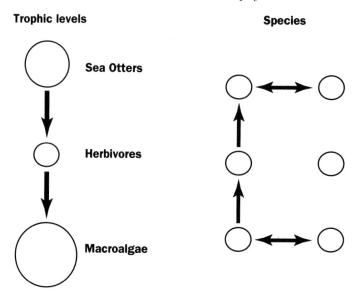

Fig. 10.3. Schematic diagram showing the breadth of the sea otter's influence in kelp forest communities. The flow chart to the left indicates that sea otter predation has strong top-down influences on the kelp forest, reducing the herbivores (and their effects) and enhancing the autotrophs (and their effects). The flow chart to the right indicates that these top-down influences may cause two classes of indirect effects, (1) enrichment (shown by the vertical arrows) and (2) competitive interactions (shown by the horizontal arrows).

is that these autotrophs, despite their diverse phylogenies, fix the stable isotopes of carbon (^{12}C and ^{13}C) in different ratios during photosynthesis. These features permitted the use of a mixing model to determine the relative importance of kelp versus phytoplankton production. This determination is based on the assumption that the net rate of microalgal production is unaffected by sea otters, and is thus similar among the various study islands. A comparison of carbon isotope ratios within species across islands with and without sea otters (using the data from Duggins *et al.* 1989) indicate respective macroalgal contributions to total organic carbon of about 75% and 25%, which translates into roughly a three-fold increase in total primary production at the otter-dominated islands.

Because most macroalgal biomass ultimately degrades into particulate organic carbon (POC), we reasoned that the hypothesized variation in production from macroalgal production could be assessed by comparing the performance of filter-feeding organisms (consumers of POC) between islands with and without otters. This was accomplished by measuring growth rates of mussels and barnacles in these two environments. To control for genetic or pre-settlement ontogenetic differences within species among treatment islands, we initiated this experiment with newly settled mussels and barnacles from a common population at San Juan Island,

Washington. We then translocated these organisms to islands with and without sea otters in the western Aleutian archipelago and measured their growth increments one year later. Growth rates in all instances were significantly (two to five times) greater at islands where otters abounded (Duggins *et al.* 1989, Fig. 10.4).

Increased production resulting from sea otter predation could affect other members of the coastal food web. Although largely unstudied, there is some evidence supporting this idea. One such example concerns the otters themselves. That the density of sea otter populations may be enhanced by kelp production, and thus by the otters' own predatory influences, is suggested for central California and Amchitka Island, Alaska. The coastal habitat in central California is a mosaic of rocky reefs and sandy-bottom communities, both of which are inhabited by otters. However, otter densities are two to three times greater in the rocky bottom (kelp bed) habitats (R. Jameson and J. Estes, unpublished data). This does not establish kelp production as the cause, although it is consistent with that possibility. At Amchitka Island, the sea otter population appears to have attained an elevated carrying capacity following an earlier post-recovery decline and the subsequent development of extensive kelp forests in that area (Estes *et al.* 1989, Estes 1990).

A second class of indirect effects from sea otter

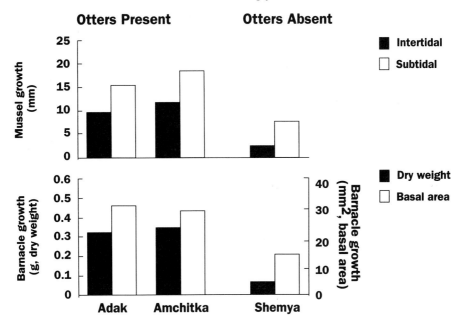

Otters Present   **Otters Absent**

Legend: ■ Intertidal □ Subtidal

■ Dry weight □ Basal area

Adak   Amchitka   Shemya

Fig. 10.4. Growth rates of suspension feeders (mussels and barnacles) at Aleutian Islands where sea otters were either abundant (Adak and Amchitka) or absent (Shemya). (Redrawn from Duggins *et al.* 1989).

predation is those extending laterally through coastal food webs (Fig. 10.3). Here again, the examples and evidence are mainly from inter-island comparisons in the Aleutian archipelago. The first of these involves competitive interactions between the two most common genera of epibenthic kelps, *Laminaria* and *Agarum*. *Laminaria* includes a group of species that are strong competitors while being poorly defended (chemically) against herbivores. *Agarum*, in contrast, is well defended chemically, but competitively subordinate to *Laminaria* (Dayton 1975, Estes & Steinberg 1988). Sea otters, by limiting the abundance of herbivorous sea urchins in shallow water, create an environment in which *Laminaria* is able to competitively exclude *Agarum*. *Agarum*, on the other hand, is more successful in habitats in which the intensity of herbivory is greater. Competitive interactions among algal species have been demonstrated at several other locations in the north-east Pacific Ocean where sea otters occur (Duggins 1980, Reed & Foster 1984).

Two further examples involve seabird species that feed on benthic invertebrates, thus being potential competitors with sea otters. One of these examples concerns the foraging behaviour of glaucous-winged gulls *Larus glaucescens*. Where otters were absent, the gulls foraged extensively on benthic invertebrates in the mid- to lower-littoral zones, and during summer at least, their diet comprised these forms almost exclus-

ively. In contrast, glaucous-winged gulls fed mostly on small fishes at Amchitka Island, where sea otters were abundant (Trapp 1979, Irons *et al.* 1986). A second example concerns the common eider *Somateria mollissima*, a benthic feeding sea duck. A remarkably strong negative correlation exists between eider density and the presence/absence of sea otters across the Aleutian Islands, and a causal role for otters is suggested by changes observed at Attu Island, where common eider density declined more than tenfold in the 1980s and 1990s, following re-establishment of the sea otter. During this same period, eider densities remained largely unchanged elsewhere in the Aleutian Islands (US Fish & Wildlife Service, unpublished data). This is unlikely to be a direct effect of predation because sea otters rarely eat seabirds (Riedman & Estes 1988).

Other likely examples of indirect food web effects from sea otter predation can be inferred from ecological and oceanographic studies in which the sea otter was not directly considered. For instance, surface canopy-forming kelps have been shown to block the shoreward movement of barnacle *Balanus* spp. larvae, thus influencing the structure of adult barnacle populations in the rocky littoral zone (Gaines & Roughgarden 1985, Roughgarden *et al.* 1988). These findings suggest that sea otters, by enhancing kelp abundance, may exert a negative effect on the abundance of adult barnacles.

The above-described cases all involve large, conspicuous, or well-studied species. Many other coastal-living species, although unstudied, also may be indirectly influenced by sea otter predation. This poorly known set of potential interactions is a rich arena for further research.

EVOLUTION

Strong species interactions, when persistent over evolutionary timescales, should have selective effects on participating species. Macroevolutionary events of this kind sometimes can be understood by contrasting species characteristics over sufficiently large scales of space or time, across which the purported agent of selection also varies. This approach has provided evidence for a selective role by consumers in a variety of species and ecosystems. Marine examples include the influence of terrestrial versus aquatic predators on the escape behaviour of arctic and antarctic pinnipeds (Stirling 1983), escalating predation through geological time on the evolution of gastropod shell form (Vermeij 1977), and escalating herbivory on the functional morphology of crustose coralline algae (Steneck 1983).

We have used this approach to study the selective influences of sea otters and their recent ancestors on marine plant/herbivore interactions (Steinberg *et al.* 1995). This was done by contrasting biotic assemblages between the temperate north-east and south-west Pacific oceans. The north-east Pacific region, as shown previously, is influenced by the effect of sea otter predation on herbivorous invertebrates, and probably has been since at least the upper Miocene (Repenning *et al.* 1979, Berta & Morgan 1986). Temperate Australasia, in contrast, while containing similar guilds of plants and herbivores, lacks a predator of comparable influence to the sea otter (Estes & Steinberg 1988). Thus we imagine that northern hemisphere kelp forests have evolved as three-tiered food webs, with sea otters and their recent ancestors at the top, herbivores in the middle, and kelps and other macroalgae as the autotrophs. In odd-numbered food webs of this sort, herbivores are limited by predation, thus protecting the autotrophs from intense and chronic disturbance via herbivory. In contrast, herbivores in even-numbered food webs are not limited by herbivory, thus leaving the autotrophs unprotected. These concepts (Hairston *et al.* 1960, Fretwell 1977, Power 1990) and patterns led us to predict that temperate latitude marine autotrophs should be defended against their herbivores in Australasia but not in the North Pacific, a prediction which has since been substantiated. Marine algae extensively use secondary metabolites as defences against herbivory (Hay & Fenical 1988), and northern hemisphere phaeophytes (brown algae) contain low

concentrations of secondary metabolites (phlorotannins) whereas those in southern hemisphere algae are much higher (Steinberg 1989). There is also evidence that herbivores from these regions have evolved differing abilities to cope with chemical defences (Steinberg & van Altena 1992, Steinberg *et al.* 1995).

These findings suggest that sea otter predation, by freeing North Pacific autotrophs from the need to defend themselves, has been a key factor in the evolution of a marine flora that is vulnerable to herbivory. This perspective may help explain why north-east Pacific kelp forests, many of which still lack sea otters, are so frequently destabilized by destructive herbivory (Harrold & Pearse 1987).

SOCIAL AND ECONOMIC IMPACTS

The influences of sea otter predation probably impact human welfare in several important ways. The most obvious and well known of these is a detrimental impact on shellfisheries. The majority of North Pacific shellfisheries would never have developed had sea otters not first been removed from the system (Estes & VanBlaricom 1985). However, otters were exterminated from most of their range, and people subsequently developed livelihoods and recreational pursuits around the abundant standing stocks of sea otter prey that formed in their absence. Otters have since been protected, and as the range of recovering populations has expanded, their prey have been returned to natural levels. These events, complex even within the realm of the natural sciences, have taken on economic, social, and political dimensions that have propelled the issues beyond the traditional considerations of resource management and conservation.

Resource conflicts revolving around sea otters and shellfisheries are complicated by significant indirect effects that emanate from the predator – prey interactions. For instance, giant kelp *Macrocystis pyrifera* is exploited in a viable industry; kelp beds provide habitat for certain fish species that are exploited in recreational or commercial fisheries, and either the physical presence of kelp, or the organic detritus derived from kelp beds probably sustains juvenile stages of fish species that as adults are exploited by humans. The social and economic consequences of these interactions are mostly unstudied, and thus largely unknown.

BEYOND OTTERS

Sea otter predation has a wide range of important impacts on coastal ecosystems. Are impacts of this nature typical for other species of aquatic birds and mammals (most of which are top-level carnivores), or are they unique to the sea otter? Although important ecological functions have been suggested for several other marine mammals and birds, the majority of species are unstudied from this perspective. How should we go about answering this question? I think it unlikely that the answers will be found in descriptions of the internal workings of static systems. Experiments involving variation in predator abundance are needed to properly assess the ecological functions of these species, although purposeful manipulations are unlikely to succeed in most cases for both logistical and ethical reasons. For most large, mobile predators, natural experiments (comparisons using unplanned spatial or temporal variation in predator abundance) can offer a reasonable compromise. When interpreted with sufficient care and forethought, these situations can provide large-scale views of systems under the varying influences of predators, without the attendant problems of having to manipulate predator populations. Unfortunately, the necessary conditions for such natural experiments do not occur often and are thus rarely available when needed. Nonetheless, if ever our present state of ignorance concerning the ecological importance of large, mobile predators is to be rectified, opportunities for the application of natural experiments must be looked for and exploited when they arise.

ACKNOWLEDGEMENTS

I thank the many friends and colleagues that have participated in the work so briefly summarized here. I also am grateful to the conference organizers, especially Dr Simon Greenstreet, for making it possible for me to attend this conference.

REFERENCES

Berta, A. & Morgan, G.S. (1986) A new sea otter (Carnivora: Mustelidae) from the late Miocene and early Pliocene (Hemphilian) of California. *Journal of Paleontology*, **59**, 809–819.

Dayton, P.K. (1975) Experimental studies of algal canopy interactions in a sea otter-dominated community at Amchitka Island, Alaska. *Fishery Bulletin*, **73**, 230–237.

Duggins, D.O. (1980) Kelp beds and sea otters: an experimental approach. *Ecology*, **61**, 447–453.

Duggins, D.O., Simenstad, C.A. & Estes, J.A. (1989) Magnification of secondary production by kelp detritus in coastal marine ecosystems. *Science*, **245**, 170–173.

Estes, J.A. (1990) Growth and equilibrium in sea otter populations. *Journal of Animal Ecology*, **59**, 385–401.

Estes, J.A. & Duggins, D.O. (1995) Sea otters and kelp forests in Alaska: generality and variation in a community ecological paradigm. *Ecological Monographs*, **65**, 75–100.

Estes, J.A. & Palmisano, J.F. (1974) Sea otters: their role in structuring nearshore communities. *Science*, **185**, 1058–1060.

Estes, J.A. & Steinberg, P.D. (1988) Predation, herbivory, and kelp evolution. *Paleobiology*, **14**, 19–36.

Estes, J.A. & VanBlaricom, G.R. (1985) Sea-otters and shell-fisheries. *Marine Mammals and Fisheries* (eds J.R. Beddington, R.J.H. Beverton & D.M. Lavigne), pp. 187–235. George Allen and Unwin, London.

Estes, J.A., Duggins, D.O. & Rathbun, G.B. (1989) The ecology of extinctions in kelp forest communities. *Conservation Biology*, **3**, 252–264.

Foster, M.S. (1990) Organization of macroalgal assemblages in the northeast Pacific: the assumption of homogeneity and the illusion of generality. *Hydrobiologia*, **192**, 21–33.

Foster, M.S. & Schiel, D.R. (1988) Kelp communities and sea otters: keystone species or just another brick in the wall? *The Community Ecology of Sea Otters* (eds G.R. VanBlaricom & J.A. Estes), pp. 92–108. Springer-Verlag, Berlin.

Fretwell, S.D. (1977) The regulation of plant communities by food chains exploiting them. *Perspectives in Biology and Medicine*, **20**, 169–185.

Gaines, S. & Roughgarden, J. (1985) Larval settlement rate: a leading determinant of structure in an ecology community of the marine intertidal zone. *Proceedings of the National Academy of Sciences*, **82**, 3707–3711.

Hairston, N., Smith, F.E. & Slobodkin, L. (1960) Community structure, population control, and competition. *American Naturalist*, **94**, 421–425.

Harrold, C. & Pearse, J.S. (1987) The ecological role of echinoderms in kelp forests. *Echinoderm Studies* (eds M. Jangoux & J.M. Lawrence), pp. 137–233. A.A. Balkema, Rotterdam.

Hay, M.E. & Fenical, W. (1988) Marine plant–herbivore interactions: the ecology of chemical defense. *Annual Review of Ecology and Systematics*, **19**, 111–145.

Hockey, P.A.R. & Branch, G.M. (1984) Oystercatchers and limpets: impact and implications. *Ardea*, **72**, 199–206.

Irons, D.B., Anthony, R.G. & Estes, J.A. (1986) Foraging strategies of glaucous-winged gulls in a rocky intertidal community. *Ecology*, **67**, 1460–1474.

Laws, R.M. (1977) Seals and whales in the Southern Ocean. *Philosophical Transactions of the Royal Society of London, Series B*, **279**, 81–96.

Mann, K.H. (1982) *Ecology of Coastal Waters*. University of California Press, Berkeley.

Marsh, C.P. (1987) Impact of avian predators on high intertidal limpet populations. *Journal of Experimental Marine Biology and Ecology*, **104**, 185–201.

Oliver, J.S., Kvitek, R.G. & Slattery, P.N. (1985) Walrus disturbance: scavenging habits and recolonization of the Bering Sea benthos. *Journal of Experimental Marine Biology and Ecology*, **91**, 233–246.

Oliver, J.S. & Slattery, P.N. (1985) Destruction and opportunity on the sea floor: effects of gray whale feeding. *Ecology*, **66**, 1965–1975.

Power, G. & Gregoire, J. (1978) Predation by fresh water seals on the fish community of Lower Seal Lake, Quebec. *Journal of the Fisheries Research Board of Canada*, **35**, 844–850.

Power, M.E. (1990) Effects of fish in river food webs. *Science*, **250**, 411–415.

Reed, D.C. & Foster, M.S. (1984) The effects of canopy shading on algal recruitment and growth in a giant kelp forest. *Ecology*, **65**, 937–948.

Repenning, C.A., Ray, C.E. & Grigorescu (1979) Pinniped biogeography. *Historical Biogeography, Plate Tectonics, and the Changing Environment* (eds J. Gray & A.J. Boucot), pp. 357–369. Oregon State University Press, Corvallis.

Riedman, M.L. & Estes, J.A. (1988) Predation on seabirds by sea otters. *Canadian Journal of Zoology*, **66**, 1396–1402.

Roughgarden, J., Gaines, S.D. & Possingham, H.P. (1988) Recruitment dynamics in complex life cycles. *Science*, **241**, 1460–1466.

Steinberg, P.D. (1989) Biogeographical variation in brown algal polyphenolics and other secondary metabolites: comparison between temperate Australasia and North America. *Oecologia*, **78**, 374–383.

Steinberg, P.D. & van Altena (1992) Tolerance of marine invertebrate herbivores to brown algal phlorotannins in temperate Australasia. *Ecological Monographs*, **62**, 189–222.

Steinberg, P.D., Estes, J.A. & Winter, F.C. (1995) Evolutionary consequences of food chain length in kelp forest communities. *Proceedings of the National Academy of Sciences USA*, **92**, 8145–8148.

Steneck, R.S. (1983) Escalating herbivory and resulting adaptive trends in calcareous algal crusts. *Paleobiology*, **9**, 44–61.

Stirling, I. (1983) The evolution of mating systems in pinnipeds. *Recent Advances in the Study of Animal Behaviour* (eds J.F. Eisenberg & D.G. Kleiman), pp. 489–527. Special Publication No.7, American Society of Mammalogists, Lawrence, Kansas.

Strong, D.R. (1992) Are trophic cascades all wet? Differentiation and donor control in speciose ecosystems. *Ecology*, **73**, 747–754.

Trapp, J.L. (1979) Variation in summer diet of glaucous-winged gulls in the western Aleutian Islands: an ecological interpretation. *Wilson Bulletin*, **91**, 412–419.

Vermeij, G.J. (1977) The mezozoic marine revolution: evidence from snails, predators, and grazers. *Paleobiology*, **3**, 245–258.

Watson, J.C. (1993) *Effects of sea otter* Enhydra lutris *foraging on rocky sublittoral communities off northwestern Vancouver Island, British Columbia*. PhD Dissertation, University of California, Santa Cruz, California.

CHAPTER 11

Modelling the responses of mussel *Mytilus edulis* and oystercatcher *Haematopus ostralegus* populations to environmental change

J.D. Goss-Custard and R.I. Willows

SUMMARY

(1) Models which attempt to predict the responses of populations to new environmental circumstances from empirical relationships acquired in the current environment may give unreliable predictions when applied to situations that lie beyond the empirical range. Indeed, practical constraints may prevent the necessary empirical relationships from being acquired in the first place. Using the interaction between mussels *Mytilus edulis* and oystercatchers *Haematopus ostralegus* as the example, this paper discusses how such difficulties may be addressed by models built from functions that are derived from fundamental properties of the organisms themselves.

(2) In the case of the mussels, the model is based on the energetics and physiological responses of individual animals to natural and anthropogenic factors. In the case of the oystercatchers, which eat the mussels, the model is based on the energetic consequences for individual birds of their behavioural responses to spatial and temporal variations in the abundance and quality of these bivalves and on their responses to each other. Both models give predictions that are qualitatively in line with field data in the main study area, the Exe estuary in south-west England.

(3) Combining these two models allows us to predict how the interaction between this predator and prey will be affected by novel combinations of environmental factors, such as those arising from changed levels of environmental contaminants and increased shell-fishing. Because the models are largely based on properties of the organisms that are unlikely to change in the new circumstances, they should provide a firm basis for prediction.

(4) The particular circumstances discussed in this paper concern increased hydrocarbon pollution combined with a reduction in oystercatcher feeding areas arising, for example, from overfishing mussel beds to extinction. The chronic presence of low levels of hydrocarbons affects the metabolism of the mussels and reduces their growth rates. Hydrocarbons, in addition to other contaminants such as the biocide TBT, may also act to increase the thickness of their shells. Both these changes reduce the rate at which the birds can feed and further intensifies the increased interference and exploitation competition amongst them resulting from the loss of feeding areas. This, in turn, affects the ability of the birds to build up the fat reserves they need for surviving the winter and for migrating to their breeding areas in spring.

(5) The models will allow the long-term consequences of such environmental changes for the dynamics of both the mussel and oystercatcher populations, and of the interaction between them, to be explored. They can also be applied to other shellfish and shorebird species, such as the cockle *Cerastoderma edule* and the knot *Calidris canutus* which eat them.

Key-words: environmental change, estuary, *Haematopus ostralegus*, model, mussel, *Mytilus edulis*, oystercatcher, population dynamics, predator–prey

---

INTRODUCTION

The many human activities that directly or indirectly influence estuaries not only affect commercial and recreational interests but also affect the specialized organisms that live there. In Britain, for example, conservationists are particularly concerned that anthropogenic factors will affect the shorebirds and wildfowl for which British estuaries are so important as feeding areas when the birds are overwintering or on passage. However, as the shellfish industry illustrates, estuaries also have great economic potential, the increasing realization of which may have an effect on these birds. Such potential conflicts of interest require predictions as to how various human activities will affect the ecological functioning of estuaries, especially those that bear upon these commercial, recreational and conservational interests. In such circumstances, vague forecasts as to what will happen if one or other compromise decision is taken are likely to be inadequate. Furthermore, inaccurate predictions could have detrimental economic and environmental consequences. Quantitative predictions are required so that the costs and benefits of particular policy options can be evaluated. To achieve this, some form of modelling is clearly needed.

The difficulty here is that predictions are often required for quite novel environmental circumstances. This raises the danger that models which attempt to make predictions from empirical relationships that have been acquired in the current environment may give unreliable predictions when applied to situations that, in effect, lie beyond the empirical range. Indeed, as Figure 11.1 also illustrates, practical constraints may make it difficult to establish necessary empirical relationships in the first place. Using the interaction between mussels *Mytilus edulis* and oystercatchers *Haematopus ostralegus* on the Exe estuary in south-west England as the example, this paper discusses how such difficulties may be addressed by models that are built from functions that are derived from basic properties of the organisms themselves. In the case of the mussels, the functions are based on the energetics and physiological responses of individual animals to natural and anthropogenic factors. In the case of the oystercatchers, which eat the mussels, the functions are based on the energetic consequences for individual birds of their behavioural responses to spatial and temporal variations in the abundance and quality of these bivalves and on

their responses to each other. Although the two models are only now being integrated to study the reciprocal interactions between this prey and predator, sufficient progress has been made to illustrate how they can be used to predict the responses of both populations to combinations of environmental changes that have not yet been observed directly.

THE MUSSEL MODEL

The processes determining the dynamics of mussel populations can be divided into two components, those acting at the level of the population and those dependent on the physiological performance of individual mussels. The population level processes are particularly difficult to resolve. They include, for example, the direct consequences of the population density and size-structure on the availability of food, and on individual physiological performance and survival (Fréchette *et al.* 1992). Furthermore there are density-dependent effects associated with other predators in addition to the birds (McGrorty *et al.* 1990). Perhaps most problematical are the factors determining the settlement and recruitment of the long-lived, widely dispersed planktonic larval stage into particular beds of adult mussels. The second component concerns the impact of the environment on the physiology and energetics of individual mussels. There exists a wealth of published information on the ecophysiology of *M. edulis*, and here we confine ourselves to this aspect of the modelling effort.

Model structure

The physiological model provides a description of the feeding, growth, respiration, and reproductive partitioning of cohorts of mussels from settlement. Furthermore, the model incorporates functions which allow the impact of physiological state variables, such as size and reproductive output, on survivorship to be modelled. The model is defined as a coupled set of differential equations. It is conceptually similar to that proposed by Ross & Nisbet (1990); however, it incorporates more detail of respiratory physiology, but omits any explicit consideration of a storage component.

The model equations (Table 11.1) define the rate of assimilation of energy as a function of mussel size, bed exposure and food concentration. Although detailed

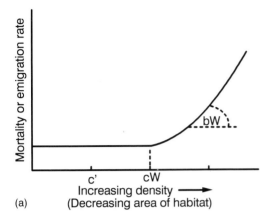

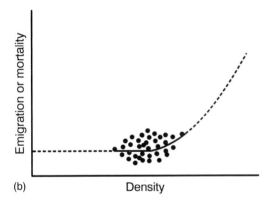

Fig. 11.1. (a) Hypothetical relationship between the mortality rate of a large shorebird, such as an oystercatcher, and their density on the feeding grounds. Below a certain density cW the rates are independent of density, but above this level an increasing proportion of birds leave the estuary or die. The slope bW of this density-dependent part of the function will have a critical effect on the response of the population to habitat loss which, of course, will increase bird density if the same number of birds attempt initially to stay in the area. (b) This shows, however, that it may be extremely difficult to estimate both cW and bW from field measurements of mortality and emigration (which are difficult to separate in such mobile animals anyway). Each dot refers to one estimate for one estuary of the overwinter mortality or emigration rate in a long-lived species whose population size does not change much over several decades. Despite a lifetime of work, the researcher would still have little idea of what would happen to both rates outside the narrow range observed and would have great difficulty in estimating the critical parameters cW and bW. Yet both parameters need to be estimated if the quantitative consequences of habitat loss are to be forecast. Furthermore, this empirical relationship would only still apply if habitat of average quality were to be removed. If feeding areas of above-average quality were lost, for example, the values of both cW and bW would change as the function changed shape (Goss-Custard *et al.* 1995b).

understanding of mussel feeding behaviour and physiology exists, and has been successfully modelled (Ross & Nisbet 1990, Willows 1992), the much simpler representations used here are adequate to generate realistic growth responses to variable feeding environments. Respiration is defined as a function of size (equivalent to the empiricists' basal metabolic rate) and also a function of assimilation (equivalent to the specific dynamic action) and growth rate (equivalent to the costs of growth) (Widdows & Hawkins 1989).

By modelling both assimilation rate (A) and respiration rate (R), scope-for-growth can be calculated as SfG = A − R. Here we do not explicitly consider the energy costs associated with the excretion of nitrogenous wastes. Excretory losses, when measured as part of an energy budget study, are generally small but dependent on physiological activity. For example, Widdows & Johnson (1988) estimated the energetic equivalent of nitrogenous excretion to be 1–7% of respiration or assimilation rate, dependent on the experimental growth conditions. In the context of this model, excretory losses can be considered to be partitioned between the activity costs associated with growth, assimilation and maintenance, and subsumed, for convenience, within the total respiratory cost term.

The individual net rate of production is equal to scope-for-growth, and is equivalent to the growth rate prior to the onset of reproductive partitioning. In order to determine the start of reproductive allocation, we invoke a hypothesis that states that reproductive maturity is achieved at the age (and size) at which the rate of net production with respect to size (the accumulated production) reaches a maximum. Following reproductive maturity, the fraction of scope-for-growth allocated to reproduction is an increasing function of size, unless scope-for-growth is negative under which circumstances the fractional allocation is zero. The accumulated reproductive allocate is set to zero annually, that is spawning is triggered, with reference to the seasonal increase in the available food concentration.

Length is calculated as an allometric function of size (which excludes resources allocated to reproduction), with an exponent of 1/3. When scope-for-growth is negative the size of the mussel decreases, but the length is held constant. Additional length increase is permitted once sufficient growth has occurred such that the maximum size previously achieved is equalled. With these rules the model successfully reproduces observed seasonal changes in length/weight allometries (e.g. Bayne & Worrall 1980) although it fails to model detailed uncoupling of growth in length and weight reported for some populations (Hilbish 1986).

Scope-for-growth has been widely used as a sensitive measure of the degree of chronic sublethal toxic effect

Table 11.1. Equations used in the mussel model. Forcing variable: F, food concentration. State variables: S, size; R_c, reproductive allocate; N, cohort density. Subsidiary variables: A, assimilation rate; L, length; R, respiratory costs; BMR, basal metabolic rate; AMR, active metabolic rate; P_r, reproductive allocation fraction. Model parameters: F_{max}, F_{min}, A_{max}, R_{max}, α, β, γ, δ, ε, ϕ, η, φ, κ, μ, ω, υ, ν.

Food	F	$= F_{min} + 0.5(F_{max} - F_{min})(1 + \sin(2\pi t))$
Assimilation	A	$= A_{max}(1 - \exp^{-\kappa F})(1 - \exp^{-a\{F\}S})$
Respiration	R	$= BMR + AMR$
Scope-for-growth	SfG	$= A - R$
Growth	dS/dt	$= SfG(1 - P_r)$
Length	dL/dt	$= \beta S/L \, SfG$
Reproduction	dR_c/dt	$= SfG \, P_r$
Mortality	dN/dt	$= -(\mu P_r^{\omega} + \upsilon \exp^{-\nu S})\,N$
where	BMR	$= \alpha S$
	AMR	$= \gamma A + (1 - P_r)\delta(A(1 - \gamma) - BMR)/(1 + \delta(1 - P_r))$
If SfG > 0 and S > S_r	P_r	$= R_{max}(1 - \exp^{-\eta(S - Sr)})$ else $P_r = 0$
	a	$= \phi + \varphi \exp^{-\varepsilon F}$

resulting from the presence of anthropogenic contaminants (Widdows *et al.* 1987, Widdows & Donkin 1991). A benefit of explicitly incorporating such physiological detail is that it is then possible to model the effects of classes of environmental contaminants with differing mechanisms of sublethal toxicity. Three mechanisms of toxic action have been recognized from studies on *Mytilus*, each having distinct and separate effects on feeding and respiratory physiology (Widdows & Donkin 1991). The incorporation of these distinct sublethal mechanisms within simple models of physiology has been outlined elsewhere, and the implications for physiological performance under static environmental conditions analysed (Willows 1994). Low, environmentally relevant levels of hydrocarbons, cause a reversible non-specific narcosis which characteristically reduces the feeding activity of mussels. The proportionate reduction in feeding rate has been found to be directly proportional to the log exposure or tissue concentration of certain aromatic hydrocarbons (Donkin *et al.* 1989). Although contaminants may affect both feeding (energy intake rate) and respiration rate, under simple narcosis only feeding rate is affected. It is the consequences of a reduction in net energy intake rate for mussel growth and survival that are discussed here, within the context of a dynamic feeding environment. This allows us to model the impact of environmental changes resulting from changed levels of contaminants and concentrations of suspended particulate food on the availability of suitably sized mussel prey, but not the influence of population level feedbacks on either growth or survival.

Parameter estimation

A formal estimation of parameter values, fitting the model to the observed Exe mussel populations (McGrorty *et al.* 1990), has yet to be accomplished and awaits completion of the data analysis. Unfortunately no suitable data exist on the supply of suspended food particulates to mussel beds within the Exe estuary. Hence an arbitrary, seasonally variable food supply has been assumed, and the model parameterized to provide an approximate fit to observed variation in growth trajectories for cohorts of mussels occupying different beds, with different durations of tidal innundation, within the Exe system. In order to select parameter values to give representative variations in tissue weight, resulting from the reproductive cycle and spawning, and overwinter weight loss due to starvation (that is metabolism of tissues), reference has been made to published data on other populations within south-west England (Bayne & Worrall 1980, Bayne *et al.* 1983). Parameterization of the components of respiratory metabolism is based on Widdows & Hawkins (1989).

Model results – consequences of physiological stress

Within the Exe estuary adult mussels achieve maximum lengths of between 55 and 80 mm at about ten years of age, dependent on the particular bed (S. McGrorty, personal communication). As a result of large individual variability in growth rates they achieve 30 mm, the size at which oystercatchers start to eat mussels, at between two and five years of age. Hence factors which act to

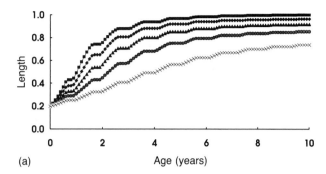

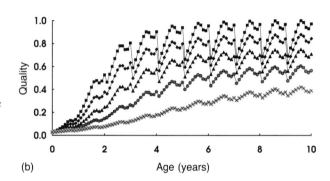

Fig. 11.2. Mussel physiological model simulations in response to increasing levels of sub-lethal toxic stress reducing feeding rate, and hence energy intake rate. (a) Mussel growth is shown as the change in length, and (b) quality expressed as total tissue weight (somatic plus reproductive tissues) per unit length, at proportionate reductions of: ■, 0%; ◆, 10%; ▲, 20%; ●, 30% and ×, 40% of the unimpacted energy intake rate. Length and quality are expressed as a proportion of their maximum unimpacted values.

reduce the maximum size mussels may achieve within the system, or reduce the rate at which they achieve the threshold size at which, under present feeding conditions, they start to be eaten, will reduce the quality of the feeding habitat for the birds. Figure 11.2(a) shows the simulated growth performance of five cohorts of mussels under hypothetical increased levels of chronic sublethal stress that act to reduce net energy intake rate by up to 40%. The effect of such stress is to have an increasing impact upon the maximum asymptotic length achieved after ten years, reducing maximum length by up to 30%. However, the effect on growth rate is more marked, with animals of age two to three years less than half the size of non-impacted individuals. By expressing mussel quality as total tissue weight (somatic plus reproductive tissues) per unit length it is possible to derive a simple standardized index of the quality of mussels as food (Fig. 11.2(b)). Note that the effects of stress on mussel quality are particularly large on the younger age groups, due to the much lower rates of growth achieved by these animals.

Currently oystercatchers only feed on mussels that have reached a size that is approximately 40% of the maximum achieved by the best performing mussels within the Exe estuary. The age at which mussels first achieve 40% of maximum size is rather sensitive to increasing levels of sublethal stress (Fig. 11.3). This increased time taken to enter the size classes at which they are consumed corresponds to a decreased rate of supply of food for the oystercatchers. If these birds are unable to maintain body condition by feeding on mussels smaller than this size for the estuary, then the model suggests that moderate levels of sublethal stress for the mussels may have a disproportionately large impact on the quality of the estuary as an overwintering site for avian predators.

Decreasing size-dependent mortality and increased mortality associated with reproduction, particularly spawning, have been recorded for some mussel populations (McGrorty *et al.* 1990, Worrall & Widdows 1984), although detailed functional forms are very difficult to elucidate from field-based studies. By hypothesizing the form of the possible dependency of survivorship on mussel size (mortality decreasing with increasing body size) and reproductive effort (mortality increasing with fractional allocation to reproduction) (see Table 11.1), we are able to simulate the possible consequences of the reduced growth potential of the environment on the physiological dependent components of survivorship. Of course, the sensitivity of survivorship to changes in sublethal stress, resulting in reduced size and reproductive output at a given age,

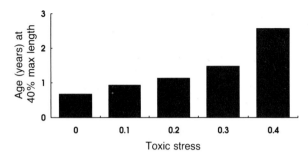

Fig. 11.3. The predicted time taken by mussels to achieve 40% of the maximum length observed for mussels growing within the Exe, at increasing levels of sublethal stress. 40% of maximum size corresponds to the minimum length of mussels upon which oystercatchers will feed.

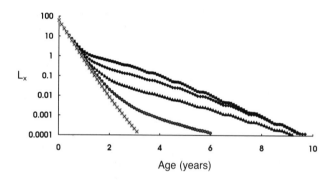

Fig. 11.4. Modelled impact of increased levels of sublethal toxic stress acting to reduce feeding rate, and hence energy intake rate, on the physiological component of survivorship (L_x) of a cohort of mussels. Reductions are: ■, 0%; ◆, 10%; ▲, 20%; ●, 30% and ×, 40% of the unimpacted energy intake rate.

depend on the particular functional forms chosen and their parameter values. In addition there may be density dependent compensations occurring at the population level which may mitigate any increased mortality. However, the model illustrates the possible consequences for the density of mussels surviving to any given age and size under conditions of increasing sublethal stress (Fig. 11.4). Clearly, with size-dependent mortality acting strongly on the smaller size groups, a reduced rate of growth may have a particularly large and increasing net effect on the density of survivors, and hence on the availability of suitably large prey items.

THE OYSTERCATCHER MODEL

The dynamics of populations are critically affected by the form and precise parameter values of their density-dependent functions (May 1981). Yet, not only may practical constraints often limit our ability to define these functions but existing functions may not apply anyway in the new circumstances for which forecasts are required (Fig. 11.1). An alternative approach to defining the form, and measuring the parameter values, of the critical feedback functions is therefore required.

On the Exe estuary, an attempt is being made to derive the overwinter mortality function from studies of variation in the behaviour of individual oystercatchers. Although the feedback from bird density to the rates of emigration and mortality might be caused by either predators or parasites, we have so far focused on competition for food. Many of the oystercatchers that die in Europe in winter have starved during severe weather conditions, those that die often having not laid down sufficient energy reserves during the earlier part of the winter prior to the onset of severe weather (Goss-Custard *et al.*, 1996c). Accordingly, attention has focused on competition for food as the most appropriate feedback process upon which to base model functions.

Feedback from oystercatcher population density to the intake rate of individual birds arises from prey depletion, both over one winter and in the long term, and from interference between foraging birds (Goss-Custard 1977, 1980, Zwarts & Drent 1981, Meire *et al.* 1994). The idea underlying the model is that, where competition occurs for resources, measurement of the distribution of competitive abilities amongst the individuals in the population will enable the proportion that fail to compete effectively – and so die or emigrate – at different population sizes to be predicted, thus allowing the parameters of the density-dependent function to be estimated (Goss-Custard 1993). Since the resources are spatially heterogenous, the proportion failing will also depend on the pattern of spatial variation in the resource, with more individuals failing as the proportion of feeding areas that are of poor quality increases. As the overwinter depletion of the food supplies of oystercatchers may occur at different rates in different places (Hancock 1971, Brown & O'Connor 1974, Horwood & Goss-Custard 1977), and as the quality of individual prey may change as they lose condition (Dare & Edwards 1975, Bayne & Worrall 1980), seasonal changes in food supply in different places must also be considered. Accordingly, the responses of individual oystercatchers to each other and to their spatially and seasonally varying mussel food supply have been studied and the results used to derive key functions in an individuals-based game theoretic model. In this, the choices made by one competing individual as to where to feed at any one time during the winter are contingent on those made by all the other survivors exploiting the estuary at that time. The model tracks the body condition and overwinter survival of each individual and, by running simulations over a range of initial population sizes, the shape and parameter values cW and bW of the winter density dependent function (Fig. 11.1) can be derived.

Model structure

The intake rates of oystercatchers feeding on mussels depend on several factors. For all individuals, intake rate is affected by the density, size-distribution, energy content and, in some birds perhaps, the average shell thickness of the mussels (Zwarts *et al.*, 1996). The values of all these parameters in mussels on the Exe vary both spatially and seasonally (Goss-Custard *et al.* 1993), and will affect all individuals in the model. However, the rates achieved by individual birds feeding in the same place at any one time will also vary for two reasons. Firstly, individuals differ in the rate at which they consume mussel flesh in the absence of interference from other birds. This 'interference-free intake rate' (IFIR) is thought to reflect the efficiency with which a bird forages. Individuals of low foraging efficiency are the birds most likely to be affected by the increased prey depletion following an increase in bird numbers. The intake rate an individual achieves depends secondly on the immediate interference effect that other foraging oystercatchers have on its rate of consumption. An individual's 'susceptibility to interference' (STI) is the amount by which its intake rate is reduced per unit increase in oystercatcher density. STI is, in turn, related to a bird's dominance and to its feeding method, being highest in sub-dominant and hammering birds (Goss-

Custard & Durell 1988). By definition, the amount by which intake rate is reduced below its IFIR by interference also depends on the density of oystercatchers; clearly, subdominants are more affected by increased levels of interference when bird densities are high than when they are low. An individual's intake rate at any one time and place thus depends on the food supply, on its foraging efficiency and on its dominance and on the density of oystercatchers where it is feeding.

The model tracks the feeding location, intake rate and body condition of each bird on each day of its presence on the Exe from mid-September to mid-March. The biomass of mussels above 30 mm long on each of the 12 mussel beds studied, averaged over eight Septembers, defines the food supply at the beginning of the winter. Although the area of each bed remains constant over the winter, the proportion of each that is accessible to oystercatchers varies with the neap–spring tide cycle. Depending on its position on the shore, 0–100% of a bed may remain covered at low water on extreme neap tides, thus causing oystercatcher densities, and so interference competition, to be very high in the reduced areas that are exposed (Goss-Custard & Durell 1987). From September onwards, the food supply declines on each bed as it is exploited by oystercatchers and as mussels lose condition and disappear from storms and other mortality agents. At present, the losses due to storms and loss of condition are included as rates empirically determined in the field, but in future versions of the model, the loss of condition will be derived from the mussel model.

At each stage during the winter, the intake rate of an oystercatcher of average foraging competence, in the absence of interference from other birds, is estimated for each mussel bed from the biomass density of mussels present, using a functional response derived from a simple, empirically-derived optimal foraging model (Goss-Custard *et al.* 1996b). The putative intake rate in the absence of interference is then calculated for each individual bird on each bed given its efficiency relative to that of a bird of average competence. The rate at which an oystercatcher would actually feed on each bed is then affected by the amount of interference it would experience on each bed, given the density of oyster-catchers present and its own dominance relative to those of the other birds present. In each daily iteration of the model, each individual is selected in random order to choose, within the 3% limits of its empirically-determined ability to discriminate (Goss-Custard *et al.* 1995a), the mussel bed on which it achieves the highest gross intake rate at that time. The assumption that they make foraging decisions on this basis has much empirical support; for example, oystercatchers select prey size-classes largely on this basis (Zwarts *et al.*, 1996). Many birds, particularly the sub-dominant ones, continually change their feeding site; because of depletion, the relative quality of the mussel beds change and competitors change their foraging location.

The age distribution of the population in the model in September matches that recorded on the Exe itself. Each bird is given a level of fat reserves in mid-September drawn at random, and currently without regard to its dominance or foraging efficiency, from the distribution of weights measured on the Exe at that time of year for each age-class. Subsequently, birds either put on fat reserves, or metabolize them, according to how well they feed each day. Birds can feed for 12 hours in every 24 and, following Hulscher (1996), at the same rate at night as during the day. The limited capacity of the oystercatcher gut to process mussel flesh (Kersten & Visser 1996), along with the known efficiency with which it assimilates mussel flesh (Kersten & Piersma 1987), limits the rate at which each bird in the model can consume food. Daily energy requirements for each day of the winter are estimated from the studies of Kersten & Piersma (1987).

Once an individual has consumed its energy requirements during a 24 h period, it lays down any surplus energy consumed as fat, deposited with an empirically measured efficiency (Kersten & Piersma 1987), up to a maximum rate of 5% of current body weight per day (Zwarts *et al.* 1990). Each bird attempts during autumn and winter to accumulate fat reserves at the rate observed by its age-class on the Exe over 15 years. The reserves are used to maintain the bird on days when it fails through foraging to meet its current temperature-related energy requirements. Since immature oyster-catchers on the Exe lose weight after mid-December, all young birds in the model draw on their energy reserves, before beginning to feed, from mid-winter onwards, at the average daily rate of weight loss measured in the field. Individuals die in the model if their fat reserves at any time fall to zero for one day. The present version only considers reserves of energy stored as fat and ignores the small, and only slowly metabolizable fraction, stored as muscle protein (Davidson & Evans 1982).

Model tests

The model was tested by comparing its outputs with behavioural and ecological empirical data from the Exe. The model predicts quite well the preference ranks of the 12 mussel beds, the tendency for birds to spread out over more of the mussel beds as the total numbers present increased, the decrease in the numbers of immature oystercatchers on the most preferred mussel beds as adult numbers increased, the higher proportion of immatures on the less preferred beds, the greater rate of movement between beds of birds with the lowest dominance scores, and the higher intake rates of dominant individuals and the similarity between the intake rates of adults and immatures with the same dominance. The model predicts quite well the observed density-dependent overwinter redistribution of oyster-catchers over the 12 mussel beds; there is a net movement of birds from the initially high-density to the initially low-density beds over the six winter months. With mussel losses due to other factors taken into account, the model also predicts quite well the observed density-dependent overwinter depletion of the mussels. Although, with a few exceptions, the qualitative trends were predicted correctly, their magnitudes were sometimes under-predicted, suggesting some parameter values will need to be refined and, perhaps, new functions added to the model (Goss-Custard *et al.* 1995b,c, 1996d).

The overall reasonable correspondence between observation and prediction is encouraging and suggests that a first attempt at deriving the density-dependent emigration and mortality functions is worthwhile. However, compared with the observed rates, the model predicts the overwinter mortality rates of young birds quite well but considerably over-predicts that of adults, as measured over five mild winters, because it under-estimates the ability of real oystercatchers to lay down fat reserves during the autumn and early winter. There are several explanations for this. For example, the model assumes that the birds obtain all their food from the mussel beds whereas, in fact, they take other prey before and after the mussel beds have been exposed and covered by the tide. That this explains the discrepancy is supported by the finding that the model accurately predicts that oystercatchers should start to collect supplementary food from fields over high tide in late October and that an increasing number need to do so as the winter progresses (Goss-Custard *et al.* 1995b). By this and other means, real oystercatchers can extend their feeding period beyond the 12 h in every 24 that the model currently assumes.

As the amount of extra food obtained this way has yet to be estimated, a rescaling coefficient was used to allow the predicted daily amounts of food ingested by each individual to be increased or decreased until the model predicted the mortality rates that were actually observed in each age-class. With these included, the winter gains and losses in weight coincide very closely with the changes observed in nature. These rescaling coefficients, which varied only between 1.2 and 1.7, provide a measure of by how much the current model falls short of its objective and force us continually to rethink our research programme. Meanwhile, they also allow us to make preliminary predictions of the density-dependent winter emigration and mortality functions.

Deriving present-day density-dependent functions

The density-dependent functions for oystercatchers feeding on the 12 mussel beds of the Exe were predicted in the model by running simulations over a wide range of September starting population sizes and seeding each mussel bed with the average September mussel biomass recorded over eight Septembers. On the assumption that birds only die and do not emigrate, the proportion of birds whose energy reserves fell to zero at any time during the winter, and so died, are plotted in Figure 11.5 against the initial numbers settling in September. However, since birds might actually emigrate before starving, the simulations were also run on the assumption that birds leave the estuary when their fat reserves fall to a point equivalent to 8% of their body weight, the point at which oystercatchers leave the Wadden Sea in severe weather (Hulscher 1989, 1990). As the mean of three simulation runs in each case illustrate, the predicted emigration function lies only slightly above that arising solely from mortality. Mortality and emigration first become density dependent when the numbers of birds arriving on the estuary reaches 1500 and then increase rapidly, although at a decelerating rate. As there are at present some

2000 birds, this implies that the population is already experiencing density-dependent mortality and/or emigration.

Deriving density-dependent functions for new circumstances

What happens to these density-dependent mortality and emigration functions if feeding areas are lost or if their average quality is reduced? Clearly, removing feeding areas of average quality simply causes mortality to rise up along the present day density-dependent function. The effect of removing feeding areas of above average quality was explored by running simulations and reducing the total area of the the mussel beds by steps of 10%, starting with the best 10%. The average quality decreases at the same time as the total area decreases. The predicted density-dependent mortality function rises steeply as an increasing proportion of the best feeding area is removed (Fig. 11.6).

The population consequences of habitat loss and change

These results are not suprising but their significance lies in the indication they give of the shape and para-

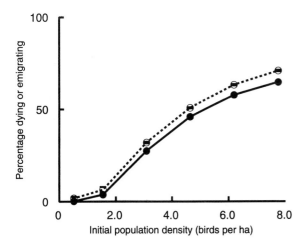

Fig. 11.5. Density-dependent emigration (⊖) and mortality (●) functions derived from the individuals-based oystercatcher model for the present-day Exe estuary (Goss-Custard *et al.* 1995b).

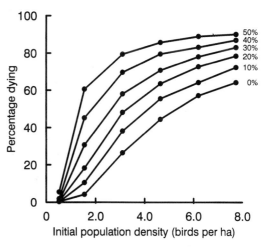

Fig. 11.6. Density-dependent mortality functions derived from the individuals-based oystercatcher model for the Exe estuary following the removal of up to the best 50% of the feeding area. The curve labelled 0% is the present-day function. The curve labelled 10% is the function obtained when the best 10% of the mussel feeding areas have been removed, and so on (Goss-Custard *et al.* 1995b).

meter values of the density-dependent functions. How the size of a population whose density-dependent winter mortality functions changed as shown in Figure 11.6 would respond to successive habitat loss was explored by population modelling. This was done using a simple demographic model of oystercatchers, the main features and parameter values of which are described in Goss-Custard *et al.* (1995d,e). Apart from winter mortality, the main source of density-dependence in the model is competition for breeding territories. Figure 11.7 shows the output from two simulations in which 10% of the feeding areas were successively removed and the appropriate density-dependent winter mortality functions, as derived in Figure 11.6, inserted at each stage. After every additional 10% loss of habitat, and the resulting change in the density-dependent function, the population quickly reduced to a new, and lower, equilibrium level (Fig. 11.7). As would be expected, the population was much more severely reduced if the best quality feeding areas were removed first.

DISCUSSION

In the case of the oystercatchers, the comparisons in Figure 11.6 show just how much a critical population

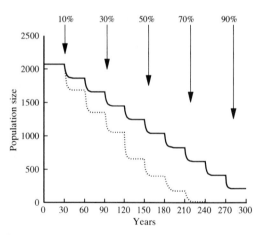

Fig. 11.7. The effect on the size of the mussel-feeding oystercatcher population of the Exe estuary of successive 10% reductions in feeding area. Solid line: habitat of average quality is removed at each step, the density-dependent mortality function shown in Figure 11.5 being used throughout the simulation. Dotted line: the best 10% to 90% of the feeding areas are removed and the appropriate function is introduced successively at each step; the functions for the removal of 10–50% are shown in Figure 11.6 but the functions for 60–90% loss were obtained the same way (Goss-Custard *et al.* 1996e).

function can change when habitat is changed. The results in Figure 11.7 show, in turn, how large the effect can be on equilibrium population size of removing an increasing proportion of the feeding area, especially when the quality of the areas that remain decline as well. It is difficult to see how the shape and parameter values of such functions in the new circumstances brought about by the reduction in the size and quality of the feeding areas could be established from conventional demographic studies which simply relate mortality rate to population density. When feeding areas are removed, the birds redistribute themselves with consequences for their body condition and survival rates that would be difficult to predict without the kind of simulation model being developed. This point can be made even more forcefully when other changes in the mussel food supply also accompany habitat loss. The simulations with the mussel model show how the growth rates, body quality and density of size classes of mussel prey would change in response to hypothetical circumstances of an increase in the levels of contaminants, such as hydrocarbons, within the estuary (Fig. 11.2). These changes would reduce the intake rates of the average oystercatcher because these depend so much on mussel size and flesh content (Zwarts *et al.*, 1996). In addition certain contaminants, for example TBT, can also increase the thickness of the mussel shells (D.S. Page, unpublished data) which, in turn, would make it more difficult for the two-thirds of the population of oystercatchers that open mussels by hammering – of which a disproportionately high number are adults – to break into their prey (Meire, 1996). Any effect that this has on the rate at which the birds can feed, and thus their chances of obtaining their daily energy requirements, can be considered in the present version of the oystercatcher model because the effect of shell thickness on intake rate can also be included (Goss-Custard *et al.* 1996b). Indeed, versions of the model currently under development include other factors that affect prey choice in oystercatchers, such as the extent to which mussels are covered by barnacles (Meire & Ervynck 1986, Ens & Alting, 1996). Although the simulations have not yet been run, it should be possible before long to explore how quite novel combinations of environmental changes, such as gradually removing the best feeding areas while hydrocarbon levels increase, would affect the condition and survival of both the mussels and their oystercatcher predators.

In both the mussel and oystercatcher models, more research remains to be done before they reach the stage at which they can be used for making confident prognostic predictions. For example, at present, any long-term impact of the oystercatchers on the mussels are not taken into account, and this could be considerable (McGrorty *et al.* 1990, Goss-Custard *et al.* 1996a). Similarly, there are additional functions that need to be added to both models. For example, within the mussel model, the difficulties of population level processes and larval recruitment need to be addressed, although evidence of strong density-dependent mortality within the first year of settlement (McGrorty *et al.* 1990) suggests that the dynamics of these benthic populations will not be sensitive to the processes controlling their recruitment dynamics when the settlement rate is high (Gaines & Roughgarden 1985, Roughgarden & Iwasa 1986). For the oystercatcher model, the decisions made by the birds may need to depend more on their body condition and the risks of being taken by a predator, or of becoming parasitized, rather than on the simple principle of maximizing current intake rate (McNamarra & Houston 1990). Furthermore, individuals in the model do not yet change their decisions with experience, for example, when they learn about the prey distribution (Bernstein *et al.* 1988, 1991). As such improvements are made, the confidence with which predictions to new circumstances can be made will gradually increase.

The models can also be extended to other predator–prey systems within estuaries. For example, it would be relatively simple to parameterize the model for the interaction between oystercatchers and cockles *Cerastoderma edule*, and this is in progress. In this case, depletion of the prey by the birds is likely to play a much larger role than for the mussels on the Exe because oystercatchers can remove very large amounts of the cockle standing crop (Brown & O'Connor 1974, Hancock 1971, Horwood & Goss-Custard 1977, Spencer 1991). Other shorebird predators can also be included; for example, knot *Calidris canutus* take small cockles in large numbers and may considerably affect cockle abundance which, in turn, affects the ability of the birds to store the energy to fuel their long migrations to their arctic breeding grounds (Piersma 1994).

Clearly, the value of the whole approach depends on how well understood are the properties of the animals which form the basis of model functions. In the case of both the oystercatchers and the mussels, this includes the physiology of ingestion, digestion and assimilation and the energetic requirements under various schedules of activity and environmental conditions. For the mussels it includes reproductive maturity, the reproductive cycle, shell deposition and physiological components contributing to survivorship. For the oystercatchers, moult and fat deposition are also included; a comprehensive review of the issues involved can be found in Piersma (1994). It also includes the behavioural decision rules used by birds in choosing on what and where to feed. Establishing such rules is the central issue in behavioural ecology. This discipline has

for a long time provided population ecology with an evolutionary framework within which hypotheses at the population level may be generated and first evaluated; for example, the kin-selection hypothesis of cycles (Charnov & Finerty 1980), now being tested in red grouse *Lagopus lagopus* (Mountford *et al.* 1989, Moss & Watson 1991). However, as illustrated here, behavioural ecology also provides a basis for estimating population parameters that may be difficult to measure directly and then only apply over the limited range of conditions encountered during the study period itself. By being based on well-established principles and on basic properties of the animals concerned, it is hoped that such functions may be used with more confidence than direct correlation measures of density-dependence to predict the response of a population to quite novel environmental changes.

ACKNOWLEDGEMENTS

We are very grateful to Dr A.J. Gray for his valuable comments on the manuscript and to Ralph Clarke, Sarah Durell and Andy West for help with building the oystercatcher models and running the simulations.

REFERENCES

Bayne, B.L. & Worrall, C.M. (1980) Growth and production of mussels *Mytilus edulis* from two populations. *Marine Ecology Progress Series*, **3**, 317–328.

Bayne, B.L., Salkeld, P. & Worrall, C.M. (1983) Reproductive effort and value in different populations of the marine mussel, *Mytilus edulis* L. *Oecologia*, **59**, 18–26.

Bernstein, C., Kacelnik, A. & Krebs, J.R. (1988) Individual decisions and the distribution of predators in a patchy environment. *Journal of Animal Ecology*, **57**, 1007–1026.

Bernstein, C., Krebs, J.R. & Kacelnik, A. (1991) Distribution of birds amongst habitats: theory and relevance to conservation. *Bird Population Studies* (eds C.M. Perrins, J.-D. Lebreton & G.J.M. Hirons), pp. 317–345. Oxford University Press, Oxford.

Brown, R.A. & O'Connor, R.J. (1974) Some observations on the relationships between oystercatchers *Haematopus ostralegus* L. and cockles *Cardium edule* L. in Strangford Lough. *Irish Naturalist*, **18**, 73–80.

Charnov, E.L. & Finerty, J.P. (1980) Vole population cycles: a case for kin selection. *Oecologia*, **45**, 1–2.

Dare, P.J. & Edwards, D.B. (1975) Seasonal changes in flesh weight and biochemical composition of mussels (*Mytilus edulis* L.) in the Conway Estuary, North Wales. *Journal of Experimental Marine Biology and Ecology*, **18**, 89–97.

Davidson, N.C. & Evans, P.R. (1982) Mortality of redshanks and oystercatchers from starvation during severe weather. *Bird Study*, **29**, 183–188.

Donkin, P., Widdows, J., Evans, S.V., Worrall, C.M. & Carr, M. (1989) Quantitative structure–activity relationships for the effect of hydrophobic organic chemicals on rate of feeding by mussels (*Mytilus edulis*). *Aquatic Toxicology*, 14, 277–294.

Ens, B.J. & Alting, D. (1996) Prey selection of a captive oystercatcher, *Haematopus ostralegus*, hammering mussels, *Mytilus edulis*, from the ventral side. *Ardea*, 83.

Fréchette, M., Aitken, A.E. & Pagé, L. (1992) Interdependence of food and space limitation of a benthic suspension feeder: consequences for self-thinning relationships. *Marine Ecology Progress Series*, 83, 55–62.

Gaines, S. & Roughgarden, J.D. (1985) Larval settlement rate: A leading determinant of structure in an ecological community of the marine intertidal zone. *Proceedings of the National Academy of Sciences*, 82, 3707–3711.

Goss-Custard, J.D. (1977) The ecology of the Wash. III. Density-related behaviour and the possible effects of a loss of feeding grounds on wading birds (Charadrii). *Journal of Applied Ecology*, 14, 721–739.

Goss-Custard, J.D. (1980) Competition for food and interference among waders. *Ardea*, 68, 31–52.

Goss-Custard, J.D. (1993) The effect of migration and scale on the study of bird populations: 1991 Witherby Lecture. *Bird Study*, 40, 81–96.

Goss-Custard, J.D. & Durell, S.E.A.leV.dit. (1987) Age-related effects in oystercatchers, *Haematopus ostralegus*, feeding on mussels, *Mytilus edulis*. 1. Foraging efficiency and interference. *Journal of Animal Ecology*, 56, 521–536.

Goss-Custard, J.D. & Durell, S.E.A.leV.dit. (1988) The effect of dominance and feeding method on the intake rates of oystercatchers, *Haematopus ostralegus*, feeding on mussels, *Mytilus edulis*. *Journal of Animal Ecology*, 57, 827–844.

Goss-Custard, J.D., West, A.D. & Durell, S.E.A.leV.dit. (1993) The availability and quality of the mussel prey (*Mytilus edulis*) of oystercatchers (*Haematopus ostralegus*). *Netherlands Journal of Sea Research*, 31, 419–439.

Goss-Custard, J.D., Caldow, R.W.G., Clarke, R.T., Durell, S.E.A.leV.dit. & Sutherland, W.J. (1995a) Deriving population parameters from individual variations in foraging behaviour. I. Empirical game theory distribution model of oystercatchers *Haematopus ostralegus* feeding on mussels *Mytilus edulis*. *Journal of Animal Ecology*, 64, 265–276.

Goss-Custard, J.D., Caldow, R.W.G., Clarke, R.T., Durell, S.E.A.leV.dit., Urfi, A.J. & West, A.D. (1995b) Consequences of habitat loss and change to populations of wintering migratory birds: predicting the local and global effects from studies of individuals. *Ibis* 137 (Suppl. 1), 56–66.

Goss-Custard, J.D., Caldow, R.W.G., Clarke, R.T. & West, A.D. (1995c) Deriving population parameters from individual variations in foraging behaviour. II. Model tests and population parameters. *Journal of Animal Ecology*, 64, 277–289.

Goss-Custard, J.D., Clarke, R.T., Briggs, K.B., Ens, B.J., Exo, K-M., Smit, C., Beintema, A.J., Caldow, R.W.G., Catt, D.C., Clark, N., Durell, S.E.A.leV.dit, Harris, M.P., Hulscher, J.B., Meininger, P.L., Picozzi, N., Prys-Jones, R., Safriel, U. & West, A.D. (1995d) Population consequences

of winter habitat loss in a migratory shorebird. I. Estimating model parameters. *Journal of Applied Ecology*, 32, 317–333.

Goss-Custard, J.D., Clarke, R.T., Durell, S.E.A.leV.dit., Caldow, R.W.G. & Ens, B.J. (1995e) Population consequences of winter habitat loss in a migratory shorebird. II. Model predictions. *Journal of Applied Ecology*, 32, 334–348.

Goss-Custard, J.D., McGrorty, S. & Durell, S.E.A.leV.dit. (1996a) The effect of the European oystercatcher on shellfish populations. *Ardea*, 83.

Goss-Custard, J.D., West, A.D., Caldow, R.W.G., Durell, S.E.A.leV.dit. & McGrorty, S. (1996b) An empirical optimality model to predict the intake rates of oystercatchers *Haematopus ostralegus* feeding on mussels *Mytilus edulis*. *Ardea*, 83.

Goss-Custard, J.D., Durell, S.E.A.leV.dit., Goater, C.P., Hulscher, J.B., Lambeck, R.H.D., Meininger, P.L. & Urfi, A.J. (1996c) How oystercatchers survive the winter. *The Oystercatcher: from Individuals to Populations* (ed. J.D. Goss-Custard). Oxford University Press, Oxford.

Goss-Custard, J.D., West, A.D. & Sutherland, W.J. (1996d) Where to feed. *Behaviour and Ecology of the Oystercatcher: from Individuals to Populations* (ed. J.D. Goss-Custard). Oxford University Press, Oxford.

Goss-Custard, J.D., Durell, S.E.A.leV.dit., Clarke, R.T., Beintema, A.J., Caldow, R.W.G., Meininger, P.L. and Smit, C. (1996e) Population dynamics: predicting the consequences of habitat change at the continental scale. In *The Oystercatcher: from Individuals to Populations* (ed. J.D. Goss-Custard), Oxford University Press, Oxford.

Hancock, D.A. (1971) The role of predators and parasites in a fishery for the mollusc *Cardium edule* L. *Dynamics of Populations*. (eds P.J. den Boer & G.R. Gradwell), pp. 419–439. Centre for Agriculture, Publicity and Documentation, Wageningen.

Hilbish, T.J. (1986) Growth trajectories of shell and soft tissues in bivalves: seasonal variation in *Mytilus edulis* L. *Journal of Experimental Biology and Ecology*, 96, 103–113.

Horwood, J.W. & Goss-Custard, J.D. (1977) Predation by the oystercatcher, *Haematopus ostralegus* (L.), in relation to the cockle, *Cerastoderma edule* (L.), fishery in the Burry Inlet, South Wales. *Journal of Applied Ecology*, 14, 139–158.

Hulscher, J.B. (1989) Mortality and survival of oystercatchers *Haematopus ostralegus* during severe winter conditions. *Limosa*, 62, 177–181.

Hulscher, J.B. (1990) Survival of oystercatchers during hard winter weather. *Ring*, 1–2, 167–172.

Hulscher, J.B. (1996) The food and feeding behaviour of oystercatchers. *The Oystercatcher: from Individuals to Populations* (ed. J.D. Goss-Custard). Oxford University Press, Oxford.

Kersten, M. & Piersma, T. (1987) High levels of energy expenditure in shorebirds; metabolic adaptations to an energetically expensive way of life. *Ardea*, 75, 175–188.

Kersten, M. & Visser, W. (1996) The rate of food processing in the oystercatcher: food intake and energy expenditure constrained by a digestive bottleneck. *Functional Ecology*, 9.

McGrorty, S., Clarke, R.T., Reading, C.J. & Goss-Custard, J.D. (1990) Population dynamics of the mussel *Mytilus edulis*: density changes and regulation of the population in the Exe estuary, Devon. *Marine Ecology Progress Series*, **67**, 157–169.

McNamara, J.M. & Houston, A.I. (1990) The value of fat reserves and the trade-off between starvation and predation. *Acta Biotheoretica*, **38**, 37–61.

May, R.M. (1981) *Theoretical Ecology: Principles and Applications*. Blackwell Scientific Publications, Oxford.

Meire, P.M. (1996) Interactions between oystercatchers (*Haematopus ostralegus*) and mussels (*Mytilus edulis*): implications from optimal foraging theory. *Ardea*, **83**.

Meire, P.M. & Ervynck, A. (1986) Are oystercatchers (*Haematopus ostralegus*) selecting the most profitable mussels (*Mytilus edulis*)? *Animal Behaviour*, **34**, 1427–1435.

Meire, P.M., Schekkerman, H. & Meininger, P.L. (1994) Consumption of benthic invertebrates by waterbirds in the Oosterschelde estuary, SW Netherlands. *Hydrobiologia*, **282/283**, 525–546.

Moss, R. & Watson, A. (1991) Population cycles and kin selection in red grouse *Lagopus lagopus scoticus*. *Ibis*, **133** (Suppl. 1), 113–120.

Mountford, M.D., Watson, A., Moss, R., Parr, R. & Rothery, P. (1989) Land inheritance and population cycles of red grouse. *Red Grouse Population Processes* (eds A.N. Lance & J.H. Lawton), pp. 78–83. Royal Society for the Protection of Birds, Sandy.

Piersma, T. (1994) *Close to the Edge: Energetic Bottlenecks and the Evolution of Migratory Pathways in Knots*. Uitgeverij Het Open Boek, Texel, The Netherlands.

Roughgarden, J.D. & Iwasa, Y. (1986) Dynamics of a metapopulation with space-limited subpopulations. *Theoretical Population Biology*, **29**, 235–261.

Ross, A.H. & Nisbet, R.M. (1990) Dynamic models of growth and reproduction of the mussel *Mytilus edulis* L. *Functional Ecology*, **4**, 777–787.

Spencer, B.E. (1991) Predators and methods of control in molluscan shellfish cultivation in north European waters. *Aquaculture and the Environment* (eds N. De Pauw & J. Joyce), pp. 309–337. European Aquaculture Society Special Publication No. 16, Gent, Belgium.

Widdows, J. & Donkin, P. (1991) Role of physiological energetics in ecotoxicology. *Comparative Biochemistry and Physiology*, **100C**, 69–75.

Widdows, J., Donkin, P. & Evans, S.V. (1987) Physiological responses of *Mytilus edulis* during chronic oil exposure and recovery. *Marine Environmental Research*, **23**, 15–32.

Widdows, J. & Hawkins, A.J.S. (1989) Partitioning rate of heat dissipation by *Mytilus edulis* into maintenance, feeding and growth components. *Physiological Zoology*, **62**, 764–784.

Widdows, J. & Johnson, D. (1988) Physiological energetics of *Mytilus edulis*: scope for growth. *Marine Ecology Progress Series*, **46**, 113–121.

Willows, R.I. (1992) Optimal digestive investment: a model for filter feeders experiencing variable diets. *Limnology and Oceanography*, **37**, 829–847.

Willows, R.I. (1994) The ecological impact of different mechanisms of chronic sub-lethal toxicity on feeding and respiratory physiology. *Water Quality and Stress Indicators: Linking Levels of Organisation* (eds J.G. Jones & D.W. Sutcliffe), pp. 88–97. Proceedings of the Joint Associations No. 1.

Worrall, C.M. & Widdows, J. (1984) Investigation of factors influencing mortality in *Mytilus edulis* L. *Marine Biology Letters*, **5**, 85–97.

Zwarts, L. & Drent, R.H. (1981) Prey depletion and the regulation of predator density: oystercatchers (*Haematopus ostralegus*) feeding on mussels (*Mytilus edulis*). *Feeding and Survival Strategies of Estuarine Organisms* (eds N.V. Jones & W.J. Wolff), pp. 193–216. Plenum Press, London.

Zwarts, L., Ens, B.J., Kersten, M. & Piersma, T. (1990) Moult, mass and flight range of waders ready to take off for long-distance migration. *Ardea*, **78**, 339–364.

Zwarts, L., Cayford, J.T., Hulscher, J.B., Kersten, M., Meire, P.M. & Triplet, P. (1996) Prey size selection and intake rate. *The Oystercatcher: from Individuals to Populations* (ed. J.D. Goss-Custard). Oxford University Press, Oxford.

CHAPTER 12

Modelling techniques and data requirements in aquatic systems management

W.S.C. Gurney, D.A.J. Middleton, A.H. Ross, R.M. Nisbet, E. McCauley,
W.W. Murdoch and A. Deroos

SUMMARY

(1) We consider some of the issues involved in producing models of aquatic systems that are sufficiently robust to be of value in addressing population management questions. In particular we consider the data requirements of such models.

(2) We consider the use of unstructured, biomass-based 'box' models. Although models of this type are generally considered unverifiable, we discuss the case of a fjordic ecosystem model where the nature of the system, and a large body of data, allowed a rigorous cycle of model falsification and improvement.

(3) To assess the information requirements of individual-based population models we examine the steady state properties of a suite of models which represent different rules governing individual energy allocation and mortality. The behavioural complexity of such models implies a requirement for a large volume of detailed individual-level data for model formulation and testing.

(4) These examples illustrate the essential interaction between data-gathering, experimentation and modelling that is required if models are to be used safely to predict the outcome of management decisions in aquatic ecosystems.

Key-words: biomass-based models, *Daphnia pulex*, data requirements, ecosystem management, equilibrium demography, individual-based models, sea-lochs

INTRODUCTION

No model which is to be capable of illuminating the practical questions involved in managing a population can avoid incorporating biological detail about the constituent individuals, the population structure and the broader context in which the population exists. However, in almost all cases, simply synthesizing all existing knowledge of these areas will inevitably result in a construct which violates the most fundamental rule of useful modelling; namely that any given set of predictions should be robust against realistic uncertainty in parameter values and that the mechanism by which they arise should be (at least *a posteriori*) intuitively obvious.

The earliest attempts to see populations in their ecological context were the biomass-based whole ecosystem models spawned in the early 1970s by the International Biological Programme. Interest in such models was originally driven by the success of conservation laws in physical systems. However, the difficulty of defining and measuring the fluxes of energy across the boundaries of (invariably) open biological systems, together with a recognition that most energy flows within biological systems are in bound form and thus accompanied by flows of elemental nutrients, led to alternative formulations using carbon or nitrogen as their currency. For various reasons, the boundaries of the systems which such models attempted to describe tended to be very widely drawn. This resulted in structurally complex models with a large list of parameters whose values were extremely hard to determine accurately, partly because of their sheer number and partly because of the difficulty of obtaining sufficiently precise interpretations of their meaning.

These parameterization difficulties were often exacerbated by the sensitivity of the predicted dynamics to certain critical parameters. Where accurate system level data are available, one resolution of such difficulties is to use the critical parameters to optimize the model's fit to part of a well-defined dataset. However, most of the early ecosystem models represented systems whose boundaries were so widely defined as to render such a strategy impracticable. The models were therefore not generally subjected to the virtuous cycle of falsification and refinement, but where systematic testing was attempted the results were almost invariably discouraging.

Two reactions to these early failures, which have been very helpful scientifically but are less relevant in a management context, have been to model artificially simplified ecosystems (e.g. Andersen & Nival 1989), or to focus on short-term representation of a subset of ecosystem processes (e.g. Taylor *et al.* 1992, Tett *et al.* 1986, Cloern & Cheng 1981). Of more relevance to the construction of management models has been the movement towards the incorporation of additional biological detail at the level of individuals and populations. Such approaches, often generically known as 'individual-

based models' can use a wide variety of technical representations, from the distribution function methods pioneered by Metz & Diekmann (1986) to the explicit individual by individual computational representation of an entire population (e.g. Huston 1992, MacKay 1992).

A common feature of all individual-based approaches is that the additional detail which they incorporate is necessarily bought at the price of a considerably increased hunger for data in both the parameterization and the testing phases of any project. In this paper we seek to shed light on the precise extent of the data requirements of structured models. We also re-examine, with the aid of an example, the possibility of using unstructured, biomass-based, models for management purposes.

A BIOMASS-BASED ECOSYSTEM MODEL

Describing a fjord

In Figure 12.1 we show the biological structure of the fjord ecosystem model developed by Ross *et al.* (1993, 1994). This is essentially a very conventional linear trophic chain model with primary production controlled by the combination of irradiance and inorganic nutrient availability, secondary and tertiary production controlled by the availability of biomass at the next lower trophic

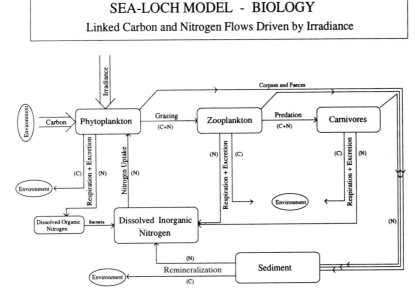

Fig. 12.1. The biological structure assumed in the fjord ecosystem model of Ross *et al.* (1993, 1994).

level, and inorganic nutrient recycled through the sediments. Its only slightly unusual feature is that it takes separate account of the flows of carbon and the primary limiting nutrient (nitrogen), since these are closely coupled over only part of the nutrient cycle.

Numerous attempts have been made to represent both marine and terrestrial ecosystems in such terms, and the record of the resulting models when tested against data is not generally encouraging (see Gurney *et al.* in press, for further discussion). However, the system we are modelling here has two very important special features, both arising from the hydrological properties of the physical system in which the biological components are operating (Fig. 12.2). The first of these is that the continuous inflow of (low density) fresh water into a sea-loch coupled with regular flushing with (high-density) water from the outside sea, implies that the system is relatively strongly stratified throughout the year. The second is that because of these water movements, all the components of the ecosystem from inorganic nutrients to carnivores are coupled to external sources and sinks.

The open nature of this system makes modelling it much simpler than describing the equivalent closed system. In terms of dissolved nutrient dynamics, the system behaves in a manner analogous to a laboratory chemostat, importing nutrient (both from terrestrial run-off and from the external sea) when the internal concentration is low and exporting nutrient to the external sea when the internal concentration is high. The result is that ambient nutrient concentrations never fall to levels low enough to inhibit the growth of primary producers (Ross *et al.* 1994). Primary productivity is thus controlled by irradiance levels, and the standing crop of primary producers is controlled by grazing.

Testing the model against data

Despite its relative simplicity, this model has 36 parameters describing the biota, eight parameters describing the physical structure and exchanges with the outside world, and 14 driving functions specifying the annual cycle in the appropriate parts of the environment. However, the physical characteristics of Scottish sea-lochs have been intensively studied and are well known. The species structure of the trophic groups is relatively simple and the principal organisms concerned are common. We were thus able to obtain good estimates of two-thirds of the biological parameters from the literature. The biological conditions in the sea off Killary Harbour have also been intensively studied, and the Firth of Lorne (external to Lochs Etive and Creran) has been recently surveyed by Scottish Office personnel.

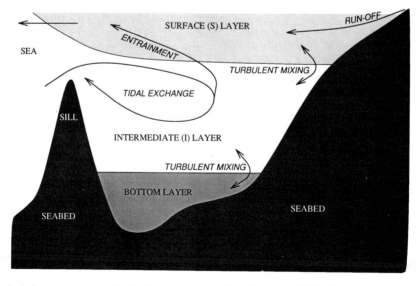

Fig. 12.2. The physical structure assumed in the fjord ecosystem model of Ross *et al.* (1993, 1994).

Numerical studies showed that the behaviour of the model depends on the parameters (including those whose values we did not know well) in a very smooth and continuous way. We were thus able to conduct an exhaustive investigation of its behaviour, with special emphasis on undertaking the mechanisms underlying all predicted effects. At the end of this exercise we were able to draw a set of conclusions about how sea-lochs worked which were highly robust against quite large changes in almost all the parameters.

Initial semi-qualitative comparisons against an ad-hoc dataset compiled from the literature for Killary Harbour (a long shallow inlet on the west coast of Ireland) showed that the model behaviour was in broad agreement with observation. This encouraged us to seek to develop it into a construct capable of quantitative prediction of sea-loch annual cycles. We first compiled a test dataset comprising partial datasets on three Scottish sea-lochs with distinct hydrological properties:

- Loch Airdbhair – a shallow rapidly flushed embayment on the north west coast
- Loch Etive – a deep two-basin system with a very slowly flushed inner basin
- Loch Creran – similar in conformation to Killary Harbour, but deeper and more slowly flushed.

We hypothesized that the speciation of the functional groups was the same in all four of our test systems so that we should be able to predict the annual cycle in all four with a single set of biological parameters by changing the (well known) physical and external environmental parameters appropriately. We determined the values of the parameters we did not know by optimizing the model fit to the Killary Harbour dataset (which had the widest trophic coverage) and then used the partial datasets from the three Scottish sea-lochs as the formal test by which the model might be falsified.

Our initial attempts at this virtuous cycle were unsurprisingly successful at falsifying early iterations of the model. However, aided by our knowledge of its internal dynamics we were able to identify the structural features responsible for these failings and remedy the inappropriate or inaccurate descriptions which underlay them. For example, we had omitted self-shading effects, which were unimportant in the turbid conditions of Killary Harbour, but which play a significant role in clearer Scottish waters. We had failed to take appropriate account of the increase in zooplankton losses due to autumn resting egg production, and we had neglected some aspects of zooplankton grazing behaviour which had important consequences on system stability. Once these deficiencies had been remedied, the large majority

of the discrepancies between observations and model predictions fell to within the (observational) noise in the data. We illustrate the quality of the final fits obtained in Figure 12.3.

Why does this model work?

Despite being conceptually similar to earlier box-models, and having a similarly intimidating parameter count, this model appears to be capable of passing tests of a severity which most biomass ecosystem models would certainly fail. This naturally leads us to ask in what way this model is different. Although there is clearly no single simple answer to this question we believe that several factors play important roles.

Firstly, the open nature of the sea-loch ecosystem removes many of the subtle modelling problems associated with system behaviour in nutrient impoverished or winter conditions and substitutes the problems of measuring fluxes across the system boundaries. However, these boundaries are well defined and are geographically limited in extent, so accurate measurement of the fluxes across them is an entirely superable problem.

Secondly, the biological system being modelled lends itself to a structurally simple representation whose dynamics depend on parameter values in a satisfactorily smooth and continuous way. Furthermore, the parts of the biological community which are important to the whole-loch nutrient-balance questions, upon which the model focuses, contains few organisms whose characteristic timescales are longer than a few weeks.

Lastly, and (we believe) most importantly, there was a very large body of available data at all levels from the individual to the ecosystem. This enabled us to use individual level data for primary parameterization, confirm the appropriateness of those parameters and determine optimum values for those we did not know by reference to population level data, and still have enough data left over to form the basis of a viable exercise in model falsification.

The first important take-home message seems to be that the data requirement for a successful ecosystem model, even of a simple box-model type, is very considerable indeed. We emphasize that without the extensive population and community level testing which has been possible with this model, the predictions of early model variants, based only on available individual level information, could have seriously mislead management discussions.

This data hunger is, in general, increased rather than decreased by the incorporation of additional individual level biological detail, and we would thus argue that where there are no clear contra-indications, a box-

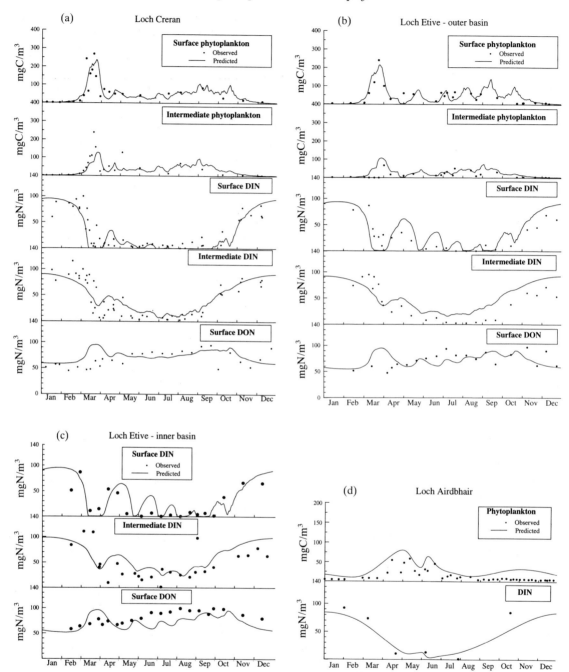

Fig. 12.3. Testing the fjord model against datasets from four characteristically different fjords: (a) Loch Creran, (b) Loch Etive – outer basin, (c) Loch Etive – inner basin, (d) Loch Airdbhair, (e) Killary Harbour. The model definition and parameters are to be found in Ross *et al.* 1994.

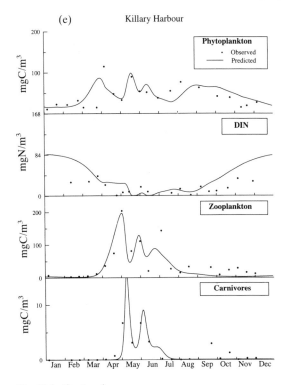

Fig. 12.3. *Continued.*

modelling approach should be given serious consideration. However, the clearest contra-indication to box-modelling is that some or all of the organisms of interest are complex, long-lived, animals whose population dynamics are significantly affected by changes in the population's physiological structure. In such cases it is inevitable that we must take a physiologically structured view of some parts of our management model, and in the next section of this paper we consider what this implies.

STRUCTURED POPULATION MODELS

Introduction

We now seek to assess the information requirements of models which incorporate more biological detail than simple biomass-based box models. As part of this enterprise we shall examine the properties of one rather simple component of such a model, namely a population of individuals who are distinguishable by virtue of their weight. Although we shall conduct our investigations in the context of a distribution function description of the

type made familiar by the structured population model studies of Metz & Diekmann (1986) the conclusions which we draw apply to all individual-based models including the 'collection of individuals' models discussed in DeAngelis & Gross (1992).

In order to make our discussion as general as possible we focus our attention on the steady state characteristics of the population. We shall show that the shape of the equilibrium weight distribution is a property of the demographic characteristics of the individuals who make up the population and is not influenced by the nature of the feedback processes which bring about the equilibrium state. This enables us to make a very general determination of the way quantities which might be the subject of practicable observations, such as average adult fecundity, will change in response to environmental variation. The sensitivity of this response acts as a guide to the importance of the associated individual characteristics in population control.

Theoretical background

We consider a closed population whose individuals are distinguishable only by their weight w, and who share a common resource whose abundance we write as R. The per capita fecundity β, mortality δ, and growth rate g, are thus functions only of w and R. To characterize the state of the population we define a weight distribution f(w,t) such that f(w,t)dw represents the number of individuals in the population at time t who have weights in the range w→w + dw.

The dynamics of this population are described by three equations. The McKendrick–von Foerster equation

$$\frac{\partial f(w,t)}{\partial t} + \frac{\partial}{\partial w}(g(w,R)f(w,t)) + \delta(w,R)f(w,t) = 0 \quad (1)$$

represents the processes of growth and mortality for the current members of the population. The production of offspring and their recruitment into the population (at weight w_0) is described by the boundary condition

$$g(w_0,R)f(w_0,t) = \int_{w_0}^{\infty} \beta(w,R)f(w,t)dw \quad (2)$$

Finally we must have a description of the feedback mechanism which relates resource abundance to the population size and weight distribution. Although we shall show below that the details of this mechanism are irrelevant to the present discussion, we shall, for completeness, show an example of the type of feedback normally incorporated in models of this type. Suppose that the resource R is introduced into the system at a contant rate P and is consumed by individuals with weight w at a rate U(w,R). In this case simple book-keeping demands that

$$\frac{dR}{dt} = P - \int_{w_0}^{\infty} U(w,R)f(w,t)dw \qquad (3)$$

A steady-state solution of the continuity equation (Equation (1)) which occurs at a resource abundance R, must obey the requirement that

$$f^*(w) = f^*(w_0)\xi(w,R) \qquad (4)$$

where

$$\xi(w,R) \equiv \exp\left[-\int_{w_0}^{w} \frac{\delta(x,R) + g'(x,R)}{d(x,R)}dx\right] \qquad (5)$$

with g'(w,R) being defined as the partial derivative of the growth rate with respect to weight. Back substitution of Equation (4) in the renewal condition (Equation (2)) now tells us that the value of resource availability required to establish equilibrium R*, is the solution of

$$g(w_0,R^*) = \int_{w_0}^{\infty} \beta(w,R^*)\xi(w,R^*)dw \qquad (6)$$

Equation (5), with the resource abundance set to the value satisfying Equation (6), defines the shape of the equilibrium weight distribution. The final step in determining the equilibrium distribution is to substitute this equilibrium shape factor into the feedback equation and hence determine the value of the weight distribution at the recruitment boundary which brings the resource dynamics into equilibrium. For the special case shown as Equation (3), this would yield

$$f^*(w_0) = \frac{P}{\int_{w_0}^{\infty} U(w,R^*)\xi(w,R^*)dw} \qquad (7)$$

Observable demographic parameters

Even if the approximation that individuals are distinguishable only by their weight is a good one, direct experimental observation of a population weight distribution is an intimidating exercise in laboratory conditions and a practical impossibility in the field. In practice, the kind of information likely to be available will take the form of average demographic quantities; for example, the mean time taken for a new recruit to reach sexual maturity, the mean juvenile mortality, or the mean adult fecundity. We now relate these observable demographic parameters to the equilibrium state of our weight structured population model.

If an individual reaches sexual maturity when it reaches weight w_M and the equilibrium resource abundance is R* then, at equilibrium, the average adult fecundity is clearly the ratio of the total number of offspring produced per unit time to the total adult population:

$$\bar{\beta} = \frac{\int_{w_M}^{\infty} \beta(w,R^*)\xi(w,R^*)dw}{\int_{w_M}^{\infty} \xi(w,R^*)dw} \qquad (8)$$

To evaluate the juvenile to adult delay (τ) we need first to obtain the weight, w(a,R), of an individual of age a growing in conditions of constant resource availability R. This is given by the solution of

$$\frac{dw(a,R)}{da} = g(w(a,R),R) \qquad w(0) = w_0 \qquad (9)$$

The development delay, τ is then the solution of

$$w(\tau,R^*) = w_M \qquad (10)$$

Once we know τ and the equilibrium shape factor $\xi(w,R^*)$ we can evaluate the average juvenile death rate as

$$\bar{\delta} = \frac{1}{\tau}\ln\left[\frac{1}{\xi(w_M,R^*)}\right] \qquad (11)$$

Individual demography

Although it is sometimes possible to evaluate the integral in Equation (6), it is almost never possible to proceed further and obtain a closed form solution for the resource abundance required to produce equilibrium. In order to make progress we must make some explicit assumptions about the functions (β, δ, g) which describe the demography of the individuals who make up the population and hence ultimately determine its dynamics.

We shall discuss individual growth and fecundity together, since both are intimately linked to the assimilation of energy and to the rules by which the organism decides what should be allocated to growth and what to reproduction. To explore the influence such internal allocation rules may exercise on population dynamics we analyse two possible combinations of rules, one developed over many years by S.A.L.M. Kooijman and J.A.J. Metz and stated in its most complete form by Kooijman (1993) and the other adapted from the ideas of Paloheimo *et al.* (1982). Both models assume that all individuals in the population feed from a single food supply with a type II functional response.

The key assumptions of the Kooijman–Metz allocation model are as follows:

- maximum assimilation rate scales with weight to the power 2/3 (a surrogate for length^2)
- search volume scales with A_{max} so half-saturation food density is weight independent
- metabolic costs scale with weight
- a fraction κ of gross assimilate is used for growth and maintenance, the remainder being used for reproduction in mature ($w > w_M$) animals and for building gonads in immatures.

Defining the constants of proportionality which relate metabolic cost to weight and maximum assimilation rate to $w^{2/3}$ as m and α respectively, and writing the type II functional response as H(R), we find that the Kooijman–Metz allocation model implies that

$$g(w,R) = \kappa\alpha H(R)w^{2/3} - mw \quad (12)$$

and

$$\beta(w,R) = \begin{cases} 0 & w < w_M \\ (1 - \kappa)\alpha H(R)w^{2/3}/w_E & \text{otherwise} \end{cases} \quad (13)$$

As Figure 12.4(a) shows, these assumptions lead to an individual growth curve in which weight tends to an asymptotic value which depends on resource abundance. By contrast, the model which we have adapted from the work of Paloheimo *et al.* (1982), whose key assumptions are

- maximum assimilation rate and metabolic costs both scale with weight

- the proportion of gross assimilate allocated to reproduction is propotional to $w - w_M$

implies that

$$g(w,R) = [\gamma H(R)w - mw][1 - \eta(w - w_M)^+] \quad (14)$$

and

$$\beta(w,R) = [\gamma H(R)w - mw][\eta(w - w_M)^+] \quad (15)$$

where γ and η are the constants of proportionality relating maximum uptake rate and allocation to reproduction to weight and excess weight respectively. Clearly, in this case the proportion of gross assimilate routed to growth goes to zero when weight reaches the (resource independent) value $w_M + 1/\eta$. The resulting growth curves are illustrated in Figure 12.4(b).

Model behaviour

Although it is possible to make some progress with analytical investigation of some of the models discussed here, in no case has a wholly analytical solution proved possible. We have therefore adopted a primarily numerical approach, using as the basis of our investigation a set of parameters (Table 12.1) which are loosely designed to represent *Daphnia pulex*.

In this paper we have stated the model structure in its most intuitive form. However, it is a generic property of all models with dynamics underpinned by a saturating individual growth curve, that at low values of total death rate the equilibrium weight distribution has a singularity at the asymptotic weight (Fig. 12.5). This can give rise

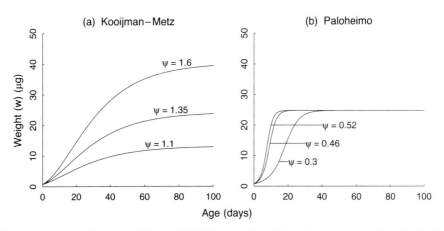

Fig. 12.4. Growth curves for the Kooijman–Metz and Paloheimo models with *D. pulex* parameters at three levels of resource abundance. Kooijman–Metz model parameters are $\kappa = 0.3$, $m = 0.14\,\text{day}^{-1}$, $w_R = 0.77\,\mu g$, $\alpha H(R) = 1.1, 1.35, 1.6$. Paloheimo model parameters are $m = 0.14\,\text{day}^{-1}$, $\eta = 0.035\,\text{mg}^{-1}$, $w_R = 0.77\,\mu g$, $w_M = 8.1\,\mu g$, $\gamma H(R) = 0.3, 0.46, 0.52$.

Table 12.1. *Daphnia pulex parameters.*

All models	$W_R = 0.77\,\mu g$	($\Rightarrow L_R = 0.6\,mm$)
	$W_M = 8.1\,\mu g$	($\Rightarrow L_M = 1.6\,mm$)
	$W_E = 1.23\,\mu g$	($\Rightarrow W_R = 0.63\,W_E$)
	$m = 0.14\,day^{-1}$	($\Rightarrow$ starvation time = 7 days)
Kooijman–Metz	$\kappa = 0.3$	
Paloheimo	$\eta = 0.035\,\mu g^{-1}$	($\Rightarrow W_\infty = 25\,\mu g$)

to serious numerical problems which can most easily be circumvented by restating the problem in terms of an age-distribution and a weight-at-age function. We have therefore recast the problem in these terms (see Gurney *et al.* 1995) and used standard numerical procedures from the NAG (1994) library to obtain solutions.

In Figure 12.6 we show how the observable demographic characteristics of the equilibrium population vary with background mortality, for two models in which the only source of mortality is the environmental (or background) mortality δ_B. One model uses a Kooijman–Metz (K-M) growth and fecundity model, while the other uses the Paloheimo, Crabtree and Taylor (PCT) model. It is immediately clear that while the predictions of the two models differ only quantitatively at high background death rates, they imply qualitatively different patterns of behaviour at low death rate. This difference in behaviour is shown even more clearly by the accompanying frames of Figure 12.6, where we have shown the relative sensitivities (R_τ and R_β), which we define as

$$R_\tau \equiv (d\tau/\tau)/(d\delta_B/\delta_B) \qquad (16a)$$

and

$$R_\beta \equiv (d\bar\beta/\bar\beta)/(d\delta_B/\delta_B) \qquad (16b)$$

so that they represent the proportional change in β or τ which is brought about by a given proportional change in δ_B. From these frames we see that at low death rates the adult fecundity predicted by the PCT model goes smoothly to zero (with the relative sensitivity remaining constant), while that predicted by the K-M model becomes entirely insensitive to changes in background death rate. We can conclude from this that population exhibiting PCT allocation and constant death rate is controlled by a combination of fecundity and development time under all conditions, but one exhibiting K-M allocation will come to be controlled by development time alone if the background death rate falls low enough.

To see how general this effect is, we shall now examine two models which differ from the two discussed above only in the fact that mortality has a component which depends on the ratio of uptake to costs. To make the behaviour as straightforward as possible, we envisage that uptake dependent component as a proportional enhancement of the background death rate, and write

$$\delta = \delta_B\left[1 + \frac{\gamma}{\dfrac{A(w,\psi)}{mw} - \rho}\right] \qquad (17)$$

This implies that the death rate is essentially the back-

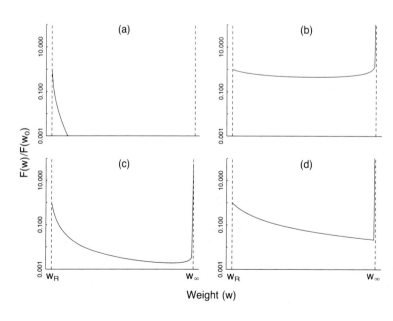

Fig. 12.5. Equilibrium weight distributions for the Kooijman–Metz and Paloheimo models: (a) K–M model with *D. pulex* parameters and constant death rate of 0.5 day^{-1}; (b) K–M model with *D. pulex* parameters and constant death rate of 0.03 day^{-1}; (c) Paloheimo model with *D. pulex* parameters and δ of 0.5 day^{-1}; (d) as (c) except that η is 100.

Weight (w)

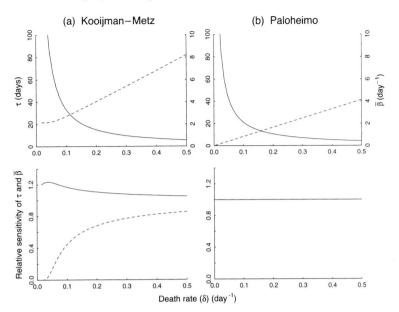

Fig. 12.6. Variation of time to sexual maturity (continuous line) and average adult fecundity (dashed line) with mortality for models with constant death rate and (a) Kooijman–Metz and (b) Paloheimo allocation. The model parameters used are those given for *D. pulex* in Table 12.1.

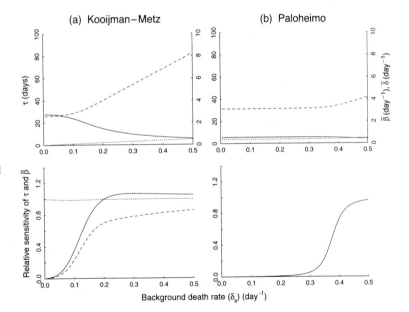

Fig. 12.7. Variation of time to sexual maturity (continuous line), average adult fecundity (dashed line) and average juvenile death rate (dotted line) with background mortality for models with assimilation dependent death rate ($\gamma = 0.01$, $\rho = 4$) and (a) Kooijman–Metz and (b) Paloheimo allocation. The model parameters used are those given for *D. pulex* in Table 12.1.

ground rate when the ratio of assimilation to maintenance costs (A/mw) is large compared to the threshold value (ρ) and rises sharply as this ratio falls towards the critical value.

The effects of this modification are shown in Figure 12.7, from which it is clear that the behaviour of both models is radically different from that which they exhibit under the assumption of constant death rate,

and strongly dependent on the underlying allocation model. The model with Paloheimo allocation shows that, except at the very highest value of background death rate, the assimilation dependent component of mortality becomes such a sensitive function of assimilation rate that it can almost completely compensate for changes in background mortality, and all the observable demographic characteristics become essentially independent of background mortality. By contrast, the model with Kooijman–Metz allocation shows a similar compensation effect operating on fecundity and development time, but average juvenile death rate continuing to change in sympathy with the background value.

In the case of Paloheimo allocation and assimilation dependent mortality it is relatively straightforward to diagnose the 'density-dependent' mortality as being the main factor in controlling the population dyamics. In the case of Kooijman–Metz allocation it would be very tempting to interpret the sensitivity of average juvenile mortality as implying that this is the main control mechanism. However, more careful examination shows this to be in serious error. The fact that the K-M model assumes different allometry for assimilation and costs, implies that with this growth and reproduction model adult costs can approach the critical proportion of assimilation, and hence drive adult mortality to high levels, while juvenile death rate remains essentially at the background level. Thus the correct interpretation of the observed model behaviour is that the controlling factor is assimilation dependent *adult* mortality. The observed sensitivity of juvenile mortality to changes in background death rate happens because the juvenile death rate is essentially identical to the background level.

DISCUSSION

The work surveyed in this paper has emphasized the central role played by high quality data in formulating, parameterizing and testing models of all kinds, and most especially those models which are designed to assist in the consideration of management matters. The paramount necessity to understand individuals and populations in their community context if they are to be managed properly, necessarily implies that management oriented models will always be complex objects. However, the key to keeping the cost and complexity of gathering the necessary data within acceptable bounds seems to be careful focusing on a compact group of questions, and elimination of detail extraneous to answering those questions.

Our biomass ecosystem model example illustrated that when the prime focus is on questions of system-wide energy and material balance, the key biological and physical processes can often be represented successfully by very crude caricatures. Although this approach has a number of salient advantages, particularly that it can often result in a model of sufficient simplicity to yield good qualitative insights into the mechanisms of causation with the system, it by no means removes the need for data to drive the exercise.

Rather, it changes the kind of data needed from exhaustive observations of individual and population performance under a wide variety of environmental conditions, to long time-series of simultaneous high resolution observations of system averaged properties.

The more complex or long-lived are the keystone species in the system or sub-system of interest, the less likely is it that such a broad-brush approach will be successful. Although some of the prevailing fashion for individual-based models seems to be driven as much by technological possibility as by dispassionate examination of options, there is little doubt that where additional biological detail is required, an individual-based perspective is invariably helpful. However, our study of the equilibrium dynamics of structured population models has clearly shown that such an approach is in no sense an easy option.

Far from reducing the data requirements of model construction, the behavioural complexity of which individual-based models are capable, and the subtlety of the causal linkages upon which this behaviour is based, both imply that very large quantities of detailed individual level data are essential. Even more intimidating is the implication, which emerges very clearly from a comparison of Figures 12.6 and 12.7, that assumptions traditionally made in the absence of knowledge in unstructured population modelling (for example a constant death rate) can have unexpectedly radical effects on the behaviour of structured models.

Thus May's (1981) conclusion that the behaviour of (unstructured) population models depends critically on the functional form of all its density-dependent functions, applies with redoubled vigour to structured or individual-based models. While, at least for some organisms, the direct observations of these density-dependences seems to be a practical possibility, the difficulty of making observations in the appropriate environmental context can be insurmountable. It has been argued that the most appropriate reaction to this difficulty is to formulate causal models of individual demography (often based on energetic) considerations, in the hope that such descriptions will be more robust than empirical characterizations. While there is undoubted scientific gain in formulating such models, their utility as extrapolators of behaviour outside the region in which they have been tested is still unproven. Moreover, the present work has shown that there can be subtle interactions between rules formulated at the individual level and population demography which, while they are by no means hard to understand *a posteriori*, may well provide unwelcome surprises to the incautious modeller.

We believe the overall conclusion must be that while modelling will almost invariably add considerably to the value of any empirical or experimental study, it cannot in any sense act as a substitute for gathering data or doing experiments. While strategic (i.e. untested and probably untestable) models can play an important role in refining ideas and defining experimental strategies they cannot safely be used as quantitative (or even qualitative) predictors of management outcomes until after the failure of a sufficient volume of attempts to falsify them.

ACKNOWLEDGEMENTS

This work was supported by CEC MAST Contract No. MAS2-CT920033 and MAFF Contract CSG-FC0105. Niall Broekhuizen assisted in the preparation of the figures.

REFERENCES

Andersen, V. & Nival, P. (1989) Modelling of phytoplankton population dynamics in an enclosed water column. *Journal of the Marine Biological Association, UK,* **69**, 625–646.

Cloern, J.E. & Cheng, R.T. (1981) Simulation model of *Skeletonema costatum* population dynamics in northern San Fransisco bay. *Estuarine Coastal and Shelf Science,* **12**, 83–100.

DeAngelis, D.L. & Gross, L.J. (1992) *Individual-based Models and Approaches in Ecology.* Chapman and Hall, London.

Gurney, W.S.C., Middleton, D.A.J., Nisbet, R.M., McCauley, E., Murdoch, W.W. & DeRoos, A. (in press) Individual energetics and the equilibrium demography of structured populations. *Theoretical Population Biology.*

Gurney, W.S.C., Ross, A.H. & Broekhuizen, N. (1995) Coupling the dynamics of species and materials. *Linking Species and Ecosystems* (eds C.G. Jones & J.H. Lawton), pp. 176–193. Chapman and Hall, London.

Huston, M.A. (1992) Individual based forest succession models and the theory of plant competition. *Individual Based Models and Approaches in Ecology* (eds D.L. DeAngelis & L.J. Gross), pp. 408–420. Chapman and Hall, London.

Kooijman, S.A.L.M. (1993) *Dynamical Energy Budgets in Biological Systems.* Cambridge University Press, New York.

MacKay, N. (1992) Evaluating the size effects of lampreys and their prey: an application of an individual based model. *Individual Based Models and Approaches in Ecology* (eds D.L. DeAngelis & L.J. Gross), pp. 408–420. Chapman and Hall, London.

May, R.M. (1981) *Theoretical Ecology: Principals and Applications.* Blackwell Scientfic Publications, Oxford.

Metz, J.A.J. & Diekmann, O. (1986) The dynamics of physiologically structured populations. *Lecture Notes in BioMathematics,* **68**. Springer-Verlag, Berlin.

NAG (1994) *NAG Fortran Library Mk16.* Numerical Algorithms Group Ltd., Oxford.

Paloheimo, J.E., Crabtree, S.J. & Taylor, W.D. (1982) Growth model of *Daphnia. Canadian Journal of Fisheries and Aquatic Science,* **39**, 598–606.

Ross, A.H., Gurney, W.S.C., Heath, M.R., Hay, S.J. & Henderson, E.W. (1993) A strategic simulation model of a fjord ecosystem. *Limnology and Oceanography,* **38**, 128–153.

Ross, A.H., Gurney, W.S.C. & Heath, M.R. (1994) A comparative study of the ecosystem dynamics of four fjordic ecosystems. *Limnology and Oceanography,* **39**, 318–343.

Taylor, A.H., Watson, A.J. & Robertson, J.E. (1992) The influence of the spring phytoplankton bloom on carbon dioxide and oxygen concentrations in the surface waters of the north-east Atlantic during 1989. *Deep Sea Research,* **39**, 137–152.

Tett, P.A., Edwards, A. & Jones, K. (1986) A model for the growth of shelf-sea phytoplankton in summer. *Estuarine Coastal and Shelf Science,* **23**, 641–672.

CHAPTER 13

The impact of introductions of new fish species on predator–prey relationships in freshwater lakes

C.E. Adams

SUMMARY

(1) Rapid change frequently accompanies the establishment of a new species introduced into a freshwater fish community. The few studies that have examined in detail the mechanisms of such change, have often identified predator–prey dynamics as the key factor responsible.

(2) Where a piscivorous fish species has been introduced, change may occur by direct predation of the introduced species upon one or more native fish species. Cases where this has occurred are typified by very rapid establishment and population expansion of the introduced species, followed by catastrophic decline often to near extirpation of native prey species and consequent instability in the dynamics of both predator and prey populations.

(3) If the introduced species is not piscivorous, disruption of predator–prey dynamics may occur through indirect predation effects if the introduced species and one or more native species share a common predator.

(4) In this paper it is demonstrated that such indirect effects of an introduced species are occurring in Loch Lomond, Scotland.

(5) Following the introduction and establishment of a large population of ruffe, the predation pressure of three of the principal predators (pike, herons and cormorants) was deflected from their native species on to ruffe.

(6) The effect on native species is not known, but will depend upon the response of the predators. If predator population sizes remain stable, predation pressure on native species will have been reduced. If however, the predator population size increases in response to greater food availability then, because the ruffe population is extremely large and expanding, predator–prey dynamics may become unstable if the ruffe population size exceeds the long-term carrying capacity of the loch.

Key-words: direct predation, fish invasions, indirect predation, Loch Lomond

INTRODUCTION

The successful invasion by a new species is frequently a powerful force for change in animal communities. This may be particularly true of closed communities normally protected from invasion by a physical barrier (Roughgarten 1986). Evidence suggests that the species composition of obligate freshwater fish communities, isolated from each other by land barriers and by virtue of their limited powers of dispersal, remains relatively stable over long time periods (Maitland 1977). When invasions of new fish species do occur they are usually attributable to deliberate or accidental introductions by humans (Maitland 1987).

The impact of invading species on native fish communities is rarely positive and often catastrophic. Although changes in fish communities have been observed and documented frequently, it is often difficult to determine the exact nature of the mechanisms reshaping community structure (Moyle 1986). In studies where the mechanisms of change have been determined, disruption of predator–prey relationships has been shown to be a route through which very rapid community change may occur (Moyle 1986, Arthrington 1991). Changes in predator–prey relationships following the introduction of new species may be of a direct or indirect nature.

Direct predation effects

Probably the most widely reported impact on a freshwater fish community following the invasion and establishment of a new species is the direct predation of the introduced species upon one or more native species. One of the best documented examples of this, is that of the introduction of the piscivorous Nile perch *Lates niloticus* into Lake Victoria in East Africa.

Nile perch were introduced to Lake Victoria in the early 1960s; after an initial delay they became established and spread throughout the northern part of the lake. Initially Nile perch fed mostly on haplochromines, an abundant fish group including more than 300 endemic species. However, as a result of the rapidly expanding population and a high predation rate, in addition to other pressures such as overexploitation and eutrophication, these species very quickly became depleted (Fig. 13.1) and are now rare in many areas of

the lake. Following the rapid decline in prey species, small Nile perch changed to preying mainly upon a small prawn species (*Carina*) while larger individuals shifted to feeding on juvenile Nile perch. Thus a highly unstable relationship between predator and its prey species developed when the predator population size exceeded that which prey populations were able to sustain (for reviews see Barel *et al.* 1985, Achieng 1990, Oguto-Ohwayo & Hecky 1991, Craig 1992).

A second example of the potential effect of predation by an introduction on a native species is that of the rainbowfish *Melanotaenia eachamensis*. The rainbowfish was a species endemic to a single crater lake in tropical rainforest in northern Queensland, Australia, Lake Eacham. In 1984 rainbowfish were still abundant in this lake, but by 1987 surveys suggested that it had become extinct in the wild. It is thought that its demise was primarily the result of predation by the aptly named mouth almighty *Glossamia aprion*. This species was introduced into the lake in the mid 1980s and increased in population size, presumably in response to an abundant food supply, resulting in a high predation rate and rapid apparent extirpation of the rainbowfish (Barlow *et al.* 1987, Arthrington 1991).

At least two studies have examined the effects of the introduction of zander *Stizostedion lucioperca* on native fish communities. In 1976 the first zander to be noted in Coombe Abbey Lake, southern England was captured. In the following years the population increased dramatically, but started to stabilize by 1984. During this period of expansion, zander fed almost exclusively on small roach *Rutilus rutilus*, (1+ and 2+ age classes). As a consequence the roach population declined, until by 1984 relative abundance of predator and prey was such that there was one zander for every 3.7 roach prey. This ratio of predator to prey is much lower than that required to sustain the predator population (for review see Hickley 1986).

In 1955, zander were deliberately introduced to Lake Egirdir, south-western Turkey. At this time the lake contained 11 native species. Zander quickly became established and supported a sizable local fishery. However by the early 1970s the growth rate of zander began to decline. By the late 1980s, nine of the original 11 native species were absent from the lake, and although two of these retained small populations in tributaries, the other seven had apparently become extinct. Campbell (1992) examined the diet of zander between 1988 and 1989. He found that zander fed exclusively on invertebrates until they had reached at least 300 mm (total length), a size much larger than the size at which zander have been reported as switching to piscivory in more stable populations (typically 100 mm). Fish in excess of 300 mm fed almost exclusively on

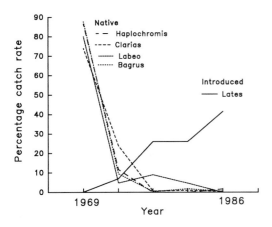

Fig. 13.1. Changes in catch rates in the Nyanze Gulf, Lake Victoria of native fish species and *Lates niloticus* following its introduction (after Achieng 1990).

juvenile zander; however, as 0+ age juveniles grew they became too large for fish under 450 mm to consume. Thus zander between 300 and 450 mm were forced to revert to feeding on invertebrates until the following year-class became available as prey. Only fish in excess of 450 mm were completely piscivorous. These findings suggest that a relationship developed where the only available fish prey species supporting the zander population in Lake Egirdir was the zander itself (for review see Campbell 1992).

Indirect predation effects

Direct predation by an introduced piscivorous species upon native species may not be the only mechanism through which changing predator–prey interactions may affect native species. For example, very little is known about the more subtle adjustments in relationships between predators and prey within fish communities when one or more populations may be growing or contracting rapidly, as is often the case when an introduced species is becoming established. In particular, little is known about the role of indirect predation interactions during such periods. Such interactions may take a number of forms but are generally characterized by one species (the donor) modifying the direct effect of a second species (the transmitter) on a third (the receiver) (Miller & Kerfoot 1987).

The role that such interactions can play in the dynamics of the species concerned has been investigated both theoretically and by manipulation experiments where species are added to, or removed from, small manageable habitats or artificial enclosures (Lubchenco 1978, Davidson et al. 1984, Kerfoot 1987, Perfecto 1990). Despite this, there is a paucity of data on the effects of indirect predation on large natural ecosystems and no reports of their impact following the accidental introduction of new fish species. However, evidence is reported here that indirect predation effects are occurring in one freshwater community (Loch Lomond, Scotland) that has been disrupted by a series of new fish species introductions over the last ten years.

FISH INTRODUCTIONS TO LOCH LOMOND

After a long period of stability in species composition, lasting at least two centuries (Maitland 1972), five species of fish previously unknown in the catchment were introduced to Loch Lomond and at least four of these have become well established. In 1982 the first ruffe *Gymnocephalus cernuus* in Scotland was recorded

there (Maitland *et al.* 1983). At around the same time gudgeon *Gobio gobio*, was also recorded (Maitland *et al.* 1983). Following this, the presence of both dace *Leuciscus leuciscus* and chub *Leuciscus cephalus* was confirmed in 1987 (Adams *et al.* 1990). In 1991 one further new species, crucian carp *Carassius carassius* was discovered (Adams & Mitchell 1992). Until the early 1980s all of these species had a highly restricted distribution in Scotland. For one species, the gudgeon, the mechanism of its arrival is well documented, as there is a record of attempts to establish a population of this species in a small lake in the catchment (Burkel 1971). For the other species evidence is entirely circumstantial, although it is highly likely that these populations emanated from discarded live-bait brought into the catchment by anglers.

Recent work has shown that all species (with the exception of crucian carp for which there are no data) have become established (Adams & Tippett 1991, Adams 1994). In the case of the ruffe, population expansion over the years following its first discovery has been rapid and exponential (Maitland & East 1989). By 1989 ruffe was probably the most common fish in Loch Lomond (Adams 1994), and there is evidence that the population has doubled at least twice since then (P.S. Maitland, personal communication).

To examine the possibility that the relationships between predators of fish and their native prey have changed since the establishment of ruffe and other non-native fish species populations in Loch Lomond, the diet of three of the principal resident predators of fish, cormorants *Phalacrocorax carbo*, pike *Esox lucius* and herons *Ardea cinerea*, was examined between 1989 and 1991 after the establishment of large populations of introduced species (Maitland & East 1989, Adams 1994). In the case of both pike and herons, it was possible to compare the diet from these studies with data on diet collected before the establishment of introduced species.

Cormorant diet was examined between April 1990 and March 1991 from recoveries of fish otoliths from pellets deposited at a cormorant roost site in the south of Loch Lomond. Approximately 20 pellets were collected every 14 days and dissolved using detergent. All fish otoliths were recovered and identified by reference to a collection of known origin (see Adams *et al.* 1994 for full details). Unfortunately there are no comparable data on cormorant diet prior to the arrival of ruffe.

The diet of herons was estimated from an analysis of regurgitated remains and pellets recovered from beneath nests in 1978 during the breeding season at one large colony in the south-east of Loch Lomond by Giles (1981). This study was repeated during the 1990 breeding season, using identical techniques (Adams & Mitchell 1995).

The food of pike was assessed on the basis of gut contents determined by dissection of fish collected from sites throughout Loch Lomond between 1989 and 1990 (full details in Adams 1991). These data were compared with similar data for pike collected between 1955 and 1967, presented in Shafi & Maitland (1971).

Figure 13.2 shows the occurrence of fish in the diet of each of the three predator species after the establishment of ruffe and for herons and pike also before the establishment of ruffe in Loch Lomond. For all three species there has been a marked shift in the species preyed upon since the arrival of ruffe. For cormorants feeding on Loch Lomond, we do not know the prey species that comprised the diet before the establishment of ruffe; however, by 1990/91 ruffe was the principal prey species making up about 85% of all fish (by number) consumed. The diet of herons during the breeding season of 1978 consisted primarily of roach

(63% by number) with brown trout *Salmo trutta* making 13% and pike a further 3%. By 1990 the diet of the same colony, at the same time of year, consisted mainly of ruffe (61%) with roach, brown trout and pike making up 5%, 3% and 1%, respectively. The diet of pike shows a similar change in species composition. Before the arrival of ruffe (1955–1967) powan was the dominant species in the diet making up 57% by number, with *Salmo* sp. (discrimination between *S. trutta* and *S. salar* was not always possible), perch *Perca fluviatilis* and three-spined stickleback *Gasterosteus aculeatus* making up 27%, 7% and 5%, respectively. By 1989/90 ruffe was the dominant prey species making up 44% of the diet with minnow *Phoxinus phoxinus*, roach and *Salmo* sp. making up 17%, 14% and 10% of the remainder of the species preyed upon.

As a result of this switch in diet of these three predators from native species to introduced ruffe, it would appear that ruffe have caused an indirect predation interaction by deflecting the predation press-

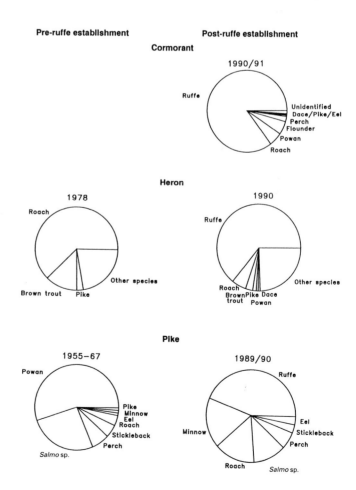

Fig. 13.2. The diet of cormorants, herons and pike in Loch Lomond, following the establishment of introduced ruffe and for herons and pike prior to the establishment of ruffe (derived from Shafi & Maitland 1971, Giles 1981, Adams 1991, Adams *et al.* 1994).

ure exerted by these predators from native species on to the newly established ruffe population. Thus at least superficially it would appear that native fish have benefitted from the indirect effect of relief of predation pressure resulting from the establishment of a new ruffe population. However, to assess fully the resultant impact of this indirect predation effect on the native fish populations it is necessary to answer three questions:

(1) What is the magnitude of the predation pressure that has been deflected on to the new ruffe population?
(2) What will be the effect of reduced predation pressure on the prey population?
(3) What will be the long-term response of the predator population?

Given our current lack of detailed quantitative knowledge of the role that predator–prey interactions play in regulating prey or predator population size and the difficulties inherent in quantifying population size in large lake-dwelling fish communities, it is not possible to fully answer any of these questions. However, by making some relatively simple calculations using published data from a number of sources it is possible to gain some insight into the possible impact of these indirect effects.

Predator population size

For both cormorants and herons, fairly accurate data on population size are available. Between 1984, a year when the ruffe population reached a substantial size (Maitland & East 1989), and 1992, there were six approximately annual, systematic counts of the cormorant population on Loch Lomondside giving a mean count of 48 ± 1.72 (SE) individuals (J. Mitchell, personal communication). For herons, regular annual counts have been made at breeding colonies throughout Loch Lomondside since 1967 (Mitchell 1993). Over the period 1984–1992, the mean count of herons on Loch Lomondside was 58 ± 7.6 (SE) (range 36 to 98). For pike there are no data of any kind that would allow a direct estimate of the Loch Lomond population size. The best estimates of pike population size from a large lake are those of Kipling and Frost (1970) for Lake Windermere, using mark–recapture techniques between 1944 and 1962. Mean fish density over this period was 4.84 ± 0.35 fish $\cdot$ ha^{-1} (range 2.6 to 8.3). Lake Windermere is similar in topography and production to the south and mid basins of Loch Lomond. Transposing Kipling and Frost's mean density to give a working estimate of pike population size in Loch Lomond's mid and south basins would give a rough population estimate of 23 000 individuals (Table 13.1).

Daily ration of predators

A number of studies have made estimates of prey consumption rates for these predators (Table 13.1). For cormorants the daily intake has been estimated at between 425 and 700 g per bird (van Dobben 1952, Mills 1965, Rae 1969). For herons estimates range from 330 to 500 g per bird per day (see Cramp & Simmons 1980). The most comprehensive estimates of consumption of pike comes from Kipling and Frost's

Table 13.1. Estimates of predator number, annual consumption and the predation deflected by establishment of ruffe for three resident predators of fish on Loch Lomond.

Predator	Predator numbers	Ration	Annual consumption (tonnes)	% ruffe in diet by biomass	Deflected predation (tonnes)	Ruffe consumption (N $\cdot$ yr^{-1})
Cormorant	48.0 ± 1.72 (mean $\pm$ 1 SE 1984–92; N = 6)	425–700 g $\cdot$ day^{-1} (van Dobben 1952, Mills 1965)	7.5–12.3	21	1.6–2.6	0.27–0.43×10^6
Heron	58.0 ± 7.6 (mean $\pm$ 1 SE 1984–92; N = 7)	330–500 g $\cdot$ day^{-1} (Cramp & Simmons 1980)	10.4	45	3.2–4.8	0.35–0.53×10^6
Pike	23 000[a]	1.46 kg $\cdot$ fish^{-1} $\cdot$ yr^{-1} (Kipling & Frost 1970)	33.5	38	12.8	1.6×10^6

[a] Pike numbers estimated from mean density in Lake Windermere (4.84 ± 0.35 (SE) fish $\cdot$ ha^{-1} (Kipling & Frost 1970)), see text for more details.

(1970) study. These authors estimate annual consumption rate in Lake Windermere as varying from 1.14 to 1.76 kg per fish per annum with a mean of 1.46 ± 0.35 (SE) kg per fish per annum (mean fish size 0.47 kg).

Predation pressure

By multiplying the estimates of predator population size and annual consumption rate it is possible to calculate a rough working estimate of the predation pressure exerted on the fish populations of Loch Lomond by each of these predator species. From the calculations as presented (Table 13.1) it would appear that the predation pressure exerted by pike exceeds that of herons and cormorants combined. It is likely that this difference may be even greater than suggested here, as it is likely that annual consumption for herons and cormorants is overestimated. For herons, this is because there is some evidence that, outside the breeding period some individuals may disperse to forage outside the catchment. For cormorants it is known that for two to three months of each year the foraging population declines when breeding birds are absent from the catchment (Adams *et al.* 1994). In contrast it is likely that the estimate of pike predation pressure is a conservative one, as these estimates take no account of the small but significant population of pike in the north basin.

To provide an estimate of the potential magnitude of the predation deflected to the new ruffe population, the proportion of the annual intake of ruffe by biomass was estimated for each predator. For cormorants the mean weight of approximately 100 individual prey, chosen at random, for each of the three commonest species in the diet was estimated from otolith size–fish size relationships. The mean size of ruffe consumed was smaller (6.0 g) than that of powan (176.8 g) and roach (194.3 g) (all other species made up less than 2% by number) (Table 13.2). Thus ruffe made up around 21% of the annual intake of cormorants by biomass.

For herons, prey fork-length was determined for all undamaged prey recovered. This was converted to weight using length–weight regressions for each species. Mean size of ruffe consumed was 9.1 g; this was smaller than the mean size of eels (77.9 g), trout (12.0 g) and the only recovered powan (200 g) but larger than the mean size of roach (0.7 g), three-spined stickleback (1.0 g), stoneloach (5.5 g) and minnow (0.4 g). Overall ruffe made up 45% of the diet of herons by biomass.

For pike, recovered prey items were weighed directly or where digestion was well advanced, length measures were converted to weights from known length–weight regressions for each species. Mean ruffe size was 8.0 g, which is larger than the mean size of minnow (0.3 g), three-spined stickleback (0.5 g) and roach (0.8 g) but smaller than mean trout size (39.7 g) and the only recovered eel (23 g) (Table 13.2).

Although the figures for predation rate by each of these predators can only be regarded as speculative, these estimates suggest that ruffe are deflecting a predation pressure equivalent to between 17.6 and 20.2 tonnes of fish per annum. This represents between 2.2 and 2.6×10^6 individual ruffe per annum. The effect that a deflection of predation pressure of this magnitude may have on native species is impossible to predict accurately. There is no information on the role that predation plays in regulating population size or even on the population size of the principal native prey species. However, even given that this predation pressure would be spread amongst several native species, it is difficult to believe that predation at this rate is inconsequential.

Table 13.2. Mean size (wt) of fish prey taken by cormorants, herons and pike feeding on Loch Lomond.

	Cormorants		Herons		Pike	
	Mean (g)	SE	Mean (g)	SE	Mean (g)	SE
Ruffe	6	0.41	9.1	0.63	8	3.04
Roach	194	27.9	0.7	0.12	0.8	0.3
Powan	177	9.27	200[a]			
Eel			77.9	6.83	23.0[a]	
Trout			12.0	2.18	39.7	7.35
Minnow					0.3	0.11
Perch			10.3	11.2	3.3	2.5
Stickleback			1.0	0.06	0.5[a]	

[a] One item only.

Effects on predator population size

An assumption implicit in the above calculations is that predation pressure has remained constant since the arrival and establishment of ruffe. However, that may not be the case. Clearly if food availability is regulating predator population size and if the establishment of a ruffe population represents an increase in available food supply, then the predator population and hence the predation pressure may increase. In the one species for which there are good data over a long period, the heron, there is no evidence that this has occurred, but it is not inconceivable that this may occur (Adams & Mitchell 1995).

Conclusion

It is impossible with our current understanding of interspecific interactions within communities to predict the outcome of the introduction of a new fish species to any fish community (Moyle 1986, Crowl *et al.* 1992). In communities where introduced fish species have become established it has been shown that changes in the dynamics of predator and prey interactions are powerful forces for change. Communities where direct predation by the introduced species upon native fish has occurred are characterized by a period of rapid change and instability in predator and/or prey populations following establishment of the invading species.

Where the invading species is not preying directly upon native fish, indirect predation effects may still occur. Although this potential mechanism of community change has received little attention by field biologists, there is a good theoretical framework for suggesting that these effects may be important in reshaping community structure (Holt 1977, Abrahams 1987, Yodzis 1988). Here it is demonstrated that indirect effects are occurring in one community where a number of new fish species have been introduced recently, the most significant of these being ruffe. The indirect predation effects described here are the deflection of the predation pressure exerted by three of the commonest predators of fish (pike, cormorants and herons) by the large population of ruffe away from native fish species. Our understanding of the consequences of this effect for native fish species is incomplete, but its magnitude and nature must depend primarily upon the response of the predator populations. If predator populations remain static, and thus predation pressure remains constant, then native fish populations should benefit from relief from some predation pressure. Tentative calculations based upon predator population size and consumption rate suggest that the relief from this predation pressure may be substantial. Alternatively, if predator populations respond to an increase in food availability by expanding, predation pressure on native species could increase. Because the ruffe population is now extremely large (P.S. Maitland, personal communication) this could destabilize the relationship between predator and prey to a point where predation pressure may temporarily exceed that which is sustainable over a medium to long time scale.

ACKNOWLEDGEMENTS

Thanks to John Mitchell for permission to quote unpublished data, to David Brown and Lorraine Keay for help in the field and to Kirsten Adams and Heather Adams for assistance with calculations of predation rate. The manuscript was much improved by comments from Felicity Huntingford and anonymous referees. Much of this work was carried out with financial support from Scottish Natural Heritage.

REFERENCES

Abrahams, P. (1987) Indirect interactions between species that share a predator: varieties of indirect effects. *Predation – Direct and Indirect Impacts on Aquatic Communities* (eds W.C. Kerfoot & A. Sih), pp. 38–55. University Press of New England, Hanover.

Achieng, A.P. (1990) The impact of the introduction of Nile perch, *Lates niloticus* (L.) on the fisheries of Lake Victoria. *Journal of Fish Biology*, 37 (Suppl. A), 17–23.

Adams, C.E. (1991) Shift in pike, *Esox lucius* L., predation pressure following the introduction of ruffe, *Gymnocephalus cernuus* (L.) to Loch Lomond. *Journal of Fish Biology*, 38, 663–667.

Adams, C.E. (1994) The fish community of Loch Lomond, Scotland: its history and rapidly changing status. *Hydrobiologia*, 290, 91–103.

Adams, C.E. & Mitchell, J. (1992) Introduction of another non-native fish species to Loch Lomond: Crucian carp (*Carassius carassius* (L.)). *Glasgow Naturalist*, 22, 165–168.

Adams, C.E. & Mitchell, J. (1995) The response of a Grey Heron *Ardea cinerea* breeding colony to rapid change in prey species. *Bird Study*, 42, 44–49.

Adams, C.E. & Tippett, R. (1991) Powan *Coregonus lavaretus* (L.) ova predation by newly introduced ruffe *Gymnocephalus cernuus* (L.) in Loch Lomond, Scotland. *Aquaculture and Fisheries Management*, 22, 239–246.

Adams, C.E., Brown, B.W. & Tippett, R. (1990) Dace (*Leuciscus leuciscus*) and chub (*Leuciscus cephalus*) new introductions to Loch Lomond. *Glasgow Naturalist*, 21, 509–513.

Adams, C.E., Brown, D.W. & Keay, L. (1994) Elevated predation risk associated with inshore migrations of fish in a large lake, Loch Lomond, Scotland. *Hydrobiologia*, 290, 91–103.

Arthrington, A.H. (1991) Ecological and genetic impacts of introduced and translocated freshwater fishes in Australia. *Canadian Journal of Fisheries and Aquatic Science*, **48** (Suppl. 1), 33–43.

Barel, C.D.N., Dorit, R., Greenwood, P.H., Fryer, G., Hughes, N., Jackson, P.B.N., Kawanabe, H., Lowe-McConnell, R.H., Nagoshi, M., Ribbink, A.J., Trewavas, E., Witte, F. & Yamaomk, K. (1985) Destruction of fisheries in Africa's lakes. *Nature*, **315**, 19–20.

Barlow, G.C., Hogan, A.E. & Rodgers, L.J. (1987) Implication of translocated fishes on the apparent extinction in the wild of the Lake Eachan rainbowfish *Melanataenia eachamensis*. *Australian Journal of Marine and Freshwater Research*, **37**, 897–902.

Burkel, D.L. (1971) Introduction of fish to a new water. *Glasgow Naturalist*, **18**, 574–575.

Campbell, R.N.B. (1992) Food of an introduced population of pikeperch *Stizostedion lucioperca* L., in Lake Egirdir, Turkey. *Aquaculture and Fisheries Management*, **23**, 71–85.

Craig, J.F. (1992) Human-induced changes in the composition of fish communities in African Great Lakes. *Reviews in Fish Biology and Fisheries*, **2**, 93–124.

Cramp, S. & Simmons, K.E.L. (eds) (1980) *The Birds of the Western Palearctic Vol. II*. Oxford University Press, Oxford.

Crowl, T.A., Townsend, C.R. & McIntosh, A.R. (1992) The impact of introduced brown and rainbow trout on native fish: the case of Australasia. *Reviews in Fish Biology and Fisheries*, **2**, 217–241.

Davidson, D.W., Inouye, R.S. & Brown, J.H. (1984) Granivory in desert ecosystems: experimental evidence for indirect facilitation of ants by rodents. *Ecology*, **65**, 1780–1786.

Giles, N. (1981) Summer diet of the grey heron. *Scottish Birds*, **11**, 153–159.

Hickley, P. (1986) Invasion by zander and the management of fish stocks. *Philosophical Transactions of the Royal Society of London B*, **314**, 571–582.

Holt, R.D. (1977) Predation, apparent competition and the structure of prey communities. *Theoretical Population Biology*, **12**, 197–229.

Kerfoot, W.C. (1987) Cascading effects and indirect pathways. *Predation – Direct and Indirect Impacts on Aquatic Communities* (eds W.C. Kerfoot & A. Sih), pp. 57–71. University Press of New England, Hanover.

Kipling, C. & Frost, W.E. (1970) A study of the mortality, population numbers, year class strengths, production and food consumption of pike, *Esox lucius* L. in Windermere from 1944 to 1962. *Journal of Animal Ecology*, **39**, 115–157.

Lubchenco, J. (1978) Plant species diversity in a marine rocky intertidal community: importance of herbivore food preference and algal competitive abilities. *American Naturalist*, **112**, 23–39.

Maitland, P.S. (1972) Loch Lomond: man's effect on the salmonid community. *Journal of the Fisheries Research Board of Canada*, **29**, 849–850.

Maitland, P.S. (1977) Freshwater fish in Scotland in the 18th, 19th and 20th centuries. *Biological Conservation*, **12**, 265–277.

Maitland, P.S. (1987) Fish introductions and translocations – their impact in the British Isles. *Angling and Wildlife in Freshwaters* (eds P.S. Maitland & A.K. Turner), pp. 57–65. ITE Symposium No. 19, Institute of Terrestrial Ecology, Grange-over-Sands, Cumbria.

Maitland, P.S. & East, K. (1989) An increase in the numbers of ruffe *Gymnocephalus cernua* (L.) in a Scottish loch from 1982 to 1987. *Aquaculture and Fisheries Management*, **20**, 227–228.

Maitland, P.S., East, K. & Morris, K.H. (1983) Ruffe, *Gymnocephalus cernua* (L.) new to Scotland in Loch Lomond. *Scottish Naturalist*, 7–9.

Mills, D.H. (1965) The Distribution and Food of the Cormorant in Scottish Inland Waters. *DAFS Freshwater and Salmon Fisheries Research Series No. 35*. Department of Agriculture and Fisheries for Scothand, Pitlochry.

Miller, T.E. & Kerfoot, W.C. (1987) Redefining indirect effects. *Predation – Direct and Indirect Impacts on Aquatic Communities* (eds K.C. Kerfoot & A. Sih), pp. 33–38. University Press of New England, Hanover.

Mitchell, J. (1993) The heronry at Gartfairn wood, Loch Lomondside – an update. *Forth Naturalist and Historian*, **16**, 58.

Moyle, P.B. (1986) Fish introductions into North America: patterns and ecological impact. *Ecology of Biological Invasions of North America and Hawaii* (eds H.A. Mooney & J.A. Drake), pp. 27–43. Springer-Verlag, New York.

Ogutu-Ohwaya, R. & Hecky, R.E. (1991) Fish introductions in Africa and some of their implications. *Canadian Journal of Fisheries and Aquatic Sciences*, **48** (Suppl. 1), 8–12.

Perfecto, I. (1990) Indirect and direct effects in a tropical agroecosystem, the maze–pest–ant system in Nicaragua. *Ecology*, **71**, 2125–2134.

Rae, B.B. (1969) The Food of Cormorants and Shags in Scottish Estuaries and Coastal Waters. *DAFS Marine Series No. 1*. Department of Agriculture and Fisheries for Scotland, Aberdeen.

Roughgarten, J. (1986) Predicting invasions and rates of spread. *Ecology of Biological Invasions of North America and Hawaii* (eds H.A. Mooney & J.A. Drake), pp. 179–188. Springer-Verlag, New York.

Shaffi, M. & Maitland, P.S. (1971) Comparative aspects of the biology of pike *Esox lucius* L. in two Scottish lochs. *Proceedings of the Royal Society of Edinburgh (B)*, **71**, 41–60.

van Dobben, W.H. (1952) The food of the cormorant in the Netherlands. *Ardea*, **40**, 1–63.

Yodzis, P. (1988) The indeterminacy of ecological interactions as perceived through perturbation experiments. *Ecology*, **69**, 508–515.

CHAPTER 14

The diet of wintering cormorants *Phalacrocorax carbo* in relation to angling catches on a coarse river fishery in north-west England

J.M. Davies and M.J. Feltham

SUMMARY

(1) Piscivorous birds are often held responsible for damaging commercial and sport fisheries. Cormorants, in particular, have been accused of reducing angling catches at some freshwater fisheries. It is, however, important to quantify the effects of these birds on fish populations.

(2) The extent of overlap between cormorant diet and angling catches on the lower River Ribble fishery in Lancashire during the winter was studied.

(3) Data on cormorant feeding behaviour and diet, and the species and sizes of fish caught by anglers, were collected. Coarse fish comprised between 46% and 72% of cormorant diet, depending on feeding site and year. The sizes of coarse fish caught by cormorants were similar to those taken by anglers, and the proportion of coarse fish in the diet of cormorants was greatest when anglers were catching most fish. This suggests that the potential for competition between anglers and cormorants may be strong.

Key-words: angling catches, coarse fish, cormorant, diet, *Phalacrocorax carbo*

INTRODUCTION

In Britain, the growing population of cormorants *Phalacrocorax carbo*, and the increasing tendency for these birds to use inland rather than coastal sites in winter, has given rise to concern, both by anglers and fisheries' managers, regarding possible damage to fisheries (Sellers 1991, Pickering *et al.* 1992, Davies & Feltham 1995). The current UK wintering population of cormorants is estimated at about 20 000 birds (Feltham & Davies 1995a) of which between 40 and 45% may be found on inland waters. Quantifying the impact of these birds on fish populations is necessary if implications for future fisheries' management are to be correctly assessed. To do this, information on the numbers of birds using a fishery, their diet, daily food intake and the availability of their prey is required (Draulans 1988, Feltham & Davies 1995a).

Studies on the foraging behaviour and diet of cormorants in Great Britain and Ireland have generally been restricted to salmonid fisheries (Mills 1965, Rae 1969, McIntosh 1978, MacDonald 1987, Kennedy & Greer 1988); few comparable data from coarse fisheries are available, particularly from river systems (Suter 1991, Feltham & Davies 1995b). A study to determine the effect of cormorants on coarse angling catches on the lower River Ribble, Lancashire, began in the winter of 1992. At present, fish stock assessment data from this river are limited to a single tributary, Boyces Beck (NGR SD651353) (Davies *et al.*, in press). No data on fish populations in either the main river or any other tributaries are available. Since 1992, the National Rivers Authority (NRA) have, however, collected catch data from coarse fishermen on the lower Ribble. Cormorants were observed feeding throughout this area, but were recorded feeding in greatest numbers on a stretch of river running through Preston (Davies & Feltham 1995). This area also included the confluence of the Ribble, with one of its tributaries, the River Darwen, on which cormorants were also regularly seen feeding.

Data on the feeding behaviour and diet of cormorants were collected from both rivers, and these were compared with angling catch statistics from the same areas. Our first objective was to examine the extent to which the diet of cormorants and the catches of anglers overlapped.

STUDY AREA AND METHODS

The River Ribble flows from its source at Newby Head Moss in the Yorkshire Pennines into the Irish Sea at Preston in Lancashire. It covers a distance of 110 km and drains a catchment of 2182 km². The upper reaches of the Ribble are principally salmonid fisheries; the lower reaches (NGR SD710371 to SD527287) however, support an important coarse fishery, particularly of cyprinid species such as chub *Leuciscus cephalus*, roach *Rutilus rutilus*, dace *Leuciscus leuciscus* and barbel *Barbus barbus*. Data were collected from a 6.5 km long (NGR SD527287 to SD577301) and 25 m wide stretch of the Ribble around the junction with the River Darwen tributary with a mean flow rate of 30 m³·s⁻¹. Data from the River Darwen were collected on a stretch 300 m long and 10 m wide, extending upstream from its confluence with the Ribble (NGR SD547283) (Fig. 14.1). Both rivers are subject to tidal influence.

Feeding behaviour and diet of cormorants

Field observations

Observations on the feeding behaviour of cormorants on the Ribble were made throughout the day in February and March of 1993 and 1994. Similar observations were made on the Darwen in February and March in 1993 and from December 1993 to March 1994. An observer walked along the river bank until a bird was located and the number of dives and the number of prey items caught during a 30 min sampling period were recorded. Observations were carried out using an Optolyth telescope with 20–60× zoom lens facility. Discrimination between different species of coarse fish caught by birds proved unreliable; therefore they were categorized simply as coarse fish, 'flatfish' (almost certainly flounder *Platichthys flesus*), or eel *Anguilla anguilla*. Prey size was estimated in relation to the bird's bill-length and converted to centimetres using bill measurements of birds shot locally (n = 15).

Analysis of stomach contents

Shooting birds was not possible on the Ribble due to the close proximity of residential areas. However, in March 1994 seven birds, for which feeding observation data had been collected, were shot under licence on the lower River Darwen. If stomach contents could not be

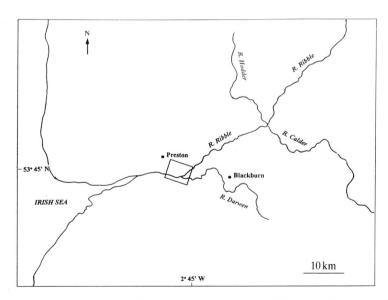

Fig. 14.1. Map of the River Ribble, in north-west England. The stretches of the Ribble and the Darwen on which data were collected are shown by inset box.

examined immediately, the whole bird was transferred to a freezer to await post-mortem examination. Fish found in the gut were identified to species level and fork-length (cm) was measured directly or estimated from bones resistant to digestion (Suter 1991, Veldkamp 1994). The proportions of different types of fish and the sizes of coarse fish found in the stomachs were then compared with data from feeding observations from the same period.

Angling catches

Angling catch data were obtained from the routine censusing of anglers by NRA bailiffs. During the winter of 1992/93, 28 anglers were censused on the Ribble and 27 on the Darwen. Equivalent figures in 1993/94 were 9 and 39, respectively. The numbers, species and approximate sizes of fish caught by anglers and the length of time fished prior to censusing were recorded. For the purposes of the census, bailiffs assigned each fish caught to one of the following size classes; 13 cm in length, 13–18 cm, 18–23 cm, 23–28 cm, 28–33 cm, 33–38 cm and 38–48 cm.

RESULTS

Feeding behaviour and diet of cormorants

The proportion of dives by cormorants which were successful, typically between 5.5% and 8% (Table

14.1), was similar, both between years and between rivers. However, the proportion of these dives resulting in the capture of coarse fish prey varied. In 1993 coarse fish constituted a greater fraction of the diet on the Darwen, while more flatfish and eels were taken on the Ribble (chi-squared test: $\chi_2^2 = 7.39$, P < 0.05), but this difference was not apparent the following year ($\chi_2^2 = 0.71$ ns) (Table 14.1). In addition, coarse fish caught by cormorants on the Darwen were larger than those caught on the Ribble in 1993 (Wilcoxon–Mann–Whitney test: $W_{79,27} = 1981.5$, p < 0.0001), but were of similar size in 1994 ($W_{19,112} = 7454.5$, ns) (Table 14.1). Consequently, the size of fish caught by birds differed between the two winters on the Darwen ($W_{27,112} = 9570.5$, p < 0.0001), but not on the Ribble ($W_{112,19} = 7545.5$, ns) (Table 14.1). The proportion (71%) and median size (18.2 cm) of coarse fish found in the stomachs of birds shot on the Darwen in 1994 did not differ significantly from estimates based on feeding observations undertaken during the same period (69% and 18.5 cm, respectively) (comparison of proportions; $\chi_1^2 = 0.26$, ns, comparison of median size; $W_{31,10} = 583.0$, ns). All ten of the coarse fish recovered from cormorant stomachs were roach.

Angling catches

The mean number of fish caught by anglers was relatively low on the Ribble in both years, being 0.44 fish per hour in 1992/93 and 0.06 in 1993/94, although the difference between years was significant (t-test; $t_{33} =$

Table 14.1. The proportions (%) of different types of fish, the median sizes (cm) of coarse fish caught by cormorants and median success of prey capture (%) of cormorants. Data were collected on the River Ribble in February and March in 1993 and 1994, and on the River Darwen in February and March 1993 and December 1993 to March 1994. Sample sizes refer to numbers of fish.

Fish type	River Ribble		River Darwen	
	1993 Feb–Mar n = 149	1994 Feb–Mar n = 41	1993 Feb–Mar n = 37	1994 Dec–Mar n = 205
Coarse (%)	53.0	46.3	73.0	54.6
Flatfish (%)	31.5	36.6	24.3	43.9
Eel (%)	15.5	13.7	2.7	1.5
Median size of coarse fish (cm) (+ quartiles)	15.0 (11.25, 18.75)	15 (15.0, 18.75)	22.5 (15.0, 30.0)	15.0 (11.25, 21.86)
Median % age of successful dives (+ quartiles)[a]	8.0 (2.7, 15.3)	5.5 (4.7, 7.9)	7.5 (0.0, 15.4)	6.5 (2.2, 10.6)

[a] Ribble 1993 versus Ribble 1994 ($W_{61,26} = 277.0$, ns), Ribble 1993 versus Darwen 1993 ($W_{61,27} = 1100.5$, ns), Ribble 1994 versus Darwen 1994 ($W_{20,108} = 7080.5$, ns), Darwen 1993 versus Darwen 1994 ($W_{27,112} = 1875.5$, ns).

2.40, P < 0.05). The most successful angler caught 6.67 fish per hour in 1992/93 and 4.16 fish per hour in 1993/94. The number of fish caught per hour on the Darwen was also significantly greater in 1992/93 than the following year ($t_{44} = 5.96$, P < 0.0001); the mean catch rates being 2.04 fish per hour and 0.37 fish per hour, respectively. The most successful angler caught 9.85 fish per hour in 1992/93 and 4.13 fish per hour the following year. The catch rate differed significantly between rivers in both years (1992/93, $t_{32} = 4.50$, P < 0.0001; 1993/94, $t_{41} = 3.00$, P < 0.05).

The species most frequently caught by anglers were chub, dace and roach (Table 14.2). In 1992/93, the proportions of these species caught were similar on both rivers ($\chi_2^2 = 5.88$, ns) (Table 14.2), but all species were significantly smaller on the Ribble compared to the Darwen (chub, $W_{81,26} = 4670.0$, p < 0.05; roach, $W_{127,30} = 10\,556.0$, p < 0.01; dace, $W_{142,22} = 12\,672.0$, p < 0.0001). Insufficient data were available on the Ribble to examine annual variation in angling catches, but these data were available from the Darwen. On this river proportionately less chub and more dace, were caught by anglers in 1992/93 than in the following year ($\chi_3^2 = 55.68$, P < 0.0001) (Table 14.2). The median size of dace and roach were, moreover, significantly greater in the winter of 1992/93 compared to the following year, but the sizes of chub were similar (chub, $W_{81,53} = 5899.0$, ns; roach, $W_{127,50} = 11\,413.0$, p < 0.0001; dace, $W_{142,7} = 10\,874.0$, p < 0.05).

DISCUSSION

The majority of fish caught by anglers on both rivers and in both years were chub, dace, and roach, which between them represented 97% to 99% of the total catch. Cormorants certainly attack all three of these

species on this fishery, since between 7–14% of chub and 2–7% of dace bore wounds consistent with cormorant attack (Davies *et al.* in press). However, the proportion that each species contributes to the diet of cormorants is not clear. All coarse fish recovered from cormorant stomachs were roach, while roach made up 33% of the angling catch in that same year ($\chi^2 = 11.06$, 1 df, P < 0.001). However, on the basis of the small number of stomachs sampled, it would be imprudent to draw any firm conclusions regarding prey selection among the different coarse fish species. This is also made difficult by the absence of any true measure of prey availability; we have, after all, no reason to presume that anglers are representitively sampling the fish stocks in the river.

The success rate of anglers was relatively low in both years on the Ribble. This was also the case on the Darwen in 1993/94, but in 1992/93, anglers on the Darwen were significantly more successful. At the times when anglers were catching relatively few fish, coarse fish made up between 46% and 54% of the diet of cormorants (Table 14.1), but in 1992/93, when anglers were more successful on the Darwen, the diet of cormorants contained a significantly higher proportion of coarse fish (72%, Table 14.1). The size of fish taken by anglers was significantly larger on the Darwen than on the Ribble at this time (Table 14.2), and this was also the case for cormorants (Table 14.1). In 1992/93 it would seem that more fish were available to both anglers and cormorants on the Darwen, and that these fish were generally larger than in 1993/94. The coarse fish caught by cormorants on the Darwen in the following year were, however, smaller than those caught by anglers. The reasons for this are unclear, although sizes of roach and dace caught by anglers were also significantly smaller in 1993/94 than in the previous year.

Table 14.2. The proportion (%) and median size classes (cm) of coarse fish species caught by anglers on the River Ribble and River Darwen in winter 1992/93 and winter 1993/94. Sample sizes refer to numbers of fish.

| Species | River Ribble | | River Darwen | | | |
| | 1992/93 n = 79 | | 1992/93 n = 360 | | 1993/94 n = 92 | |
	% of catch	Size class	% of catch	Size class	% of catch	Size class
Chub	33.0	18–23	22.5	23–28	58.7	18–23
Roach	37.9	18–23	35.3	18–23	32.6	13–18
Dace	27.8	<13	39.4	13–18	7.6	<13
Others[a]	1.3	–	2.8	–	1.1	–

[a] Others include barbel and gudgeon *Gobio gobio* L.

The degree of overlap between anglers and cormorants could not be compared directly for several reasons. Firstly, the proportions of chub, dace and roach in the diet of cormorants are not yet known with any great degree of confidence. Secondly, the angling catch statistics are constrained by (i) when and where anglers chose to fish, and (ii) when and where bailiffs were present to collect these data. Thirdly, fish stock assessment data are not currently available with which to compare angling catch statistics. Despite the above limitations, however, data from our study strongly suggest that the potential for competition between anglers and cormorants is likely to be significant. The sizes of coarse fish taken by cormorants were similar to those taken by anglers, and the proportion of coarse fish in the diet of cormorants was greatest when anglers were catching most fish. The use of angling catch data in such studies, though not ideal, is not, however, unique (Suter 1991, 1995). Damage to fisheries by fish-eating birds is often claimed simply on the basis of a perceived reduction in catch, and these data are, in many cases, the only index of fish abundance (e.g. Suter 1995, Feltham & Davies 1995a).

A recently proposed study aimed at obtaining the much needed fish stock assessment data, on the lower Ribble, and newly obtained licences to sample more birds should further improve our understanding of the relationship between cormorant diet and angling catches on these rivers.

ACKNOWLEDGEMENTS

We would like to thank the Cormorant Steering Group, in particular Peter Fox, the bailiffs for collection of creel census data and people who have helped count cormorants. The project is funded by Liverpool John Moores University and the National Rivers Authority North-West Region.

REFERENCES

Davies, J.M., Feltham, M.J. & Walsingham, M.V. (in press) Fish wounding by cormorants *Phalacrocorax carbo*. L. *Fisheries Management and Ecology*.

Davies, J.M. & Feltham, M.J. (1995) Do cormorants and anglers compete for the same resource? *Proceedings of the Institute of Fisheries Management Annual Study Course*, **25**, 167–188.

Draulans, D. (1988) Effects of fish-eating birds on freshwater fish stocks: an evaluation: *Biological Conservation*, **44**, 251–263.

Feltham, M.J. & Davies, J.M. (1995a) How much do cormorants and goosanders eat? *Proceedings of the Institute of Fisheries Management Annual Study Course*, **25**, 143–166.

Feltham, M.J. & Davies, J.M. (1995b) The diet of cormorants *Phalacrocorax carbo* L. at two fisheries in north west England. *Fisheries Management and Ecology*, **2**, 157–159.

Kennedy, G.J.A. & Greer, J.E. (1988) Predation by cormorants *Phalacrocorax carbo* (L) on the salmonid populations of an Irish river. *Aquaculture and Fisheries Management*, **19**, 159–170.

MacDonald, R.A. (1987) Cormorants and fisheries. *BTO News*, **150**, 12.

McIntosh, R. (1978) Distribution and diet of the cormorant on the lower reaches of the river Tweed. *Fisheries Management*, **9**, 107–113.

Mills, D.H. (1965) The distribution and food of the cormorant *Phalacrocorax carbo* in Scottish inland waters. *Freshwater and Salmon Fisheries Research*, **35**, HMSO, Edinburgh.

Pickering, S.P.C., Kirby, J.S. & Fox, P. (1992) Winter status of cormorants, goosanders and red-breasted mergansers in Great Britain with specific reference to the north-west National Rivers Authority region. *NRA Internal Report*. National Rivers Authority, Preston.

Rae, B.B. (1969) The food of cormorants and shags in Scottish estuaries and coastal waters. *Department of Agriculture and Fisheries of Scotland Marine Research*, **1**. HMSO, Edinburgh.

Sellers, R.M. (1991) Breeding and wintering status of the cormorant in the British Isles. *Proceedings of the Workshop 1989 on Cormorants* Phalacrocorax carbo (eds M.R. van Eerden & M. Zijlstra), pp. 30–35. Rijkswaterstaat Directorate Flevoland, Lelystad.

Suter, W. (1991) Food and feeding of cormorants *Phalacrocorax carbo* wintering in Switzerland. *Proceedings of the Workshop 1989 on Cormorants* Phalacrocorax carbo (eds M.R. van Eerden & M. Zijlstra), pp. 156–165. Rijkswaterstaat Directorate Flevoland, Lelystad.

Suter, W. (1995) The effect of predation by wintering cormorants *Phalacrocorax carbo* on grayling *Thymallus thymalus* and trout (Salmonidae) populations: two case studies from Swiss rivers. *Journal of Applied Ecology*, **32**, 29–46.

Veldkamp, R. (1994) Food choice of cormorants *Phalacrocorax carbo sinensis* in NW Overijssel, The Netherlands. *Rapport Bureau Veldkamp*. RIZA, Steenwijk.

CHAPTER 15

Estimation of by-catch composition and the numbers of by-catch animals killed annually on Manx scallop fishing grounds

A.S. Hill, A.R. Brand, U.A.W. Wilson, L.O. Veale and S.J. Hawkins

SUMMARY

(1) Scallops have been intensively fished around the Isle of Man for approximately 50 years. Scallop dredging employs gear which also removes other non-target species (by-catch). By-catch species are often severely damaged or killed during dredging, after which they are subsequently discarded.

(2) Surveys of by-catch composition and damage were undertaken on selected Manx commercial scallop grounds employing commercial scallop fishing techniques. By-catch composition varied between grounds. Some species appeared to be more susceptible than others to dredge damage with no mortality occurring amongst the more robust species.

(3) Commercial scallop fishing effort data from log books were used to estimate the total numbers of by-catch animals killed annually by the whole commercial fleet on these grounds. These calculations suggest that large numbers of the more easily damaged species may be killed each year.

(4) The significance of the removal and damage of large numbers of by-catch is discussed. Suggestions for improvements to these estimates of by-catch removal are also made.

Key-words: by-catch, dredging disturbance, scallop fishing

INTRODUCTION

The great scallop *Pecten maximus* has been actively fished for more than 50 years around the Isle of Man (see Brand *et al*. 1991 for review). Since 1969 there has also been a large fishery for the queen scallop *Aequipecten opercularis*. For some time there has been concern about the damage to benthic communities caused by trawling and scallop dredging (de Groot 1984) and some preliminary investigations of the impact of scallop dredging have been undertaken (e.g. Caddy 1973, Eleftheriou & Robertson 1992). By-catch records can underestimate the number of organisms affected by dredging (Dare *et al*. 1993) but are a readily available source of data that can be used to determine the kind of damage suffered by larger non-target species.

This study describes the common by-catch species retained within scallop dredges and assesses the damage they incur. Fishing effort exerted by a sample of the commercial fleet is estimated and used to calculate preliminary estimates of the annual numbers killed by the whole commercial fleet on three major scallop grounds.

MATERIALS AND METHODS

By-catch estimates and damage scoring

Three 5 × 5 nautical mile squares were sampled during June 1994 by the *RV Roagan* using fishing gear and methods (e.g. towing speed and haul duration)

111

similar to those used by Manx commercial fishing boats. These squares were within three frequently fished commercial grounds, Bradda Inshore, Targets and H/I Sector (Fig. 15.1). Sampling consisted of three replicate hauls of 2 nautical miles made with a gang of four spring-toothed dredges (individual dredge width 0.76 m; tooth length about 11.0 cm, 8.6 cm apart; belly rings 7.0 cm diameter) towed at a speed of 2.5 knots. By-catch was counted and scored for external damage (see Table 15.1). Counts were expressed per metre dredge width per hour of dredging (metre·hours). Damage scores are presented where five or more animals of a by-catch species were collected.

Fishing effort

Estimates of fishing effort by commercial boats have been calculated for 5 × 5 nautical mile squares around the Isle of Man (Fig. 15.1) using log book data returned over the past 15 years by, on average, 17 boats (min. = 11, max. = 24), which represents 33% of the current total commercial fleet. Information supplied includes total dredge width and tow time for each 5 × 5 nautical mile square, allowing dredging effort for each square to

be calculated in the same units as the research vessel data (i.e. metre·hours).

Commercial scallop fishery by-catch mortality calculations

By-catch animals are damaged by the dredge teeth when they first come into contact with the dredge, by the steel rings in the dredge belly and by the grinding action of stones caught in the dredge. The number and size of stones caught in dredges, and consequently the degree of damage, varies between grounds (A.S. Hill & L.O. Veale, unpublished data). By-catch damage and the subsequent mortality was therefore estimated separately for each ground. Numbers caught during the research survey were multiplied by the proportion suffering grade 3 damage, which was shown in survival experiments to result in approximately 100% mortality for most species (A.S. Hill & L.O. Veale, unpublished data). This gave an estimate of numbers killed during the survey. Multiplying by the mean annual sample fleet effort and scaling up by a factor of 3.3 gave a crude estimate of the numbers killed by the Manx commercial fleet annually within each 5 × 5 nautical mile square.

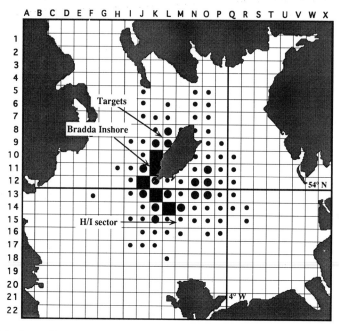

Fig. 15.1. Mean annual dredging effort per season (metre·hours) of the sample fleet between 1982 and 1992. Grid consists of 5 × 5 nautical mile squares. Squares sampled during the June 1994 by-catch survey are indicated. Fishing effort is divided into three ranges: ●, <1000 metre·hours; ●, 1000 to 4000 metre·hours; ■, >4000 metre·hours.

Table 15.1. Damage scores for biota retained as by-catch in scallop dredges.

	Score		
	1	2	3
Crabs	In good condition	Legs missing	Carapace cracked/dead
Starfish	In good condition	Arms missing	Crushed/disc damaged/dead
Urchins	In good condition	<50% of spines lost	>50% of spines lost/ crushed/ dead
Molluscs	In good condition	Chipped shell	Crushed/dead

RESULTS

Effort data (Fig. 15.1) revealed Bradda Inshore to be one of the most intensively fished grounds around the island (8388 metre·hours per year); Targets was less heavily fished (1422 metre·hours) and H/I Sector was fished at a relatively low intensity (623 metre·hours).

By-catch composition varied considerably between grounds, but *Asterias rubens* was present in relatively high numbers on all grounds (Fig. 15.2) especially Bradda Inshore. *Echinus esculentus* dominated samples from Targets and *Aequipecten opercularis* was the most

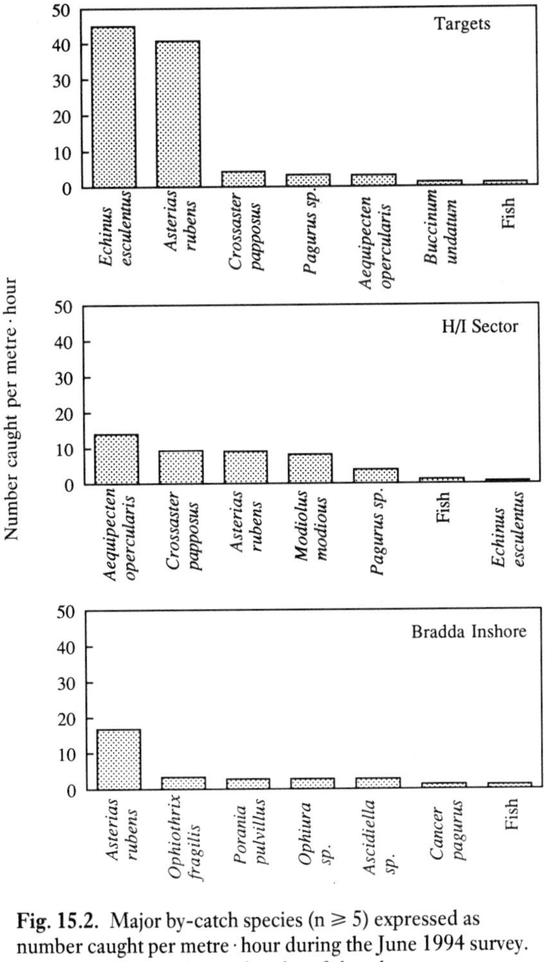

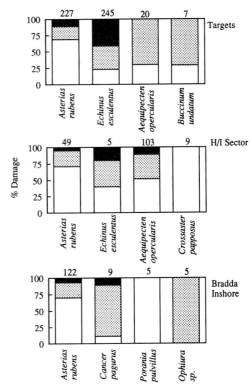

Fig. 15.3. Proportion of each damage level recorded for each species (n ≥ 5) during the June 1994 by-catch survey using the scoring system of Table 15.1 (numbers indicate sample sizes): □, 1 – in good condition; ▨, 2 – minor damage; ■, 3 – major damage.

Fig. 15.2. Major by-catch species (n ≥ 5) expressed as number caught per metre·hour during the June 1994 survey. Species are arranged in rank order of abundance.

abundant species on the H/I Sector ground. *Aequipecten* was regarded as a non-target species as the gear was set up for fishing scallops rather than queenies (i.e. larger and more widely spaced teeth on the tooth bar and larger belly rings).

Damage varied considerably between species (Fig. 15.3). *Crossaster papposus* never exhibited grade 3 damage while a consistent proportion of *Asterias* did. *Asterias* and *Echinus* occurred on more than one ground and tended to exhibit a similar range of damage scores on each ground. No grade 3 damaged *Aequipecten* were recorded from Targets but 11% of H/I Sector *Aequipecten* were fatally damaged, although this probably reflects the greater sample size from H/I Sector (103) compared with Targets (20).

Large numbers of *Asterias* and *Echinus* are taken by commercial fishing gears and killed during the season (Table 15.2). Some species appeared to be less susceptible to fatal external damage; there were no dredge-induced mortalities amongst *Crossaster*, *Buccinum undatum*, *Porania pulvillus* and *Ophiura* sp.

DISCUSSION

Each fishing ground was characterized by a unique suite of rarer species (A.S. Hill & L.O. Veale, unpublished data), although a few species, such as *Asterias*, were common to more than one ground. *Asterias* has a

considerable capacity for regeneration and *Crossaster*, *Buccinum*, *Porania* and *Ophiura* appeared to be particularly robust. Dredge disturbance may therefore favour these species and promote communities dominated by them. The return of by-catch to the sea bed may represent an important food source to some scavengers and predators. Predation studies have demonstrated the responses of predators to gear damaged animals (e.g. beam trawls: Kaiser & Spencer 1994) but further assessment of the importance of discarded by-catch to benthic food chains is required. Since these communities have been subject to dredging for many years they may already be in equilibrium with the effects of the removal and relocation of large numbers of individuals; dredging may be acting to maintain community structure rather than alter it. Further work is required to get good estimates of the population density of each species. These data will make it possible to assess whether the commercial removal and/or relocation of these species is likely to bring about further changes in community structure.

There are limitations to the accuracy of the mortality estimations. These estimates are based only on the by-catch animals incurring major damage (damage score 3), whereas a proportion of animals with lower damage scores will also die as a result of dredging. Commercial boats return the by-catch to the sea bed. These calculations take no account of increased risk of predation or disease after the by-catch has been discarded, so

Table 15.2. Estimates of total number of by-catch species ($n \geq 5$) killed annually by the Isle of Man commercial scallop fleet in 5×5 nautical mile squares on three commercial fishing grounds. Annual fishing season (1 November–31 May).

	Number caught during survey (per metre·hour)	% incurring major damage (damage score 3)	Number killed during survey (per metre·hour)	Mean annual effort by sample fleet (metre·hour)	Annual numbers killed by sample fleet	Annual numbers killed by whole fleet
TARGETS						
Asterias rubens	40.74	11	4.48	1 422	6 373	21 242
Echinus esculentus	44.97	41	18.44		26 218	87 395
Aequipecten opercularis	3.14	0	0		0	0
Buccinium undatum	1.23	0	0		0	0
H/I SECTOR						
Asterias rubens	9.02	5	0.45	623	281	937
Echinus esculentus	0.68	20	0.14		85	282
Aequipecten opercularis	14.08	11	1.55		965	3 216
Crossaster papposus	9.3	0	0		0	0
BRADDA INSHORE						
Asterias rubens	16.68	7	1.17	8 388	9 794	32 646
Cancer pagurus	1.23	11	0.14		1 135	3 783
Porania pulvillus	2.73	0	0		0	0
Ophiura sp.	2.73	0	0		0	0

mortality is underestimated. Low gear efficiency will also lead to underestimates of mortality as some animals are damaged but fail to enter the dredge while others are sufficiently small to pass through the belly rings of the dredge. Fishing effort is also likely to be underestimated as we have little knowledge of the activities of non-Manx boats or of the fishing effort for queens on these grounds. The mortality estimates will also be biased if the fishing performance of the *RV Roagan* (e.g. towing speed, gear efficiency, etc.) varied greatly from the mean values for the commercial fleet. However, this research vessel is of a similar size to the commercial fishing boats and the fishing operation was standardized, as far as possible, to imitate normal commercial practice by a crew with commercial fishing experience. Therefore any inaccuracy from this source is likely to be small in relation to other sources of error. Further work is underway to quantify gear efficiencies and to obtain more detailed fishing effort data.

By-catch samples were taken during June, at the end of the scallop fishing season, when densities are likely to be lowest due to mortality throughout the season. Most commercial grounds are subject to repeated visits during the season so estimates based on June data would tend to underestimate mortality. More detailed information on by-catch numbers throughout the season is therefore required to allow for change in the population density during the season.

Nearly all the limitations in the available data therefore appear to lead to underestimates of mortality. However, large numbers of non-target animals are killed annually and this number varies considerably between grounds. Whether these mortalities are significant in terms of the whole community is uncertain but is under investigation. We hypothesize that fishing disturbance may be a contributory factor in determining community structure and dynamics.

REFERENCES

Brand, A.R., Allison, E.H. & Murphy, E.J. (1991) North Irish Sea scallop fisheries: a review of changes. *An International Compendium of Scallop Biology and Culture* (eds S.E. Shumway & P.A. Sandifer), pp. 204–218. World Aquaculture Society, Baton Rouge.

Caddy, J.F. (1973) Underwater observations on tracks of dredges and trawls and some effects of dredging on a scallop ground. *Journal of the Fisheries Research Board of Canada*, 30, 173–180.

Dare, P.J., Key, D. & Connor, P.M. (1993) The efficiency of spring-loaded dredges used in the western English Channel fishery for scallops, *Pecten maximus* (L.). ICES CM 1993/B:15 (Ref.K).

de Groot, S.J. (1984) The impact of bottom trawling on benthic fauna of the North Sea. *Ocean Management*, 9, 177–190.

Eleftheriou, A. & Robertson, M.R. (1992) The effects of experimental scallop dredging on the fauna and physical environment of a shallow sandy community. *Netherlands Journal of Sea Research*, 30, 289–299.

Kaiser, M.J. & Spencer, B. (1994) Fish scavenging behaviour in recently trawled areas. *Marine Ecology Progress Series*, **112**, 41–49.

CHAPTER 16

Behavioural responses of scavengers to beam trawl disturbance

M.J. Kaiser and B.E. Spencer

SUMMARY

(1) Beam trawling may contribute to long-term changes in benthic communities. Most studies have concentrated on the direct effects of fishing on animals intimately associated with the seabed. However, the role of scavengers of animals damaged or disturbed by trawling is poorly understood.

(2) We investigated the behaviour of potential scavengers, at time intervals before and after fishing an area with a 4 m beam trawl, using a combination of replicate 2.8 m beam trawl tows, diver operated video surveys and extended camera observations of bait.

(3) After fishing with the commercial beam trawl, the density of dabs and gurnards increased significantly. Dabs dispersed within 48 h, whereas gurnard numbers remained high. Although the density of hermit crabs was lower immediately after fishing, they increased to the prefishing level after 24 h. Diver observations indicated that some scavengers aggregated on the trawled area within 1 h and were patchily distributed. After 24 h, common starfish and whelks were observed in greater numbers on the trawl track and were feeding on animals that had been damaged by the beam trawl.

(4) Within 30 min, dabs and whiting were attracted to a baited bag attached to a camera frame located in close proximity to the trawled area. Hermit crabs arrived after 40 min, with peak numbers occurring between 3 to 14 h after the baited camera reached the seabed. Whelks started to arrive after 7 h, peaked at 12 h and then began to disperse. Starfish continued to arrive at the bait bag for up to 17 h.

(5) Beam trawling seems to provide a food supply for a variety of scavenging species. It is conceivable that, in some areas, scavenger abundance could be related to trawling intensity and frequency, and may indicate the scale of intensity. In heavily trawled areas, communities may eventually become dominated by high abundances of a few scavenging species.

Key-words: beam trawl, disturbance, predation, scavengers

INTRODUCTION

The sediments and animal inhabitants of shallow subtidal areas are commonly subjected to natural and anthropogenic disturbance (Hall 1994). Both forms of disturbance can lead to the death of whole or parts of benthic communities (e.g. Rees *et al.* 1977, Bergman & Hup 1992). Immediately after these events, scavengers and predators have an important role in the recycling of dead or damaged animals (Dayton & Hessler 1972, Stockton & DeLaca 1982, Kaiser & Spencer 1994, Hall 1994). When disturbance events are frequent and regular they may provide an important source of food for some predatory species and consequently will affect community structure.

Demersal fishing methods, such as beam trawling, are thought to contribute to long-term changes in benthic community structure (de Groot 1984, Messieh *et al*. 1991, Witbaard & Klein 1994). Commercial beam trawls vary in width from 4 m to 12 m and are typically fitted with tickler chains or a chain mat which are designed to disturb sole or plaice that are buried in the surface sediment. In common with other demersal fishing gear, the beam trawl's design inevitably produces a by-catch of non-commercial fishes and invertebrates (Cruetzberg *et al*. 1987, Kaiser *et al*. 1994). In addition, many of the fragile infauna, such as bivalves and sea urchins, are damaged by the tickler chains or beam shoes as they pass through the sediment (Rumohr & Krost 1991). These damaged animals are presumably eaten by various predatory species that quickly migrate into areas within minutes to hours of a trawling disturbance occurring (Kaiser & Spencer 1994). Predator pressure is an important factor determining the structure of benthic communities (e.g. Paine 1980, Summerson & Peterson 1984, Thrush *et al*. 1994). Hence, if fishing activity increases the vulnerability or availability of prey species, we might predict a shift towards those predatory species best adapted to make use of this food source. In

this paper, we investigate which predators respond to disturbance created by a commercial 4 m beam trawl, and the timescale of predator arrival and departure.

METHODS AND MATERIALS

The experiment was carried out in April 1994 at a site approximately 1.5 km off the east coast of Anglesey, North Wales (Fig. 16.1). The substratum consisted of compact medium sand with ripples approximately 3–5 cm high and water depth varied between 14 and 20 m. Because of the need to synchronize various activities, two vessels (*R.V. Corystes* and *M.V. Crackers*) were used during the study. Divers operated independently from an inflatable boat.

Trawling protocol

A 4 m commercial beam trawl was fished three times along a line approximately 6 km long at a speed of about 4 knots (over the ground), to create a beam trawl disturbance effect. All three tows were within a 30 m wide area (trawl corridor) of the seabed (for full details of gear and navigation system see Kaiser & Spencer 1994).

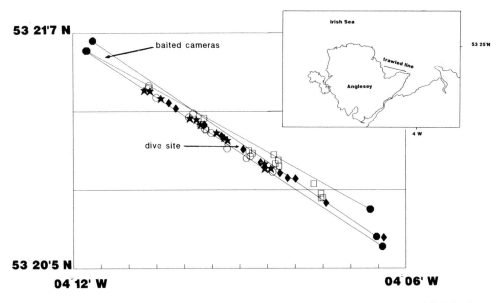

Fig. 16.1. The start and end points of the 4 m (●) and 2.8 m beam trawl tows. The start and end points of the 2.8 m beam trawl tows are not connected to save confusion, however they all fall within the area trawled by the 4 m beam trawl. ★, 2.8 m beam trawl tows 24 h before 4 m beam trawling; □, 1 h after trawling; ○, 24 h after trawling; ◆, 48 h after trawling.

Epibenthic survey

A 2.8 m wide beam trawl, fitted with six tickler chains and a 4 cm mesh codend, fished at a speed of approximately 2 knots, was used to sample the epifaunal predators in the trawl corridor. The trawl corridor was sampled using six 10 min tows on each of four different occasions: 24 h before, and 1 h, 24 h and 48 h after fishing with the 4 m beam trawl (Fig. 16.1). It took 2 h to fish each set of six tows. All trawling was in daylight. Only predatory or scavenging macrofaunal species were sorted and counted from each tow.

The area surveyed by each 10 min 2.8 m beam trawl tow did not vary significantly between sampling periods (Table 16.1). Standardizing the data to numbers per 1000 m^2 had no effect on the results; hence the raw data are presented. Significant differences in the mean number of each animal between consecutive sampling periods were analysed using one-way ANOVA on log$_e$ transformed data. When significant differences were detected ($P < 0.05$), a Tukey–Kramer multiple comparison test was used to determine differences between sampling occasions. An estimate of changes in the variation between sampling occasions was determined by calculating the coefficient of variation for the mean number of each species per sampling occasion. Differences between the coefficients of variation on each sampling occasion were then determined using a Mann–Whitney test.

Baited camera

A camera with a flash unit, mounted on a steel frame, was deployed, at the northern end of the trawl corridor (Fig. 16.1) 1 h after the last tow was hauled (12.27 GMT), with the aim of attracting scavengers in the vicinity. The camera was set to take one exposure every 10 min for 24 h. A net bag, containing 5 kg of freshly crushed queen scallops *Aequipecten opercularis* was attached to the frame such that it was 60 cm beneath the camera lens and on the seabed. Each exposure of the developed film was analysed by counting the number of animals of each species attracted to the bait. A measure of the change in the number of animals of each species was determined by comparing their number per exposure in 2 h blocks. Variation in the mean number of a species observed every 2 h, were compared using one-way ANOVA on log$_e$ transformed data. Differences between means were determined using Tukey's multiple comparison test.

Diver survey

Divers obtained video records of the appearance of the seabed and large epibenthic macrofauna using an Osprey SIT (silicon intensified tube) camera connected to a Panasonic SVHS time-lapse video recorder. The divers provided a recorded running commentary of animals observed and any food items that were in the process of being consumed. Dive surveys along a 100 m leadline, marked at 5 m intervals, were carried out 24 h before (two replicate dives on area to be trawled), 1 h (one dive parallel to trawl track, one dive over trawled area), 6 h (one dive on trawled area) and 24 h (one dive on trawled area) after fishing with the commercial trawl. The divers surveyed the same site on each occasion during light and at slack water (Fig. 16.1). The occurrence of each species was expressed as a percentage of the total number of all animals observed per dive survey, as the visible area of seabed was not consistent between dives due to variation in turbidity. The number of animals observed on the first three dives were considered to represent the undisturbed epifaunal composition. The total numbers of each species observed on these dives were pooled. The pooled numbers for each species were then calculated as the percentage of the total number of animals observed. Changes in the percentage of each species observed on consecutive dives after fishing (dives 4, 5 and 6) were tested for significant differences using the G-test (Sokal & Rohlf 1981).

Table 16.1. The mean ± SE tow length (m) estimated from the shoot and haul positions as determined from the DGPS satellite navigation system and calculated using Geoplot software.

	Time			
	24 h before	1 h after	24 h after	48 h after
Tow length	922.4 ± 27.9	921.2 ± 36.1	918.3 ± 32.3	855.9 ± 56.9

RESULTS

Small beam trawl samples

The mean number per tow of only three of the species sampled from the 2.8 m beam trawl catches varied significantly over the whole experiment (Table 16.2). The number of hermit crabs *Eupagurus bernhardus* caught per tow was significantly lower immediately after fishing (ANOVA, $F_{3,20} = 4.33$, $P < 0.02$); however, the number caught per tow 24 h prior to and 24 h after fishing were not significantly different. The number of red gurnards *Aspitrigla cuculus* caught per tow increased steadily after fishing the area, and were significantly higher after 48 h than the number caught prior to fishing (ANOVA, $F_{3,20} = 4.16$, $P < 0.02$). The mean number of dabs *Limanda limanda* caught 1 h and 24 h after fishing was significantly greater than before fishing (ANOVA, $F_{3,20} = 5.5$, $P < 0.007$), but the mean number caught after 48 h was not significantly different compared with 24 h before fishing (Table 16.2). Analysis of the coefficient of variation for each species on each sampling occasion revealed that the variation between samples was lowest 24 h before fishing, highest immediately after fishing and then gradually decreased over the next 48 h (Table 16.3).

Table 16.3. Analysis of the change in the coefficient of variation for all the animals sampled on each occasion with the 2.8 m beam trawl. Pair-wise comparisons, indicated by icons, were performed using Mann–Whitney test.

Tow	Mean ± SE	Median		P
24 h before fishing	0.45 ± 0.07	0.38	✪	✪ = 0.006
1 h after fishing	0.75 ± 0.05	0.72	✪★☆	★ = 0.003
24 h after fishing	0.48 ± 0.05	0.52	★	☆ = 0.006
48 h after fishing	0.52 ± 0.07	0.48	☆	

Baited camera

Whiting *Merlangius merlangus* and plaice *Pleuronectes platessa* were recorded at the baited bag 20–30 min after it reached the seabed. The first hermit crab was observed after 40 min, followed by starfish *Asterias rubens* after 50 min and lastly by one whelk *Buccinum undatum* after 110 min (Fig. 16.2). Other animals that were observed occasionally were swimming crabs *Liocarcinus depurator*, dabs and brittlestars *Ophiura ophiura*. One spider crab *Hyas areneus* was observed for 12 h on the bait bag. In general, the peak number of animals were observed between 8 and 14 h after the

Table 16.2. The mean ± SE number per tow (six replicates) of each of the commonest species caught in the 2.8 m beam trawl 24 h before and 1 h, 24 h and 48 h after fishing with the commercial 4 m beam trawl. For *Eupagurus bernhardus*, *Limanda limanda* and *Aspitrigla gurnardus* figures indicated with similar icons are not significantly different from each other (Tukey–Kramer multiple comparison on $\log_e$ transformed data, $P > 0.05$), none of the samples differed significantly for the other species.

Species	Time			
	24 h before	1 h after	24 h after	48 h after
Eupagurus bernhardus	30.7 ± 2.9 ✪	14.0 ± 4.1	31.3 ± 2.3 ✪	27.3 ± 6.6 ✪
Limanda limanda	12.0 ± 0.6 ★	38.5 ± 15.6 ✪	31.8 ± 4.2 ✪	24.2 ± 4.5 ✪★
Aspitrigla gurnardus	10.5 ± 2.7 ★	17.7 ± 3.1 ✪★	20.0 ± 4.8 ✪★	25.2 ± 2.1 ✪
Asterias rubens	283.5 ± 38.9	187.0 ± 49.6	249.5 ± 21.4	319.2 ± 50.5
Ophiura ophiura	175.2 ± 53.5	99.2 ± 26.1	124.7 ± 42.1	73.2 ± 13.1
Astropecten irregularis	21.7 ± 3.0	21.0 ± 5.9	17.0 ± 1.9	19.3 ± 2.0
Aphrodite aculeata	77.5 ± 3.6	66.3 ± 24.7	133.7 ± 35.1	77.8 ± 23.5
Buccinum undatum	24.0 ± 6.3	19.3 ± 13.3	31.0 ± 3.9	21.0 ± 3.2
Macropodia spp.	13.3 ± 5.2	19.5 ± 7.6	39.3 ± 12.0	28.7 ± 8.1
Liocarcinus depurator	42.5 ± 11.1	54.7 ± 14.7	74.3 ± 16.1	64.0 ± 11.4
Liocarcinus holsatus	17.7 ± 5.1	37.2 ± 10.8	28.3 ± 5.3	34.7 ± 5.2
Corystes cassivelaunus	95.7 ± 25.2	49.3 ± 12.1	112.2 ± 26.5	49.7 ± 8.8
Pleuronectes platessa	14.0 ± 4.4	14.7 ± 4.5	23.5 ± 3.4	14.7 ± 1.8
Callionymus lyra	2.3 ± 0.9	1.5 ± 0.6	3.3 ± 0.8	2.7 ± 0.9

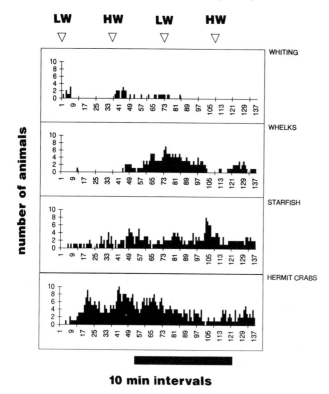

10 min intervals

Fig. 16.2. The number of whiting, whelks, starfish and hermit crabs observed every exposure (10 min interval between each exposure) over a 23 h period. The black bar indicates the period of darkness and the arrows indicate the times of low (LW) and high (HW) water.

camera frame reached the seabed; the highest diversity of animals on the bait bag, comprising three fish species and six different invertebrates, occured after 8 h. Whiting were most numerous towards sunset (Fig. 16.2), 7–8 h after the baited bag reached the seabed (Table 16.4) (ANOVA, $F_{10,121} = 8.5$, P < 0.001), while hermit crab numbers on the bait bag were highest after only 2–10 h (ANOVA, $F_{10,121} = 26.4$, P < 0.001) (Table 16.4). Hermit crab abundance did not coincide with either tide or darkness (Fig. 16.2). After 10 h, the number of hermit crabs at the bait bag began to diminish (Fig. 16.2). The number of starfish at the bait bag rose steadily to a peak after 18 h (ANOVA, $F_{10,121} = 10.41$, P < 0.001) (Table 16.4). Again, there was no obvious correlation between tide and diurnal variation (Fig. 16.2). Most whelks arrived at the bait bag after 7–17 h, with greatest numbers after 12 h (ANOVA, $F_{10,121} = 49.7$, P < 0.001) (Table 16.4). Whelk numbers at the bait bag diminished after 13–14 h (Fig. 16.2, Table 16.4). The highest numbers of whelks coincided with low water and most whelks were observed between sunset and sunrise (Fig. 16.2).

Diver observations

The frequency of occurrence of the five most common species observed differed significantly 6 h and 24 h after fishing as compared with that observed before fishing occurred (Table 16.5). After fishing, the observed number of brittlestars *Ophiura ophiura* was consistently lower than the number expected. Similarly, fewer than expected *Corystes cassivelaunus* were observed 6 h after fishing. Conversely, the observed number of starfish continually exceeded the expected value. Whelks were only observed on the trawled area.

Feeding animals were only found on the trawled area (Table 16.6). Solitary starfish were observed eating damaged *Echinocardium cordatum*, which were sometimes still buried under the surface of the sediment (confirmed when dug up by the diver). Groups of scavengers, comprising whelks, starfish and hermit crabs, were observed feeding on large animals such as *Arctica islandica*. In particular, one hermit crab, three whelks and three starfish were observed feeding on an *A. islandica*.

Table 16.4. The mean ± SE number of main predators per exposure on the baited bag. Mean number was calculated from the observations of 12 exposures, i.e. 2 h blocks. Figures accompanied by similar superscripts are not significantly different from each other (Tukey–Kramer multiple comparison test, P > 0.05).

Time interval (h)	Predators					
	Merlangius merlangus	*Pleuronectes platessa*	*Limanda limanda*	*Eupagurus bernhardus*	*Asterius rubens*	*Buccinum undatum*
2	0.67 ± 0.28 °	0.25 ± 0.17 °	–	0.58 ± 0.19 °	0.42 ± 0.14 ●	0.08 ± 0.08 *
4	–	0.17 ± 0.16 °	–	4.83 ± 0.60 ●	0.67 ± 0.18 ●	–
6	–	0.08 ± 0.08 °	0.08 ± 0.08 °	4.16 ± 0.50 ●	1.50 ± 0.37 °	–
8	1.33 ± 0.30 *	0.33 ± 0.18 °	0.83 ± 0.20 *	6.91 ± 0.54 *	1.41 ± 0.31 °	0.58 ± 0.25 *
10	0.25 ± 0.13 °	–	–	5.58 ± 0.45 *	3.08 ± 0.35 *	1.33 ± 0.22 °
12	0.58 ± 0.15 °	–	–	6.08 ± 0.33 *	1.50 ± 0.25 °	3.50 ± 0.28 ●
14	0.33 ± 0.14 °	–	–	3.91 ± 0.19 ●	2.83 ± 0.23 *	4.92 ± 0.25 *
16	–	–	–	2.75 ± 0.30 °	2.00 ± 0.27 °	2.83 ± 0.20 ●
18	–	–	–	1.67 ± 0.33 °	4.00 ± 0.62 *	1.41 ± 0.41 °
20	–	–	–	2.08 ± 0.33 °	2.16 ± 0.34 °	0.51 ± 0.14 *
22	–	–	–	2.5 ± 0.37 °	2.08 ± 0.52 *	1.75 ± 0.24 °

Table 16.5. The total number of each species observed on the first three dives is assumed to represent the normal frequency at which these animals occur in unfished areas. This gives the expected frequency for each species, which was used to calculate significant changes in frequency after the area had been fished.

		Ophiura ophiura	*Asterias rubens*	*Corystes cassivelaunus*	*Eupagurus bernhardus*	*Buccinum undatum*	G	P
Total over 3 dives prior to fishing		73	27	19	13	0		
Proportion		0.55	0.2	0.14	0.1	0		
1 h after fishing	Obs.	9	11	3	5	3	3.68	ns
	Exp.	17	6.2	4.3	3.1	0		
6 h after fishing	Obs.	36	30	0	7	4	26.9	0.001
	Exp.	42.4	15.4	10.7	7.7	0		
24 h after fishing	Obs.	14	83	2	19	7	102	0.001
	Exp.	68.8	25	17.5	12.5	0		

DISCUSSION

Our results indicate that short-term (1–24 h) increases in animal density are a direct response to the damaged tissues of animals injured by the 4 m beam trawl. Whelks, starfish, swimming and hermit crabs have all been shown to move towards baits in response to their odour (Lapointe & Sainte-Marie 1992, Nickell & Moore 1992a,b), hence these species were expected to aggregate in areas subject to trawl disturbance. However, samples collected with the 2.8 m beam trawl only detected significant variation in the number of hermit crabs, gurnards and dabs. We used a small beam trawl to collect epibenthic scavengers because this gear is particularly effective at sampling this component of the community (Kaiser *et al.* 1994). Although we were unable to detect significant fluctuations in other scavenging species, it was apparent that the coefficient of variance increased after 4 m beam trawling indicating that animal distribution had become more clumped. The coefficient of variance was greatest 1 h after 4 m beam trawling and decreased after 24 h and 48 h (Table 16.3). This result is not entirely unexpected, as we could only sample within a 30 m wide corridor, of which

Table 16.6. The numbers of the three main predators observed feeding on the diver survey area on each dive, and the prey species being consumed. Dives 1 and 2 occurred 24 h before fishing with the commercial trawl. Dive 3 surveyed an unfished area parallel to the fished line 1 h after fishing, while dive 4 surveyed the fished line at the same time. Dive 5 surveyed the fished line 6 h after fishing at 18.30 GMT and dive 6 surveyed the fished line 24 h after fishing.

	Predators		
Dive	*Asterias rubens*	*Eupagurus bernhardus*	*Buccinum undatum*
1	–	–	–
2	–	–	–
3	–	–	–
4	–	–	*C. cassivelaunus* 1
5	*Liocarcinus* sp. 3	–	–
	Mollusc 1		
	Echinocardium sp. 3		
	Unidentified 2		
6	*E. bernhardus* 1	*Arctica islandica* 1	*Arctica islandica* 4
	Echinocardium sp. 6	*B. undatum* 2	
	Ophiura sp. 1	*Echinocardium* sp. 1	
	Astropecten irregularis 1		
	Arctica islandica 9		
	Dosinia sp. 1		
	Laevocardium sp. 1		
	Unidentified 2		

a maximum of 40% of the area had been disturbed by the commercial beam trawl (maximum disturbance assumes no overlap of consecutive trawl tows). Our original assumption was that scavengers would aggregate in trawled areas, such that their numbers would also be higher in the local vicinity (15 m either side of trawl track). While this appears to be true for hermit crabs, gurnards and dabs, our results suggest that aggregation of the other species, such as whelks, is on a more local scale (distances of less than 5 m). Despite being unable to detect these subtle differences using the small beam trawl, direct observations using divers and the baited camera supported our initial hypothesis. Diver observations also indicated that animals were indeed patchily distributed, with dense aggregations of several species competing for broken *Arctica islandica*.

As anticipated from other studies (Sainte-Marie & Hargrave 1987, Nickell & Moore 1992a), whelks and starfish move slowly towards animals damaged by the 4 m beam trawl, which may explain their delayed arrival at the camera and on the trawled corridor (Nickell & Moore 1992b). Surprisingly, few swimming crabs, of either species were observed on the bait bag, but this may be attributed to the high concentration of predators, or in the case of *Liocarcinus holsatus*, diel activity rhythms (Abelló *et al.* 1990). The number of hermit crabs at the bait bag peaked and began to decrease before the peak in whelk or starfish numbers. This may be linked to a

gradual reduction in scallop flesh accessible to the hermit crabs. However, although the scallop flesh contained within the bait bag was depleted by predatory activity, the remnants were still attractive to predators such as whelks and starfish. During feeding, prey tissues are wounded releasing, among other substances, adenosine 5′-triphosphate (ATP), a potent attractant to invertebrate scavengers (Zimmer-Faust 1993), which may act as an additional stimulus, prolonging the period over which predators are attracted. This may explain the observation of large numbers of starfish and whelks feeding on damaged *Arctica islandica*. These molluscs are large and presumably take considerable time to be consumed, attracting even more scavengers.

Immediately after fishing with the 4 m beam trawl the number of dabs on the trawl corridor increased significantly and remained unchanged for the next 48 h (Table 16.2). Similarly, gurnards increased in number, but this was after a delay of 24 h. This may be caused by differences in the behavioural responses or sensory capabilities of the two species. Whiting arrived at the baited camera within 30 min, although they were not sampled by the 2.8 m beam trawl. Previously, we have shown that gurnards and whiting alter their diet composition and increase their food intake as a result of beam trawl disturbance (Kaiser & Spencer 1994). We speculate that fish aggregation is partly in response to visual cues such as the sediment cloud behind the trawl

(G. Sangster, personal communication), or the presence of other predators in a particular area which is reinforced as aggregation occurs (Ryer & Olla 1992).

Our study has shown that some predatory species quickly move into areas disturbed by beam trawls. The rate of immigration and dispersion is dependent on individual species' sensory abilities, mobility and behaviour. Animals dug up or damaged by the beam trawl are probably consumed within hours to a few days, after which the predators disperse. In some areas of the North Sea trawling activity is frequent and intense and may be considered a major disturbance of the seabed (de Groot 1984, Lindeboom 1990). While the direct effects of beam trawling result in the death of some animals, the secondary effects, i.e. the provision of an additional food for predators, may have a more significant long-term effect on the lower trophic levels of the community. Increasing the food availability for some species may increase their reproductive success, and in case of species which are not killed by beam trawling, e.g. *Asterias rubens* (Kaiser & Spencer 1993), lead to their proliferation. The resulting excessive predation pressure may accelerate a reduction in diversity in communities that are repeatedly disturbed by bottom trawls.

REFERENCES

Abelló, P., Reid, D.G. & Naylor, E. (1990) Comparative locomotor activity patterns in the protunid crabs *Liocarcinus holsatus* and *L. depurator. Journal of the Marine Biological Association, UK*, **71**, 1–10.

Bergman, M.J.N. & Hup, M. (1992) Direct effects of beam-trawling on macrofauna in a sandy sediment in the southern North Sea. *ICES Journal of Marine Science*, **49**, 5–13.

Creutzberg, F., Duineveld, G.C.A. & van Noort, G.J. (1987) The effect of different numbers of tickler chains on beam-trawl catches. *ICES Journal of Marine Science*, **43**, 159–168.

Dayton, P.K. & Hessler, R.R. (1972) Role of biological disturbance in maintaining diversity in the deep sea. *Deep Sea Research*, **19**, 199–208.

de Groot, S.J. (1984) The impact of bottom trawling on benthic fauna of the North Sea. *Ocean Management*, **10**, 21–36.

Hall, S.J. (1994) Physical disturbance and marine benthic communities: life in unconsolidated sediments. *Oceanography and Marine Biology Annual Review*, **32**, 179–239.

Kaiser, M.J. & Spencer, B.E. (1993) A preliminary assessment of the immediate effects of beam trawling on a benthic community in the Irish Sea. ICES CM 1993/B:38.

Kaiser, M.J. & Spencer, B.E. (1994) Fish scavenging behaviour in recently trawled areas. *Marine Ecology Progress Series*, **112**, 41–49

Kaiser, M.J., Rogers, S.I. & McCandless, D. (1994) Improving quantitative estimates of epibenthic communities using a modified 2-m beam trawl. *Marine Ecology Progress Series*, **106**, 131–138.

Lapointe, V. & Sainte-Marie, B (1992) Currents, predators, and the aggregation of the gastropod *Buccinum undatum* around bait. *Marine Ecology Progress Series*, **85**, 245–257.

Lindeboom, H. (1990) How trawlers are raking the North Sea to death. *Daily Telegraph*, 16 March 1990.

Messieh, S.N., Rowell, T.W., Peer, D.L. & Cranford, P.J. (1991) The effects of trawling, dredging and ocean dumping on the eastern Canadian continental shelf seabed. *Continental Shelf Research*, **11**, 1237–1263.

Nickell, T.D. & Moore, P.G. (1992a) The behavioural ecology of epibenthic scavenging invertebrates in the Clyde Sea area: laboratory experiments on attractions to bait in static water. *Journal of Experimental Marine Biology and Ecology*, **156**, 217–224.

Nickell, T.D. & Moore, P.G. (1992b) The behavioural ecology of epibenthic scavenging invertebrates in the Clyde Sea area: laboratory experiments on attraction to bait in moving water, underwater TV observations *in situ* and general conclusions. *Journal of Experimental Marine Biology and Ecology*, **159**, 15–35.

Paine, R.T. (1980) Foodwebs: linkages, interaction strength and community infrastructure. *Journal of Animal Ecology*, **49**, 667–685.

Rees, E.I.S., Nicholaidou, A. & Laskaridou, P. (1977) The effects of storms on the dynamics of shallow water benthic associations. *Biology of Benthic Organisms* (eds B.F. Keegan, P.O. Ceidigh & P.J.S. Boaden), pp. 465–474. Pergamon Press, Oxford.

Rumohr, H. & Krost, P. (1991) Experimental evidence of damage to benthos by bottom trawling with special reference to *Arctica islandica. Meeresforsch*, **33**, 340–345.

Ryer, C.H. & Olla, B.L. (1992) Social mechanisms facilitating exploitation of spatially variable ephemeral food patches in a pelagic marine fish. *Animal Behaviour*, **44**, 69–74.

Sainte-Marie, B. & Hargrave, B.T. (1987) Estimation of scavenger abundance and distance of attraction to bait. *Marine Biology*, **94**, 431–443.

Sokal, R.R. & Rohlf, F.J. (1981) *Biometry: The Principles and Practice of Statistics in Biological Research*, 2nd edn. W.H. Freeman & Co., San Francisco.

Stockton, W.L. & DeLaca, E.E. (1982) Food falls in the deep sea: occurrence, quality, and significance. *Deep Sea Research*, **29**, 157–169.

Summerson, H.C. & Peterson, C.H. (1984) Role of predation in organising benthic communities of a temperate-zone seagrass bed. *Marine Ecology Progress Series*, **15**, 63–77.

Thrush, S.F., Pridmore, R.D., Hewitt, J.E. & Cummings, V.J. (1994) The importance of predators on a sandflat: interplay between seasonal changes in prey densities and predator effects. *Marine Ecology Progress Series*, **107**, 211–222.

Witbaard, R. & Klein, R. (1994) Long-term trends on the effects of the southern North Sea beamtrawl fishery on the bivalve mollusc *Arctica islandica* L. (Mollusca, bivalva). *ICES Journal of Marine Science*, **51**, 99–105.

Zimmer-Faust, R.K. (1993) ATP: a potent prey attractant evoking carnivory. *Limonology and Oceanography*, **38**, 1271–1275.

CHAPTER 17

Seals and fishery interactions: observations and models in the Firth of Clyde, Scotland

S. des Clers and J. Prime

SUMMARY

(1) In the Firth of Clyde, seals and fishermen exploit limited fish resources. The extent of interactions between harbour seals *Phoca vitulina* and the fishery in 1993 is estimated directly through a study of seal numbers, distribution and diet, and indirectly using the sealworm *Pseudoterranova decipiens* as a tracer.

(2) Sealworm, a parasitic stomach worm of seals, is transmitted to fish where it accumulates in the flesh. In cod, infection levels are related to the size, age and growth rate of the fish. Current infection levels are similar to infection levels described in the 1960s and appear smaller only because of the absence of large cod in the fishery.

(3) Grey seals *Halichoerus grypus* are only occasional visitors to the upper Firth of Clyde; harbour seals are more numerous and show much greater site fidelity to the area. We suggest therefore, that sealworms are transmitted by harbour seals. We found no evidence to indicate any major movement of cod into or out of the northern half of the Clyde Sea. The overlap in the distribution ranges of harbour seals and cod may therefore explain why cod infection levels have always been high in the Firth of Clyde, despite relatively small numbers of seals.

(4) The numbers of harbour seals sighted at the different haul-out sites in 1993 were similar to those observed in 1989. Seal numbers in the upper Firth of Clyde may be limited by the lack of suitable intertidal habitat.

(5) Both the species and size of fish in the diet of harbour seals varied during the year. Gadids were the most common prey, followed by clupeids and other species, such as mackerel *Scomber scombrus* or butterfish *Pholis gunellus*. The main fishery in the Firth of Clyde currently targets Norway lobsters *Nephrops norvegicus*; therefore the fish prey of seals are only caught incidentally.

Key-words: clupeids, cod, harbour seal, fishery, *Phoca vitulina*, *Pseudoterranova decipiens*, sealworm

INTRODUCTION

Grey seal *Halichoerus grypus* populations around the British Isles have increased in the last decade while inshore commercial fisheries have generally declined. Public interest in marine mammals has also increased, leading to greater awareness of the potential for com-

petition for fish between seals and commercial fisheries (Harwood 1987, Harwood & Croxall 1988).

In coastal areas of the North Atlantic, the sealworm *Pseudoterranova decipiens* sensu lato (Nematoda, Ascaridoidea), a stomach nematode parasite of seals, is indirectly transmitted to cod *Gadus morhua* where it accumulates in the flesh, notably in the 'fillets'.

Between 1958 and 1973, sealworm infection levels were monitored in the cod fisheries around the British Isles by the Marine Laboratory of the Scottish Office Agriculture and Fisheries Department. Infection levels in cod from the Firth of Clyde were found to be high (Rae 1972, Wootten & Waddell 1977), although seal numbers in the area were known to be low (Rae 1963).

Our study area, the northern half of the Clyde Sea, is a relatively closed area, bounded by the west coast of Scotland and the Kintyre peninsula on each side and upper Loch Fyne to the North, and extending south to the latitude of Ayr (Fig. 17.1). A one-year study was designed to describe the nature and level of seal and fishery interactions in 1993. Numbers of grey *Halichoerus grypus* and harbour *Phoca vitulina* seals were assessed and faecal samples were collected to describe seal diet. Cod caught by demersal trawl were examined to determine their size, age and sealworm infection level. The data are analysed and combined in

a model of our current understanding of seal–fishery interactions in the Firth of Clyde.

MATERIAL AND METHODS

Seal survey

Historical data from a thermal image aerial survey were made available by the Sea Mammal Research Unit (SMRU), giving the numbers of harbour seals and grey seals visible in the Firth of Clyde on 9 and 10 August 1989.

In 1993, four systematic surveys of the coastline were carried out between April and October, in Loch Fyne, the Kyles of Bute, Inner Firth of Clyde, Kilbrannan Sound and outer Firth of Clyde as far as the latitude of Ayr (Fig. 17.1). This covered all 34 sites in the northern half of the Firth of Clyde where seals had been counted

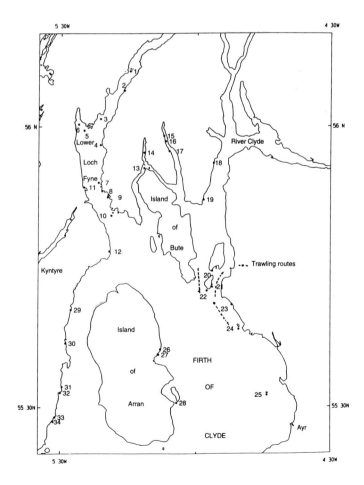

Fig. 17.1. Map of the Firth of Clyde, indicating the 34 stations where seals were sighted and seal faecal samples collected; Eilean Buidhe, where most of the faecal material was obtained, is site number 8. The three trawling routes routinely fished by University Marine Biological Station Millport are also shown.

by the SMRU survey in 1989. In April and October, seal counts were made from a dinghy launched as near as possible to seal haul-out sites. In May and August, better weather allowed the use of a chartered sailing boat, together with a small dinghy, to provide access to seal haul-out sites around Arran and off the eastern coast of the Kintyre peninsula. Each survey lasted a minimum of four days during which the distribution, numbers and species of seals sighted were recorded.

Seal diet study

During each coastline survey seal haul-out sites were visited, weather conditions permitting, to search for faecal material. Individual samples were collected and stored in polythene bags and frozen until processed. Separation of the hard, identifiable remains was facilitated by washing and brushing the samples through a nest of sieves with mesh sizes decreasing from 4.0 mm to 0.25 mm (Prime & Hammond 1990). All hard remains, including fish otoliths, bones, cephalopod beaks, eye lenses, etc., were identified and stored in 70% ethanol. Fish species were identified from the shapes of otoliths. The number of fish of each species was estimated from the number of otoliths, which were paired whenever possible. Fish otoliths were measured to provide an estimate of the original length and weight of the fish, allowing for the effect of digestion (Prime & Hammond 1990).

Cod and sealworm

Cod were caught by the Fisheries Research Vessel *Aora* trawling for an hour (rock-hopper trawl with a cod-end mesh size of 70 mm) along three different tracks at roughly monthly intervals. The fish were frozen whole on the day of capture. Once thawed, fish were

measured and weighed individually, then gutted and filleted, and their otoliths removed to determine their age assuming an official birthday on 1 January. The flesh was dissected and examined over a light table for the presence of nematodes. Worms were collected, identified and stored in 70% ethanol. Paggi *et al.* (1991) provided genetic evidence for the existence of at least three sibling species within *P. decipiens*. Worms were not examined genetically however, and sealworms are therefore referred to as *P. decipiens* sensu lato.

RESULTS

Seal species, numbers and distribution

The results of our four sighting surveys are summarized in Table 17.1. Counts from the 34 stations are grouped into five areas around the Firth. For both species of seal, the numbers counted varied greatly from one survey period to the next. The counts of both species were highest during the survey conducted in August, which is the harbour seal moulting season. Numbers were lowest in April and October, when weather conditions were poorest. Of the two seal species, harbour seals were much more abundant than grey seals.

Although the survey techniques were different, the spatial distributions of seals in the Firth of Clyde were similar in August 1993 and in August 1989 (Fig. 17.2). Harbour seal numbers were highest in Loch Fyne, followed by Kilbrannan Sound (Fig. 17.2(a)). Small numbers of grey seals were sighted in the Outer Firth and the Kyles of Bute (Fig. 17.2(b)). Most grey seals counted by SMRU in 1989 were outside our study area, in the more open and exposed southern part of the Firth, at the southern end of Arran, and on Sanda off the Mull of Kintyre.

Table 17.1. Sightings of seals in the Firth of Clyde by boat surveys in 1993: Pv, harbour seals *Phoca vitulina*; Hg, grey seals *Halichoerus grypus*.

Area	Site numbers	April		May		August		October	
		Pv	Hg	Pv	Hg	Pv	Hg	Pv	Hg
Loch Fyne	1–12	15	2	50	11	99	6	11	2
Kyles of Bute	13–17	4	8	6	4	5	9	5	5
Inner Firth of Clyde	18–22	4	3	7	2	6	4	7	8
Outer Firth of Clyde	23–28			8	9	18	13		
Kilbrannan Sound	29–34					61	0		

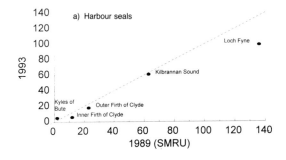

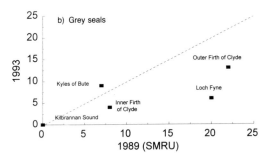

Fig. 17.2. Comparison of the numbers of (a) harbour seals and (b) grey seals, sighted during the August 1993 survey and numbers counted by SMRU using aerial thermal imagery in August 1989 in each of the five main regions.

Seal diet study

The 34 haul-out sites identified during the seal survey were all visited to collect faecal samples, but material was only found at one site (number 8) on Eilean Buidhe in Lower Loch Fyne. This site held the highest number of harbour seals counted during our study. Samples that could be confidently attributed to harbour seals were collected during May, August and October. No faecal samples were collected in April due to adverse weather conditions. The numbers of samples, otoliths, and different fish species identified for each survey are given in Table 17.2. More samples were found in

August when harbour seals spent more time hauled out.

The estimated numbers and percentages of fish prey in non-empty samples are given in Table 17.3 by survey, species group and individual species. Gadids, particularly whiting and poor cod, were the most common prey in May and October, while clupeids, mainly herring, were more prevalent in August. Other fish species, such as mackerel, dragonet, and butterfish represent more than 20% of all individual prey items identified, while flatfish featured much less. These observations agree well with studies in the Moray Firth, where harbour seals were described as 'catholic feeders' (Thompson & Miller 1990, Pierce *et al.* 1990, Thompson *et al.* 1991). All the prey species identified inhabit the Inner Firth of Clyde and Loch Fyne, in the immediate vicinity of Eilean Buidhe (site number 8, Fig. 17.1) (Millport University Marine Biological station, personal communication).

The estimated maximum length of fish eaten by harbour seals was 36 cm (Fig. 17.3). Prey size varied between surveys for all four species groups, and was smaller in May and October when gadids were the most common prey. Prey size was larger in August, around 30 cm, when herring and mackerel were more frequent in the diet. Differences in seal diet between surveys were also obvious from the estimated average weight of individual fish in each species group, and from the relative weight contribution of the different species groups (Table 17.4). In August large herring (Clupeids) and mackerel (Others), weighing more than 100 g, formed more than 80% of the diet. In October, gadids formed the bulk (58%) of the diet, consisting of numerous small, young fish of the year, weighing less than 30 g.

Cod

A total of 428 cod were caught by demersal trawl during this study. Two year old fish were most abundant in the first quarter, while the catch was dominated by one year old fish during the rest of the year, with a few young-of-the-year (0+ fish) caught in the last quarter

Table 17.2. Harbour seal faecal sample collection in Lower Loch Fyne in 1993.

	April	May	August	October
Total number of samples	0	11	31	17
Number with otoliths	0	10	28	12
Total number of otoliths	0	81	231	152
Number of indentified fish prey species	0	12	11	10

Table 17.3. Estimated occurrence of fish species (numbers and % by number) found in harbour seal faecal samples from Lower Loch Fyne in 1993.

Fish species		May		August		October	
		n	%	n	%	n	%
Gadidae	Total	23	43	43	30	59	67
Whiting	*Merlangius merlangus*	9	17	39	27	44	50
Poor cod	*Trisopterus minutus*	9	17			10	11
Cod	*Gadus morhua*			4	2.8	5	5.7
Norway pout	*Trisopterus esmarkii*	3	5.6				
4-bearded rockling	*Enchelyopus cimbrius*	2	3.8				
Clupeidae	Total	10	19	51	35	12	14
Herring	*Clupea harengus*	4	7.5	50	34	2	2.3
Sprat	*Sprattus sprattus*	6	11	1	0.7	10	11
Others	Total	15	28	44	30	13	16
Dragonet	*Callionymus lyra*	4	7.5	19	13	8	9
Mackerel	*Scomber scombrus*	1	1.9	24	17	3	3.4
Butterfish	*Phollis gunnellus*	10	19	1	0.7		
Sandeel	*Ammodytes* spp.					2	2.3
Eelpout	*Zoarces viviparus*					1	1.1
Flatfish	Total	5	9.5	7	4.9	3	3.4
Dab	*Limanda limanda*			2	1.4	3	3.4
Lemon sole	*Microstomus kitt*	2	3.8	1	0.7		
Turbot	*Psetta maxima*	2	3.8				
Witch	*Glyptocephalus cynoglossus*	1	1.9	2	1.4		
Flounder	*Platichthys flesus*			2	1.4		

(Table 17.5). Nearly three quarters (69.8%) of all cod were at or below the current minimum legal landing size of 35 cm. Market-size fish were caught mostly in autumn and winter (Fig. 17.4), a typical pattern of the demersal fishery in the area (Hislop 1986).

Seal predation on cod

Only 15 cod otoliths (15/464 = 3%) were found in harbour seal faecal samples in August and October, estimated to correspond to nine different fish. Individual sizes estimated for cod are given in Figure 17.4, with the length frequency distributions of cod caught in trawls by quarter. Most cod prey were estimated to be young of the year and were generally smaller than the cod caught in the research vessel trawls with a 70 mm mesh.

Sealworm in cod

In a previous study, large annual variations in individual worm burdens in cod from the Firth of Clyde were shown to reflect differences in the age composition of landed catch (des Clers 1989, 1990a). Fish accumulate worms from their food. In young fish of a given size, the total amount of food eaten increases with the age of the fish raised to the power three. In 1965, landings were dominated by older fish caught at the beginning of the calendar year, while in 1969 landings were mostly of younger fish caught in the autumn. Thus for a given size, fish landed in 1965 had eaten more, and hence accumulated more worms, than cod landed in 1969. These two years were therefore taken to illustrate historical variability to analyse the data collected in 1993 (Fig. 17.5).

Sealworm burdens vary greatly between individual cod; the data were therefore aggregated by 3 cm length intervals, to include at least 50 fish in each group. Leaving aside the more variable two extreme length intervals ($\leqslant$26 cm and $\geqslant$38 cm), average individual worm burdens increase monotonically with the length of host (Fig. 17.5) as expected, and sealworm burdens observed in 1993 are within the range of historical observations. Sealworm infection levels in cod in 1993 were apparently no different to levels observed almost 30 years earlier.

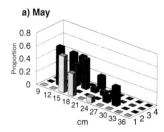

a) May

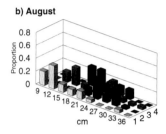

b) August

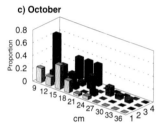

c) October

Fig. 17.3. Length frequency distributions (proportions) of the four species groups (1, gadid; 2, clupeid; 3, others; 4, flatfish; see Table 17.4) determined from otoliths found in harbour seal faecal samples collected in Lower Loch Fyne in 1993.

Table 17.4. Estimated average individual fish weight in each species group (w) and the percentage contribution (by weight) of each species group to the total diet (%) in harbour seal faecal samples from Lower Loch Fyne in 1993 (see Table 17.3 for species composition).

Fish species group	May		August		October	
	w	%	w	%	w	%
Gadidae	37	36.3	41	11.4	28	58
Clupeidae	86	36.9	162	53.9	25	10.7
Others	32	20.5	106	30.6	57	26.2
Flatfish	30	6.3	91	4.1	48	5.1

It is also worth noting that no *Anisakis simplex* larvae were found in the flesh of cod. This confirms previous observations (Wootten and Waddell 1977, S. des Clers, unpublished data) that cod from areas outside the Firth of Clyde, where *Anisakis* infections were prevalent, do not migrate into the Firth. It also suggests that any emigrants from the Firth of Clyde do not return to the area. This agrees with the description of the Clyde Sea as an isolated hydrographical entity during the summer period (Edwards *et al.* 1986).

DISCUSSION

Two types of interaction between seals and fisheries were studied in the Clyde Sea: predation on fisheries resources, and sealworm transmission to demersal fish. In the light of the data presented here, and with

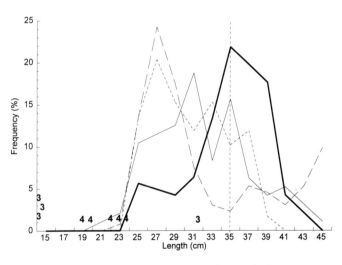

Fig. 17.4. Length frequency distribution (%) of cod caught in the Firth of Clyde in 1993, grouped by quarter. The estimated sizes of cod eaten by harbour seals in quarters 3 and 4 are indicated in bold.

Table 17.5. Numbers of cod caught by trawl near Little Cumbrae in 1993 by age group and quarter.

	Total	Jan–Mar	Apr–Jun	Jul–Sep	Oct–Dec
Total no. fish caught	428	96	131	59	142
0+	13				13
1	281	6	90	58	127
2	134	90	41	1	2

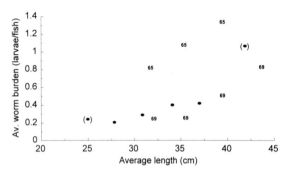

Fig. 17.5. Average sealworm burden per 3 cm length interval, in cod trawled from the Firth of Clyde in 1993. Data points for the two extreme intervals (wider than 3 cm) are shown between parentheses. Expected boundaries of natural variability are indicated by observations from Rae (1972) (see des Clers 1990a) in 1965 and 1969.

reference to related published and unpublished studies, a model is proposed to link seals and fisheries.

Seals

Seal numbers are difficult to survey, and the results from our sighting surveys cannot be directly compared with counts from the thermal image aerial survey conducted in 1989 (see Anderson 1990). Another such aerial survey would be needed to infer the current status of seal populations in the northern Firth of Clyde. Some 300 harbour seals were found dead in the Firth of Clyde as a result of the seal distemper virus epizootic in 1988 (SMRU, personal communication). As in several other locations around the British Isles and the North Sea, this level of mortality suggested that more harbour seals than were previously thought must have been present in the Firth of Clyde at the time. The great majority of haul-out sites identified from SMRU's aerial survey in 1989 and visited in 1993, are small, rocky and tidal. Many of these sites are also close to the mainland, and regularly disturbed by fishermen or others. It is possible that the lack of suitable haul-out habitat limits the number of harbour seals. Although

harbour seals can pup in the inter-tidal zone, grey seal pups spend their first few weeks on land. Thus habitat limitation is likely to be even greater for grey seals, and may explain why they have not been reported to pup in the area. However, this does not explain why they are not foraging in greater numbers in the Clyde Sea.

Fisheries

The herring fishery in the Clyde Sea has declined dramatically since the 1930s, through a combination of unrestricted catches until the late 1970s, and poor recruitment during the 1980s (Hopkins & Morrison 1987). Local demersal stocks reached low levels in the early 1970s (Hislop 1986), and have since been further depleted by by-catches of juvenile fish in the *Nephrops* fishery in the Firth of Clyde. Fishing effort for *Nephrops* increased dramatically in the 1970s and 1980s (Bailey *et al.* 1986). Very few cod were obtained in weekly demersal trawls in 1993.

Predation

Faecal samples indicated that harbour seals were feeding on pelagic and demersal fish. Although the sampling station in Loch Fyne was not in the immediate vicinity of the trawling routes (see Fig. 17.1), the diversity in prey species and prey sizes indicated a clear potential for interaction with commercial fisheries. However, the nature of possible interactions depends on the fishing activity, and these have greatly changed over the last two decades. The cod fishery and to some extent the herring fishery, which directly exploited many of the seals' prey, have been replaced by a *Nephrops* fishery, which catches demersal fish species only incidentally. Therefore, fishermen complain little of competition from seals at present. In contrast, seals' interactions with fixed gear such as salmon nets in the estuaries of salmon rivers (see Greenstreet *et al.*, 1993, S.P.R. Greenstreet, personal communication) and at fish farms, are known to cause problems locally, and should be assessed.

Sealworm transmission

The sealworm life-cycle is coastal and benthic, and the salient features of its dynamics are as follows. Eggs produced by adult worms in seals' stomachs are shed with the faeces, sink in sea water to the bottom where the hatched larvae are consumed by benthic invertebrate hosts (mostly crustaceans). These are in turn eaten by small (non-commercial) benthic living fish species. Cod may become infected early in life by eating crustaceans, but mostly become infected after their first summer by eating infected fish. Cod may not be the major source of infection in seals, as smaller demersal fish species such as long-rough dab *Hippoglossoides platessoides* or sculpin *Myoxocephalus scorpius* may carry much higher infection levels. Sealworms spend years in fish, as opposed to weeks in seal stomachs or months in benthic invertebrates, and therefore the population dynamics of the worm life-cycle are best observed in fish hosts (des Clers 1990b).

In the Firth of Clyde, sealworm burdens in cod in 1993 were not obviously different from levels observed 30 years ago. Sealworm infection levels in cod increase with the age and size of the fish. The smaller burdens observed at present reflect the predominance of small and young cod, rather than a decrease in worm transmission rates. Although historical trends in the numbers of seals are not known (Gibson 1986), the number of seals currently transmitting worms to cod is likely, therefore, to be comparable with numbers present during the 1960s.

Harbour seals, which eat more pelagic fish species than grey seals, have fewer sealworms in their stomachs (Young 1972) and were not thought to transmit sealworm infections in British waters. However, in areas where grey seals are absent, such as the Wadden Sea or the Skagerrak and Kattegat, harbour seals are known to transmit sealworms (Lick 1989, Lunneryd 1990, des Clers & Andersen 1995). Furthermore, high area fidelity in harbour seals (Härkönen 1987) may also lead to higher sealworm burdens in inshore fish than for infections transmitted by grey seals (see Möller & Klatt 1990, Jensen *et al.* 1994). Sealworm burdens in cod from the Firth of Clyde are still among the highest around the British Isles (J. Smith personal communication, SOAFD Marine Laboratory, Aberdeen unpublished survey). Therefore, given the low numbers of grey seals in the area, it is likely that sealworms are transmitted by harbour seals, as was suspected by Rae (1963).

ACKNOWLEDGEMENTS

This study was funded by the Marine Unit, World Wide Fund for Nature (WWF UK) whose support and interest are gratefully acknowledged. We are grateful to Donald Cameron and Robert Wilkie of the Specimen Supply Service at the University Marine Biological Station, Millport, for providing cod; to the Sea Mammal Research Unit (Natural Environment Research Council, Cambridge, UK) for providing the detailed seal count data from their 1989 aerial thermal imagery survey; and to Indrani Lutchman, Simon Greenstreet, John Harwood, John Hislop, John Smith, Paul Thompson and Dominic Tollitt for helpful discussions. Comments from two anonymous referees greatly improved an earlier version of the text.

REFERENCES

Anderson, S. (1990) *Seals*. Whittet Books Ltd, London.

Bailey, N., Howard, F.G. & Chapman, C.J. (1986) Clyde *Nephrops*: biology and fisheries. *Proceedings of the Royal Society of Edinburgh*, **90B**, 501–518.

des Clers, S. (1989) Modelling regional differences in 'sealworm' *Pseudoterranova decipiens* (Nematoda, Ascaridoidea) infections in some North Atlantic cod, *Gadus morhua*, stocks. *Journal of Fish Biology*, **35** (Suppl. A), 187–192.

des Clers, S. (1990a) Functional relationship between sealworm (*Pseudoterranova decipiens*, Nematoda, Ascaridoidea) burden and host size in Atlantic cod (*Gadus morhua*). *Proceedings of the Royal Society, London Series B*, **245**, 85–89.

des Clers, S. (1990b) Modelling the life-cycle of the sealworm (*Pseudoterranova decipiens*) in Scottish waters. *Population Biology of Sealworm* (Pseudoterranova decipiens) *in Relation to its Intermediate and Final Hosts* (ed. W.D. Bowen). *Canadian Bulletin of Fisheries and Aquatic Sciences*, **222**, 273–288.

des Clers, S. & Andersen, K. (1995) Sealworm (*Pseudoterranova decipiens*) transmission to fish trawled from Hvaler, Oslofjord, Norway. *Journal of Fish Biology*, **46**, 8–17.

Edwards, A., Baxter, M.S., Ellett, D.J., Martin, J.H.A., Meldrum, D.T. & Griffiths, C.R. (1986) Clyde Sea hydrography. *Proceedings of the Royal Society of Edinburgh*, **90B**, 67–83.

Gibson, J.A. (1986) Recent changes in the status of some Clyde vertebrates. *Proceedings of the Royal Society of Edinburgh*, **90B**, 453–453.

Greenstreet, S.P.R., Morgan, R.I.G., Barnett, S. & Redhead, P. (1993) Variation in the numbers of shags *Phalacrocorax aristotelis* and common seals *Phoca vitulina* near the mouth of an Atlantic salmon *Salmo salar* river at the time of the smolt run. *Journal of Animal Ecology*, **62**, 565–576.

Härkönen, T.J. (1987) Influence of feeding on haul-out patterns and sizes of sub-populations in Harbour seals. *Netherlands Journal of Sea Research*, **21**, 331–339.

Harwood, J. (1987) Competition between seals and fisheries.

Scientific Progress, Oxford, **71**, 429–437.

Harwood, J. & Croxall, J.P. (1988) The assessment of competition between seals and commercial fisheries in the North Sea and the Antarctic. *Marine Mammal Science*, **4**, 13–33.

Hislop, J.R.G. (1986) The demersal fishery in the Clyde Sea area. *Proceedings of the Royal Society of Edinburgh*, **90B**, 423–437.

Hopkins, P.J. & Morrison, J.A. (1987) Herring to the west of Scotland. *Fishing Prospects 1987*, 38 pp. MAFF Fisheries Research, Lowestoft and DAFS, Marine Laboratory, Aberdeen.

Jensen, T., Andersen, K. & des Clers, S. (1994) Sealworm (*Pseudoterranova decipiens*) infections in demersal fish from two areas in Norway. *Canadian Journal of Zoology*, **72**, 598–608.

Lick, R.R. (1989) Stomach nematodes of Harbour seals (*Phoca vitulina*) from the German and Danish Wadden Sea. ICES CM 1989/N:7

Lunneryd, S.G. (1990) Anisakine nematodes in the Harbour seal *Phoca vitulina* in the Skaggerak–Kattegat and the Baltic. *Working Paper*. Joint meeting of the International Council for the Exploration of the Sea Working Group on Baltic Seals and the Study Group on the Effects of Contaminants on Marine Mammals. ICES, Copenhagen.

Möller, H. & Klatt, S. (1990) Smelt as host of the sealworm (*Pseudoterranova decipiens*) in the Elbe estuary. *Population Biology of Sealworm* (Pseudoterranova decipiens) *in Relation to its Intermediate and Final Hosts* (ed. W.D. Bowen) *Canadian Bulletin of Fisheries and Aquatic Sciences*, **222**, 129–138.

Paggi, L., Nascetti, G., Cianchi, R., Orecchia, P., Mattiuci, S., D'Amelio, S., Berland, B., Brattey, J., Smith, J.W. & Bullini, L. (1991) Genetic evidence for three species within *Pseudoterranova decipiens* (Nematoda, Ascaridida, Ascaridoidea) in the North Atlantic and Norwegian and Barents Seas. *International Journal for Parasitology*, **21**, 195–212.

Pierce, G.J., Boyle, P.R. & Thompson, P.M. (1990) Diet selection by seals. *Trophic Relationships in the Marine Environment* (eds M. Barnes & R.N. Gibson), pp. 22–38. Aberdeen University Press, Aberdeen.

Prime, J.H. & Hammond, P.S. (1990) The diet of grey seals from the south-western North Sea assessed from the analyses of hard parts found in faeces. *Journal of Applied Ecology*, **27**, 435–447.

Rae, B.B. (1963) The incidence of larvae of *Porrocaecum decipiens* in the flesh of cod. *Marine Research 1963*, 2, 27 pp. Department of Agriculture and Fisheries for Scotland, HMSO, Edinburgh.

Rae, B.B. (1972) A review of the cod–worm problem in the North Sea and in Western Scottish Waters. *Marine Research 1972*, 2, 24 pp. Department of Agriculture and Fisheries for Scotland, HMSO, Edinburgh.

Thompson, P.M. & Miller, D. (1990) Summer foraging activity and movements of radio-tagged common seals (*Phoca vitulina* L.) in the Moray Firth, Scotland. *Journal of Applied Ecology*, **27**, 492–501.

Thompson, P.M., Pierce, G.J., Hislop, J.R.G., Miller, D. & Diack, J.S.W. (1991) Winter foraging by common seals (*Phoca vitulina*) in relation to food availability in the inner Moray Firth, NE Scotland. *Journal of Applied Ecology*, **60**, 283–294.

Wootten, R. & Waddell, I.F. (1977) Studies on the biology of larval nematodes from the musculature of cod and whiting in Scottish waters. *Journal du Conseil International pour l'Exploration de la Mer*, **37**, 266–273.

Young, P.C. (1972) The relationship between the presence of larval anisakine nematodes in cod and marine mammals in British Home waters. *Journal of Applied Ecology*, **9**, 459–485.

CHAPTER 18

The influence of a fish farm on grey heron *Ardea cinerea* breeding performance

D.N. Carss and M. Marquiss

SUMMARY

(1) Regurgitations (n = 62) and spilled food samples (n = 97), containing a total of 630 items, were used to assess the diet of grey herons *Ardea cinerea* in terms of species composition by number and mass of fishes recorded in samples. Work was carried out at six colonies in a study area in Argyll, western Scotland, during the 1986 and 1987 breeding seasons. Dates of laying the first egg, nestling age and brood survival were monitored at four of these colonies where nests could be visited. At others, dates of laying were estimated from fallen eggshells collected below nests.

(2) Grey herons at five colonies fed almost exclusively on natural, free-living prey. Small seashore fishes were the most commonly recorded items whilst salmonids were the most important group by mass. However, the diet was diverse and varied both between and during the two breeding seasons.

(3) The remaining colony was close to a rainbow trout farm where birds had easy access to these fish, weighing 5–300 g. The diet of grey herons there was dominated by rainbow trout from the farm and remained unchanged throughout both breeding seasons.

(4) Regurgitations from two natural habitat colonies and one near the fish farm contained similar numbers of fish; however, those from nestlings at the latter colony were four times heavier.

(5) In both years, grey herons at the fish farm colony started laying earlier and produced more fledged young per egg laid than did birds at other colonies, although this was only significant in 1986 when breeding was probably affected by the poor weather. Such increased productivity was, however, probably outweighed by the widespread killing of herons at fish farms.

Key-words: aquaculture, *Ardea cinerea*, breeding success, heron, predation, supplementary food

INTRODUCTION

Food is known to influence the laying date, clutch size and fledging success of some birds. In many species clutch size is inversely correlated with laying date. Experimental provision of extra food prior to the breeding season has led to earlier egg laying in the great tit *Parus major* (Kallander 1974), carrion crow *Corvus corone* (Yom-Tov 1974), sparrowhawk *Accipter nisus* (Newton & Marquiss 1981) and magpie *Pica pica* (Hogstedt 1981). However, only the last two studies provided evidence that clutch sizes were increased independently of laying date, and only for crows and magpies did supplementary food result in increased nestling survival.

The grey heron *Ardea cinerea* is a predator at Scottish fish farms (Carss 1993a). The past 15 years have seen much expansion of fish farming in Scotland (e.g. NCC 1990) and unsubstantiated claims have been made that grey heron numbers have increased due to improved breeding success, probably as a result of this expansion (Mills 1989).

Powell (1983) found that additional food increased both the clutch size and fledging success of the great white heron *Ardea herodias*, a close relative of the grey heron living in North and Central America. Unexpectedly in the light of the studies mentioned above, provision of additional food did not significantly alter the timing of nest initiation which was assumed by Powell to be correlated with laying date, and was used as a measure of the timing of breeding seasons. The author concluded that either factors other than food availability were controlling the onset of nesting or there was no strong selective pressure to nest 'earlier'. As the study population nested successfully throughout the year, the latter was certainly a possibility. In Europe the grey heron is a seasonal breeder and fish farms may act as predictable food sources which could influence the breeding of herons feeding at them.

The aims of the present study were to compare (1) nestling diet, and (2) heron breeding performance, at a large colony, close to a fish farm (where the birds appeared to take much of their food) with diet and breeding at five other colonies (where birds fed almost exclusively on wild fishes).

STUDY AREA AND METHODS

A general description of the study area is given in Carss (1993a) and the locations of the six occupied heron colonies within it (Seil Island, Lochs Feochan, Creran, Etive and Awe, and Ardconnel) are shown in Figure 18.1. Five colonies (Seil, Feochan, Creran, Etive and Ardconnel) were categorized as 'natural habitat', being either on, or near, the coast; the sixth (Awe) was 4 km from a large cage trout farm. These colonies will be referred to as S, F, C, E, Ard, and Awe, respectively, throughout the text. All were visited every 7–14 days during the 1985 and 1986 breeding seasons except Ard which was only visited during 1986.

The Loch Awe fish farm comprised about 50 cages, each holding rainbow trout *Oncorhynchus mykiss* and covered with an anti-predator top net. Many herons fed regularly here where the mesh size of the top nets allowed them to remove trout up to 300 g easily (Carss 1993a). About 90% of the trout held at the farm weighed less than 300 g (Fig. 18.2) and the size composition of stock varied little throughout the study.

Weather

The beginning of the grey heron egg-laying period varies from year to year according to the weather

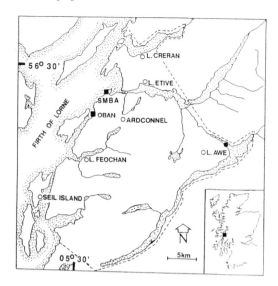

Fig. 18.1. Argyll study area showing location of SMBA Dunstaffnage Marine Laboratory and the Loch Awe fish farm (●) and the approximate locations of the six occupied heron colonies (○).

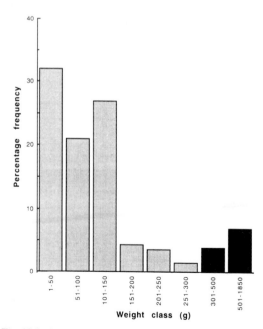

Fig. 18.2. Size distribution of the total stock of rainbow trout (n = 884 650) held at the Loch Awe fish farm, 1987. The lightly shaded portion indicates size classes easily taken by herons.

(Cruetz 1981) and so monthly rainfall and minimum temperature data were extracted from the records of the SMBA Marine Laboratory at Dunstaffnage (see Fig. 18.1).

Nestling diet

Diet was assessed from regurgitated and spilled food collected in and below nests. Nestlings often regurgitated food when observers climbed trees to examine nests. Regurgitations had clearly been voided recently and usually consisted of well-digested items. Spilled items, dropped by adults passing food to nestlings or by the nestlings themselves during manipulation, could be distinguished easily from regurgitations. They were sometimes partially digested and seldom fresh. Where trees could not be climbed, most samples were spilled items collected on the ground, although alarmed nestlings also occasionally regurgitated over the edge of the nest. Prey identification was as previously described by Carss (1993b,c).

The lengths of all fishes (see Appendix for scientific names) recorded in regurgitations and spilled food samples were similar (see Results) and so in later analyses data from both sample methods were combined. Marquiss & Leitch (1990) found that in heron regurgitations, there was a strong tendency for items of the same species to occur together and so it could not be assumed that individual items occurred independently within these samples. Therefore, each regurgitation or spill was used as the sampling unit and was assigned to one of four groups: (1) seashore species (butterfish, fifteen-spined stickleback, bull-rout, lumpsucker, wrasses, flatfishes, clupeoids, blennies and gobies), (2) gadoids (cod and saithe), (3) salmonids (salmon and trout), and (4) 'other' fishes (three-spined stickleback, minnow, common eel, perch and brook lamprey).

At three colonies (S, F, Awe), regurgitations were used to investigate the meal size (numbers and weight of fish) of nestlings to determine any differences between the fish farm colony (Awe) and elsewhere.

Breeding performance

We climbed to nests at S, F, Ard, and Awe to count and measure their contents; nests at C and E were less accessible and could be monitored only from the ground. Herons commence incubation with the first egg laid so chicks hatch at 1–2 day intervals and are therefore of disparate size. The first hatched sibling usually remains the largest and grows well so that there is little variation in the growth of the largest chicks between broods, in contrast to the very large variation between siblings within a brood (Owen 1960). The size (maximum wing chord) of the largest sibling can thus be used to estimate the age of the brood in days from the hatch of the first egg (Fig. 18.3). If food supply is poor, smaller siblings grow poorly and many die, some at an advanced age, so it cannot be assumed that the number of well-grown nestlings seen in a nest will be the number fledging. Fledged brood size can be estimated more accurately by taking into consideration the relative difference in size of siblings. From a sample of 145 nestlings whose ultimate fate was known, we calculated that those with a greater than 50% chance of dying had a wing length of less than 70% of their largest sibling (dotted line in Fig. 18.3). Thus in the present study, for the purposes of estimating fledge brood size for those nests that could not be viewed from the ground, measured nestlings that plotted above this dotted line were assumed to have survived to fledge. Those plotting below the line were assumed to have died. Some of these were found dead in nests or nearby.

Hatching dates at C and S were determined (to the nearest week) by finding hatched eggshells below nests. These eggshells are dropped from the nest by adults and are a good indication of recent hatching (Milstein *et al.* 1970). Laying dates (the date of laying the first egg in a clutch) were calculated by deducting an incubation period of 28 days (Milstein *et al.* 1970).

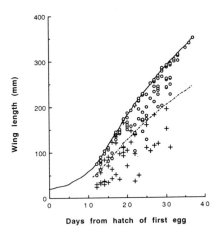

Fig. 18.3. The relationship between brood age and wing length in heron nestlings. The solid line represents the average increase in wing length with age for the first hatched siblings in 80 broods of known age. Symbols represent the wing lengths of nestlings that were known to have survived to fledge (○) or died prior to fledge (+). The dotted line separates nestlings that have a more than 50% chance of surviving from those with less.

RESULTS

Weather

The weather was markedly different between the two breeding seasons, the 1986 spring being persistently wet and cold (Fig. 18.4). February 1986 was an exceptionally cold month and daily minimum temperatures were 3.5°C cooler than average. Over twice the average rainfall was recorded in March, it was colder than average in April, and over three times the average rainfall fell in May. During 1987 the spring weather differed little from average, except during March which was wetter.

Nestling diet

Regurgitated and spilled items

The lengths of 258 regurgitated (mean = 12.1 cm) and 319 spilled fishes (mean = 12.2 cm), from all colonies in both years, were similar.

General diet

Fishes formed 97.5% of all the 630 prey items collected in regurgitation and spilled food samples (see Appendix) and so other prey groups were excluded from further analyses. Overall, diet differed between natural habitat and fish farm colonies ($\chi^2 = 97.7$, df = 3, P < 0.001). At natural habitat colonies, seashore fishes were the most commonly recorded group, followed by salmonids, 'other' fishes and gadoids. At the Awe colony salmonids, mainly rainbow trout, were most common with only a few other species recorded (Table 18.1). During the two breeding seasons 46 occupied nests were visited at this colony and food samples were collected from 32 (70%). Of these, 26 (81%) included rainbow trout. All 12 of those whose stomach contents could be examined contained commercial fish food, indicating that they had been taken from a fish farm.

Diet changes during the breeding season

Nestling diet at natural habitat colonies varied during the breeding season ($\chi^2 = 19.3$, df = 3, P < 0.001) with 3.5% of samples (n = 57) from April to June containing gadoids, compared with 34.2% (n = 38) in July–August (Fig. 18.5(a)). Nestling diet at the Awe colony remained unchanged throughout the breeding

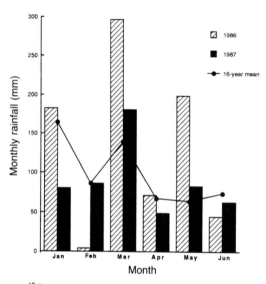

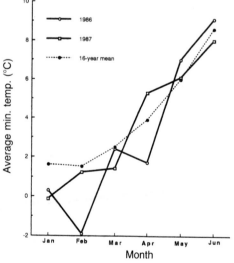

Fig. 18.4. Monthly rainfall (mm) and average minimum monthly temperature (°C) recorded at SMBA, Dunstaffnage, Jan–June 1986 and 1987. Figures also show 16 year (1971–87) means.

Table 18.1. Percentage of food samples collected from the six study colonies assigned to one of four prey categories. Colonies divided into natural habitat or fish farm on the basis of their location.

Prey category	Colony name and type	
	S, F, E, C, Ard Natural habitat	Awe Fish farm
Seashore	41.1	1.6
Gadoids	15.8	0
Salmonids	22.1	90.6
'Others'	21	7.8
No. samples (= 100%)	95	64

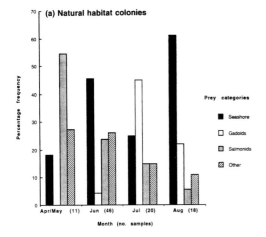

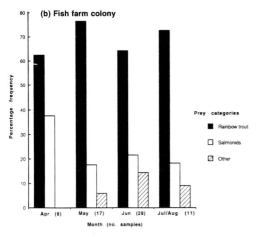

Fig. 18.5. Changes in heron nestling diet at (a) five natural habitat colonies and (b) the Loch Awe colony; for each time period the number of samples = 100%. Fish classification details are given in the text.

season, rainbow trout being the commonest prey (April–May 72%, n = 25; June–August 67%, n = 39; χ^2 = 1.41, df = 2, NS) (Fig. 18.5(b)).

Diet changes between breeding seasons

53 food samples collected on similar dates in both seasons from colonies S and F (representing 93% of all samples collected there) showed that nestling diet varied between years (χ^2 = 7.16, df = 1, P < 0.01) with higher proportions of samples in 1986 containing gadoids (28.6%, n = 21) than in 1987 (3.1%, n = 32).

All 64 food samples from the Awe colony were collected on similar dates in both seasons, and there was no difference between years (χ^2 = 3.37, df = 1, NS). Samples from both years were dominated by

rainbow trout, 55.6% (n = 18) in 1986 and 71.6% (n = 46) in 1987.

Assessing nestling meals

Regurgitations from colonies S (n = 19), F (n = 13), and Awe (n = 23) comprised mostly fish, with small proportions by number and mass of crustaceans and mammals at S and F. Samples from colonies C (n = 4), E (n = 1) and Ard (n = 2) were too small to include in any comparison between colonies, but the contents of regurgitations from the Awe colony differed markedly from those at S and F. At these last two colonies, regurgitations contained a large proportion by number and mass of small fishes and few large salmonids, so the median weight of fish per regurgitation was less than at Awe (ANOVA on log-transformed data, $F_{2,52}$ = 8.55, P < 0.001). The median weight of fish per regurgitation for the two natural habitat colonies was 32 g (range 2.5–139 g), significantly less than at the Awe colony (Mann–Whitney U-test, W = 865.5, p < 0.001), where nestling diet was dominated by rainbow trout and corresponding figures were 127 g (range 6–412 g).

Despite this four-fold difference in estimated median fish weight, regurgitations from all three colonies contained similar numbers of fish (ANOVA on log-transformed data, $F_{2,52}$ = 1.88, NS); median numbers of fish per sample being 2 (S), 4 (F), and 3 (Awe).

Breeding Performance

The timing of breeding

During 1986 47 nests were visited at S, F and Awe colonies and median laying dates varied between colonies, being 27 April, 1–2 May and 16 April, respectively. The first egg of the year was laid at Awe on 4 February, over two months earlier than the first egg of the year at the other two colonies (10 April). Laying dates (n = 17 nests) estimated by fallen eggshells at colonies E and C were similar to those at the natural habitat sites where chicks had been measured (Fig. 18.6).

In the following year, 42 nests were visited at S, F, and Awe and median laying dates again spanned a 17 day period. However, the dates (28 March, 9 April and 23 March, respectively), were 22–30 days earlier than the previous year. The first egg of the year was laid at Awe on 11 February, 34 days earlier than the first egg of the year at colonies S and F (17 March). At Ard all six nests were visited and the first-lay date (5 March) was intermediate between that at Awe and the other sites. Laying dates (n = 21) at E and C were again similar to those at natural habitat sites where nestlings had been measured (Fig. 18.6).

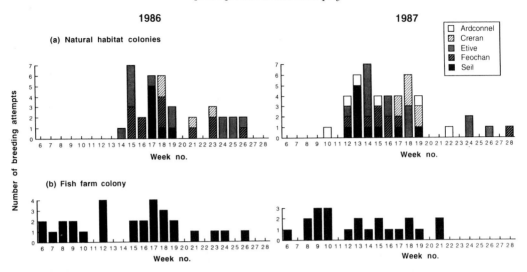

Fig. 18.6. Laying dates for each breeding attempt at five study colonies in 1986 and six in 1987. Week 6 starts 3 February in 1986 and 2 February in 1987, week 26 ends 29 June in 1986 and week 28 ends 12 July in 1987.

Clutch size

There were more large clutches at the Awe colony than at the two natural habitat sites (Table 18.2). At the natural habitat colonies S and F, 5.4% of the 37 nesting attempts (pooled from both years) produced no fledged young; the corresponding figure at the Awe colony was 9.8% of 51 attempts. Nest failures were a result of either the desertion or predation of eggs in the early stages of breeding. At the Awe colony nest failures were closely associated with the number of adult herons killed at the nearby fish farm. On one occasion in

Table 18.2. The number of failed nests and the number of clutches of various sizes recorded at Seil Island and Loch Feochan, and Loch Awe colonies. Data from 1986 and 1987 combined. At Awe there was a greater proportion of large clutches than expected ($\chi^2 = 10.2$, df = 3, P < 0.005, by combining clutches of 1 and 2, and 5 and 6).

	Colony name and type	
	S and F Natural habitat	Awe Fish farm
Failed nests	2	5
Clutch size = 1	0	0
Clutch size = 2	3	9
Clutch size = 3	18	10
Clutch size = 4	12	17
Clutch size = 5	1	9
Clutch size = 6	1	1
Breeding attempts	37	51

particular, the failure of six nests followed immediately on the death of ten herons at the farm.

Fledging success

The circumstantial evidence that failed nests were the result of breeding birds being killed at the fish farm meant we could not assume that all failures were the result of poor food supply. To relate diet to fledging success we therefore used the estimated fledged brood size of successful nests only. In 1986, herons at S and F fledged similar numbers of young per egg laid in successful nests (t = 0.26, df = 14, NS) whilst those at Awe fledged significantly more (t = 2.45, df = 34, P < 0.002) (Table 18.3).

In the following year birds at S and F again fledged similar numbers of young per egg laid (t = 1.82, df = 14, NS) whilst those at Awe had a higher productivity in successful nests, although not significantly so (t = 1.57, df = 35, P = 0.13). Production of young in successful nests at all three colonies was higher in 1987 than in the previous year, although not significantly so (S, F: t = 1.82, df = 30, NS; Awe: t = 0.38, df = 39, NS) (Table 18.3).

DISCUSSION

Nestling diet during the breeding season

Food collected from the ground below nests was expected to produce a biased picture of diet as nestlings may discard the larger items they are unable to swallow

Table 18.3. The number of nesting attempts (N) and the mean number of fledged young per egg laid for successful nests at Seil Island, Loch Feochan and Loch Awe colonies in 1986 and 1987.

	Year					
	1986			1987		
Colony	N	Fledged/egg	SE	N	Fledged/egg	SE
Seil (S)	7	0.57	0.09	9	0.58	0.05
Feochan (F)	9	0.57	0.06	7	0.75	0.10
S + F combined	16	0.57	0.05	16	0.65	0.06
Awe	20	0.74	0.05	21	0.77	0.05

or drop them during manipulation. However, no difference was found between the sizes of these prey and those regurgitated. At natural habitat colonies the majority of prey items were marine species common to the shallow waters of the Scottish west coast (e.g. Gordon & de Silva 1980, Gordon 1981). Salmonids, often salmon parr and brown trout probably caught in adjacent freshwaters, were larger than other fishes and constituted over 40% of the recorded diet by mass. Other fresh- and brackish-water species were also taken and the two rainbow trout collected may have come from a fish farm.

Heron diet varies both between and during breeding seasons (Owen 1955, 1960, Marquiss & Leitch 1990). In the present study the importance of salmonids in the diet at natural habitat colonies decreased during the breeding season whilst that of marine species (the most commonly taken items overall) increased. This may have been caused by an increase in the availability of shore fishes rather than a decrease in that of salmonids. Kislalioglu (1975) found a seasonal pattern in the feeding intensity, and hence by implication the general level of activity, of inshore fishes. Feeding levels were higher in summer than winter, being highest during July and August. Furthermore many of these species breed during the winter and spring (Gordon & de Silva 1980, Gordon 1981) and so their well-grown young are usually abundant in inshore waters during the summer. Although juvenile (0 group) saithe and cod are found close inshore in the summer and autumn (Gordon 1990), abundance is known to vary greatly between years depending on the strength of the year-class. In 1986 a high proportion of samples contained gadoids whilst only a single specimen was found in the following year. These differences were not influenced by the date of sampling and were thought to reflect natural variations in the abundance of gadoids in the natural habitat, as previously suggested in a study of shag *Phalacrocorax aristotelis* diet in the same area (Carss 1993b).

Herons from the Loch Awe colony caught most of their food at the nearby fish farm, taking advantage of the predictable and super-abundant supply of rainbow trout. In contrast to other colonies in the study area, their diet was unlikely to be influenced by fluctuations in the abundance of wild prey species because it was dominated by rainbow trout, which was readily available, throughout the study.

Breeding performance

Lack (1968) argued that most birds time their breeding to coincide with an abundance of food for their young and it is possible that the breeding season of herons feeding on natural food within the study area, and presumably elsewhere on the Scottish west coast, is timed to coincide with the increased abundance of small seashore fishes. The food supply of adults is a proximate factor in determining when they breed, but this can be modified by such things as weather. In the present study, the median dates of laying the first egg in 1986, following a cold winter and a cold and wet spring, were 22–30 days later than those in 1987 when the winter had been mild and the spring warmer and drier.

Birds overwintering in a particular area are likely to have learned the location of good foraging sites there before breeding (van Vessem & Draulans 1987). In winter herons were using the fish farm and roosting at the nearby colony. If these birds were indeed subsequent breeders, this food source may have freed them from some of the constraints imposed by naturally fluctuating food supplies and weather conditions and enabled them to come into breeding condition earlier than elsewhere. Laying in 1986 began on 4 February, 65 days earlier than at the natural habitat colonies that year. In the following year laying was slightly later (11 February), but still 34 days earlier than at the natural habitat colonies. These dates were amongst the earliest recorded in Scotland. However, the farm may influence the breeding of full-time winter residents more than birds which arrive at the colony later (having wintered elsewhere) and median laying dates for the whole colony were only a few days (11 and 5 in each year, respectively) earlier than at other colonies.

There were more large clutches at the Loch Awe colony than at two natural habitat colonies (S and F) despite the similarities in median laying dates, suggesting that clutch size had been increased at Awe independently of laying date. Here, successful pairs also fledged more young per egg laid in both breeding seasons (although not significantly so in 1987), presumably as a result of the abundant food supply at the fish farm. The growth and survival of heron nestlings is very largely determined by the amount of food brought to the nest (Junor 1972, van Vessem & Draulans 1986a). However, the lack of a significant difference in breeding success between the herons at the fish farm colony and elsewhere in 1987 suggested that under poor weather conditions (i.e. 1986) herons at Awe had an advantage over the others but that this advantage was reduced when natural conditions were better (i.e. 1987).

Although regurgitations from study colonies contained similar numbers of fish, those fish from Loch Awe were four times heavier. Food quality may also be important and it appeared that the heron chicks at the Loch Awe colony were fed a more energy-rich diet than those at other colonies. The calorific value of rainbow trout ($7.5 \, \text{kJ} \cdot \text{g}^{-1}$ wet weight) is considerably higher than that of brown trout ($3.7 \, \text{kJ} \cdot \text{g}^{-1}$ wet wight) (Prevost 1982) and other commonly caught species, e.g. butterfish ($5.0 \, \text{kJ} \cdot \text{g}^{-1}$ wet weight), saithe ($4.3 \, \text{kJ} \cdot \text{g}^{-1}$ wet weight), sea scorpion ($3.8 \, \text{kJ} \cdot \text{g}^{-1}$ wet weight) and flounder ($2.4 \, \text{kJ} \cdot \text{g-1}$ wet weight) (D.A.D. Grant, unpublished observations).

Herons living in north and west Scotland could be affected by persecution at fish farms (Marquiss 1989), which was widespread between 1984 and 1988 (Carss 1994). Set against the increased production at the Loch Awe colony is the increased risk of persecution at the nearby fish farm where 34 first-year birds and 24 adults were killed during the study period (D.N. Carss, unpublished observations). Using published figures for the average mortality of grey herons in Britain we could speculate whether such losses of adult and first-year birds were offset by the greater breeding production at

Appendix. Fish species recorded at each of the six study colonies in regurgitations and spilled food samples.

		Colony					
		S	F	C	E	Ard	Awe
Brook lamprey	*Lampetra planeri*					*	*
Common eel	*Anguilla anguilla*	*	*		*	*	*
Unidentified Clupeoid spp.		*					
Unidentified Salmonid spp.		*	*	*		*	*
Salmon	*Salmo salar*	*				*	*
Brown trout	*Salmo trutta*				*	*	*
Rainbow trout	*Oncorhynchus mykiss*	*				*	*
Minnow	*Phoxinus phoxinus*		*				*
Unidentified Gadoid spp.		*	*	*	*		
Saithe	*Pollachius virens*	*	*	*	*	*	
Cod	*Gadus morhua*	*			*		
Eel pout	*Zoarces viviparus*	*			*	*	*
Three-spined stickleback	*Gasterosteus aculeatus*	*	*			*	*
Fifteen-spined stickleback	*Spinachia spinachia*	*	*	*		*	
Bull-rout	*Myoxocephalus scorpius*	*	*	*	*	*	
Perch	*Perca fluviatilis*						*
Goldsinny wrasse	*Ctenolabrus rupestris*	*				*	
Corkwing wrasse	*Crenilabrus melops*	*		*			
Lumpsucker	*Cyclopterus lumpus*					*	
Shanny	*Lipophrys pholis*	*				*	
Butterfish	*Pholis gunnellus*	*	*	*	*	*	*
Two-spotted goby	*Gobiusculus flavescens*	*				*	
Painted goby	*Pomatoschistus minutus*					*	
Greater pipefish	*Syngnathus acus*	*	*				
Unidentified flatfish			*				
Plaice	*Plueronectes platessa*	*				*	
Flounder	*Platichthys flesus*		*	*			

the farm. Assuming that birds breed in their second year after hatching (van Vessem & Draulans 1986b) and using mortality rates of 56% and 47% respectively for the first two years of life (Mead *et al.* 1979), the farm killings could have been equivalent to the production of approximately 70 fledglings. An average of 48 fledglings were produced at the nearby colony each year, only 69% of the above figure. Such a calculation is undoubtedly an oversimplification as ringing recoveries have shown that once fledged, first-years fly up to 100 km from their natal colony while Cadbury & Fitzherbert-Brockholes (1984) found that surprisingly few of the chicks that they dye-marked at the nest were later observed at a nearby fish farm. Furthermore, the removal of adults from a local population may enhance the survival of the remaining birds by reducing competition for feeding sites. There is no evidence that the Scottish heron population has increased in recent years (Marquiss 1989) and it seems likely that the widespread persecution at fish farms (Carss 1994) has outweighed any local benefits they provided in term of increased food supply.

ACKNOWLEDGEMENTS

We are very grateful to the various landowners and fish farm managers who kindly allowed us free access to their properties. We would like to thank Robert Moss and Sarah Wanless for providing many constructive and critical comments on earlier drafts of this paper, the final version of which benefitted from the comments of two anonymous referees. This work was partially funded by a Research Studentship (to D.N.C.) from the Natural Environment Research Council.

REFERENCES

Cadbury, C.J. & Fitzherbert-Brockholes, J. (1984) Grey herons at trout farms in England and Wales. *Shore-Birds and Large Waterbirds Conservation* (eds P.R. Evans, H. Hafner & P. l'Hermite), pp. 166–177. Commission of the European Communities, Brussels.

Carss, D.N. (1993a) Grey heron, *Ardea cinerea* L., predation at cage fish farms in Argyll, western Scotland. *Aquaculture and Fisheries Management*, **24**, 29–45.

Carss, D.N. (1993b) Shags *Phalacrocorax aristotelis* at cage farms in Argyll, western Scotland. *Bird Study*, **40**, 203–211.

Carss, D.N. (1993c) Cormorants *Phalacrocorax carbo* at cage fish farms in Argyll, western Scotland. *Seabird*, **15**, 38–44.

Carss, D.N. (1994) Killing of piscivorous birds at Scottish fin fish farms, 1984–87. *Biological Conservation*, **68**, 181–188.

Cruetz, G. (1981) *Der Grauriher*. A Zeimsen Verlag, Wittenberg-Lutherstadt.

Gordon, J.D.M. (1981) The fish populations of the West of Scotland Shelf. Part II. *Oceanographic and Marine Biology Annual Review*, **19**, 405–441.

Gordon, J.D.M. (1990) The fish populations of Scottish sea lochs with particular reference to those of the Firth of Lorne Area. *Glasgow Naturalist*, **21**, 561–575.

Gordon, J.D.M. & De Silva, S.S. (1980) The fish populations of the West of Scotland Shelf. Part I. *Oceanographic and Marine Biology Annual Review*, **18**, 317–366.

Hogstedt, G. (1981) Effect of additional food on reproductive success in the magpie (*Pica pica*). *Journal of Animal Ecology*, **50**, 219–229.

Junor, F.J.R. (1972) Estimation of the daily food intake of piscivorous birds. *Ostrich*, **43**, 193–205.

Kallander, H. (1974) Advancement of laying of great tits by the provision of food. *Ibis*, **116**, 365–367.

Kislalioglu, M. (1975) *The Feeding Ecology and Behaviour of Inshore Fishes*. PhD Thesis, University of Stirling.

Lack, D. (1968) *Ecological Adaptions for Breeding in Birds*. Methuen, London.

Marquiss, M. (1989) Grey herons *Ardea cinerea* breeding in Scotland: numbers, distribution and census techniques. *Bird Study*, **36**, 181–191.

Marquiss, M. & Leitch, A.F. (1990) The diet of grey herons *Ardea cinerea* at Loch Leven, Scotland, and the importance of their predation on ducklings. *Ibis*, **132**, 535–549.

Mead, C.J., North, P.M. & Watmough, B.R. (1979) The mortality of British grey herons. *Bird Study*, **26**, 13–22.

Mills, D.H. (1989) *Ecology and Management of Atlantic Salmon*. Chapman and Hall Ltd, London.

Milstein, P. le S., Prestt, I. & Bell, A.A. (1970) The breeding of the grey herons. *Ardea*, **58**, 171–257.

NCC (1990) *Fish Farming and the Scottish Freshwater Environment*. Nature Conservancy Council, Edinburgh.

Newton, I. & Marquiss, M. (1981) Effect of additional food on laying dates and clutch sizes of sparrowhawks. *Ornis Scandinavica*, **12**, 224–229.

Owen, D.F. (1955) The food of the heron in the breeding season. *Ibis*, **97**, 276–295.

Owen, D.F. (1960) The nesting success of the heron in relation to the availability of food. *Proceedings of the Zoological Society of London*, **133**, 597–617.

Powell, G.V.N. (1983) Food availability and reproduction by great white herons *Ardea herodias*: a food addition study. *Colonial Waterbirds*, **6**, 139–147.

Prevost, Y.A. (1982) *The Wintering Ecology of Ospreys in Senegambia*. PhD Thesis, University of Edinburgh.

van Vessem, J. & Draulans, D. (1986a) Factors affecting the length of the breeding cycle and the frequency of nest attendance by grey herons. *Bird Study*, **33**, 98–104.

van Vessem, J. & Draulans, D. (1986b) The adaptive significance of colonial breeding in the grey heron: inter- and intra-colony variability in breeding. *Ornis Scandinavica*, **17**, 356–362.

van Vessem, J. & Draulans, D. (1987) Spatial distribution and time budget of radio-tagged grey herons, *Ardea cinerea*, during the breeding season. *Journal of Zoology, London*, **213**, 507–534.

Yom-Tov, Y. (1974) The effect of food and predation on breeding density and success, clutch-size and laying date of the crow (*Corvus corone* L.). *Journal of Animal Ecology*, **43**, 479–498.

CHAPTER 19

Fluctuations in the Bering Sea ecosystem as reflected in the reproductive ecology and diets of kittiwakes on the Pribilof Islands, 1975 to 1991

G.L. Hunt Jr, M.B. Decker and A. Kitaysky

SUMMARY

(1) Between 1975 and 1991, a complex series of changes occurred in the southeastern Bering Sea marine ecosystem. Sea surface temperatures increased from the late 1970s to the mid-1980s, and then decreased. Over the same period, there were inter-annual variations in the water masses surrounding the Pribilof Islands, as judged by zooplankton species composition. Subsequent to the mid-1970s, there were changes in the abundance of capelin *Mallotus villosus* and 1-group walleye pollock *Theragra chalcogramma* as determined by trawl samples.

(2) The use of capelin by both black-legged kittiwakes *Rissa tridactyla* and red-legged kittiwakes *R. brevirostris* decreased at the Pribilof Islands subsequent to 1978, as did the use of 1-group pollock in the late 1970s. Based on their occurrence in the diets of black-legged kittiwakes, the availability of fatty fishes such as myctophids, capelin and sandlance decreased after the late 1970s.

(3) Beginning in the late 1970s, there was a decrease in the number of chicks produced per nest for both black-legged and red-legged kittiwakes nesting on the Pribilof Islands.

(4) Inter-annual variation in the availability of fatty fish was at least in part responsible for variations in the production of chicks by red-legged and possibly by black-legged kittiwakes.

(5) We do not know why these changes in forage fish availability occurred in the vicinity of the Pribilof Islands, but they may have been related either to changes in the biomass of predatory adult pollock or to changes in the marine climate.

Key-words: Bering Sea, forage fish, Pribilof Islands, *Rissa brevirostris*, *Rissa tridactyla*, sea surface warming, seabird foraging, seabird reproduction

INTRODUCTION

Since the early 1970s, the marine ecosystems of the North Pacific Ocean and Bering Sea have undergone major changes, with sea surface temperatures increasing and then decreasing (e.g. Royer 1993). Over roughly the same period, northern sea lion *Eumetopias jubatus* populations decreased by about 80%, and over the past three decades, northern fur seal *Callorhinus ursinus* populations at the Pribilof Islands decreased by about 50%, eventually stabilizing their numbers during the

1980s (Castellini 1993). Further, harbour seal *Phoca vitulina* populations in the Kodiak region of the Gulf of Alaska decreased severely between 1976 and 1984 (Castellini 1993). Populations of black-legged kittiwakes *Rissa tridactyla*, red-legged kittiwakes *R. brevirostris*, and Brünnich's guillemots (thick-billed murres) *Uria lomvia* on the Pribilof Islands have also decreased by 22% to 54% since 1976, and the reproductive outputs of both species of kittiwakes have decreased dramatically subsequent to the late 1970s (Climo 1993, Dragoo & Sundseth 1993).

A number of hypotheses have been advanced to account for these changes in marine mammal and seabird populations (e.g. Loughlin 1987, Anon. 1993). One of these hypotheses emphasizes the role of walleye pollock *Theragra chalcogramma* and the potential effects of the pollock fishery on the marine prey base. A second set of hypotheses emphasizes the potential role of changes in the distribution and abundance of fatty fish (compared to pollock), such as herring *Clupea harengus*, capelin *Mallotus villosus*, sandlance *Ammodytes hexapterus*, and lantern fish (Myctophidae) (Alverson 1991, Anon. 1993). Springer (1992) focused on the importance of pollock as a keystone predator that may have pervasive influence on the availability of prey to upper trophic levels in the Bering Sea. In contrast, Decker *et al.* (1995) and Hunt *et al.* (in press) have investigated changes in the diets of Pribilof Islands' seabirds in relation to fluctuations in the marine climate. The present paper focuses on changes in the diets of the two species of kittiwakes breeding on the Pribilof Islands, and the possible influence that changes in the abundance and availability of species of fatty fish may have had on kittiwake reproductive ecology.

METHODS

Data on food habits of seabirds nesting at the Pribilof Islands were obtained by collecting regurgitations from nestlings when they were handled for weighing, and by shooting adults near the island early in the season before chicks had hatched. Samples were stored in 80% ethanol, and prey in samples were identified to the lowest taxon possible in the laboratory. Data on prey obtained from adult and nestling kittiwakes were combined because adults regurgitate prey carried in their foregut to young. Data on the diets of adult and nestling murres were analysed separately because adult murres provision their young with bill-loads of prey that differ from the foods that they ingest for their own use. Otoliths were measured to the nearest 0.1 mm, and pollock size-classes were determined using published regressions of fish length versus otolith size. Details on methodology are available in Hunt *et al.* (1981, in press).

MARINE SETTING

The Pribilof Islands are located near the edge of the continental shelf ('shelf-break' hereafter) of the southeastern Bering Sea (Fig. 19.1). The waters over the shelf are divided into discrete domains separated by fronts (Iverson *et al.* 1979, Kinder & Schumacher 1981). The middle domain of the shelf is dominated by a benthic food web, and the outer domain by a pelagic food web (Iverson *et al.* 1979, Walsh & McRoy 1986). Avian use of these domains reflects these differences in

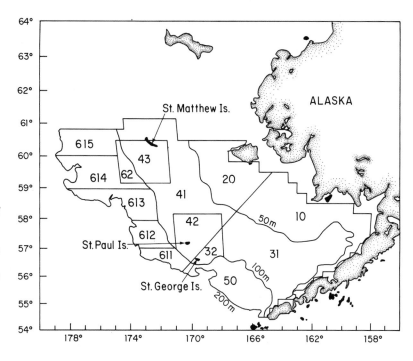

Fig. 19.1. Location of the Pribilof Islands with respect to the sampling strata used by the National Marine Fisheries Service in their bottom trawl survey of walleye pollock. The inner domain is inshore of the 50 m isobath, the middle domain between the 50 and 100 m isobaths and the outer domain between the 100 and 200 m isobaths.

food web structure (Schneider *et al.* 1986). Data from the 1970s show that the species composition of nesting seabirds and their diets on St. Paul and St. George Islands also reflect differences in the proximity of these islands to the outer domain (Hunt *et al.* 1981, Schneider & Hunt 1984).

SEA SURFACE TEMPERATURE

Sea surface temperatures vary at seasonal, annual, and longer timescales. Royer (1989) identified low-frequency fluctuations in the sea surface temperature of the northeastern North Pacific Ocean and the Bering Sea, which he estimated had a period of approximately 20 to 30 years. More recently, Miller *et al.* (1994),

Trenberth & Hurrell (1994), and Royer (1993) documented a marked shift in the climate of the northeastern North Pacific Ocean characterized by a warming of surface waters in the late 1970s. In the Bering Sea, Decker *et al.* (1995) obtained monthly mean sea surface temperatures for a 5° latitude by 10° longitude grid surrounding the Pribilof Islands that encompassed a major portion of the foraging areas used by Pribilof Islands' seabirds. These data show a similar fluctuation in surface temperatures to those found by Royer (1989) (Fig. 19.2). In the early to mid-1970s, water temperatures were considerably below the long-term mean; 1977 was the first of several years of above average sea surface temperatures. From 1979 to 1983, temperatures were near or above the average, with temperatures generally shifting to below the average after 1984.

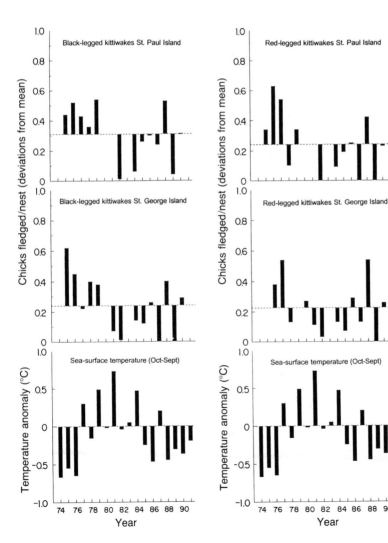

Fig. 19.2. Changes in production of young by black-legged and red-legged kittiwakes on St Paul and St. George Islands and variations in the sea surface temperature with respect to the long-term mean (1950 to 1991). After Decker *et al.* (1995), with permission.

Sea surface temperature may be a proxy variable for a variety of physical and biological changes that coincide with fluctuations in the properties of surface water. For example, Cooney & Coyle (1982) and Vidal & Smith (1986) found that in the middle domain of the Bering Sea, more of the primary production was captured in the pelagic food web during a warm-water year than in a cold-water year. The difference was determined by the rate at which the copepod grazers developed, which was temperature dependent. Similarly, atmospheric conditions that lead to sea surface warming may also displace currents and their associated fronts (T.C. Royer, personal communication), and the position of the domains may shift with respect to the Pribilof Islands. Diet data from least auklets *Aethia pusilla* show that copepod species composition varied inter-annually and support the hypothesis that St. Paul Island was within outer domain waters in 1975 and 1976, but it was surrounded by middle domain waters in 1978 and 1989 (Hunt *et al.*, in press). It is unknown if there was a long-term shift in the location of the domains subsequent to 1977, but changes in access to a particular domain or the fronts separating domains could influence the availability of prey to seabirds nesting at the Pribilof Islands.

KITTIWAKE REPRODUCTION

Both black-legged and red-legged kittiwakes showed diminished reproductive output in the 1980s as compared to the 1970s (Fig. 19.2). Reproductive output was significantly lower after 1978, but there was no significant correlation between the production of young and sea surface temperatures (Decker *et al.* 1995).

PREY AVAILABILITY

Information on the availability of prey to seabirds is difficult to obtain. In some instances, independent samples of the abundance of prey species in the vicinity of breeding colonies exist (e.g. Baird 1990, Hatch & Sanger 1992, Monaghan *et al.* 1989, in press, Wright & Bailey 1993). In the southeastern Bering Sea, National Marine Fisheries Service bottom trawl surveys provide some information on the abundance of walleye pollock and, less satisfactorily, of capelin, but these surveys do not address the question of the availability of these prey. As an alternative approach, we employed variations in the use of a prey species by a generalist predator (black-legged kittiwake) as an index of the availability of that prey relative to other prey taken. We reasoned that a generalist predator would change prey types freely

with respect to their availability, and that a specialist predator, such as the red-legged kittiwake, might be slower to switch to an alternative prey when its preferred prey became less available.

During the mid-1970s, 1-group pollock were a small but important part of the diets of kittiwakes and murres on the Pribilof Islands (Hunt *et al.*, in press). Between 1979 and 1982, the numbers of 1-group pollock estimated by trawl surveys to be present in the vicinity of the Pribilof Islands and along the shelf edge to the west of the islands dropped precipitously (Hunt *et al.*, in press) (Fig. 19.3). Similarly, west and south of the

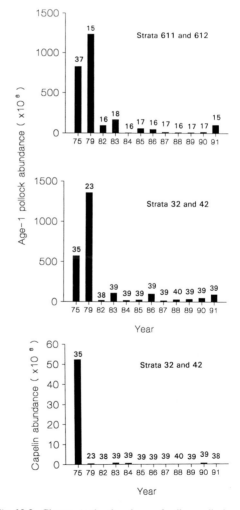

Fig. 19.3. Changes in the abundance of walleye pollock and capelin in survey strata near the Pribilof Islands. The number of trawls made in each year in the strata are indicated above the columns. Strata as in Figure 19.1. Data on pollock redrawn from Hunt *et al.* (in press), with permission.

Pribilof Islands along the shelf break, 1-group pollock numbers decreased after 1982, and remained low at least until 1991. Elsewhere on the shelf, numbers of 1-group pollock fluctuated, but showed no obvious trend, except that in most areas the unusually large 1978 year class created an obvious spike in abundance. Springer (1992) estimated population sizes for 1-group pollock for the entire eastern Bering Sea, and found no significant change in the number of 1-group pollock as a function of year when all values for the period 1973 to 1988 were used. However, there was a significant decrease when the large year classes from 1978 and 1981 were excluded. Data from both surface foraging kittiwakes and subsurface foraging guillemots breeding on the Pribilof Islands showed marked decreases in the use of 1-group pollock after the mid-1970s, with few fish of this age class taken after 1979 (Fig. 19.4). Information on the use of 1-group pollock by guillemots is included because the decrease in 1-group pollock in guillemot diets reinforces the conclusion that the change was not just in the vertical position of these fish in the water column. In sum, the available evidence shows that 1-group pollock became less available after the late 1970s, particularly in the vicinity of the Pribilof Islands. As of 1991, these fish had not returned to their former abundance near the Pribilof Islands.

During July and August, 0-group pollock are a significant component of the diets of several species of seabird breeding at the Pribilof Islands (Hunt *et al.*, in press). No measures of their abundance in the vicinity of the Pribilof Islands, independent of seabird diet samples, are available. The presence of juvenile pollock in the diets of seabirds at the Pribilof Islands decreased

in the 1980s compared to the 1970s (Fig. 19.5). Most of this change is attributable to a decrease in the amount of 0-group pollock taken, as these were the principal age class of pollock taken starting in late July (Hunt *et al.* 1981).

Capelin also decreased in abundance in the vicinity of the Pribilof Islands. Limited data from the National Marine Fisheries Service bottom trawl surveys recorded many capelin near the Pribilof Islands in 1979, but few to none in subsequent years (Fig. 19.3). Similarly, capelin disappeared from the diets of both surface foraging kittiwakes and subsurface foraging Brünnich's guillemots at the Pribilof Islands after 1978 (Fig. 19.6) (see also Decker *et al.* 1995). It is not known why capelin left the vicinity of the Pribilof Islands, but it may have been in response to increasing sea water temperatures, decreases in annual sea ice extent (Fritz *et al.* 1993), or possibly competition with young pollock for prey (Springer 1992).

There are no data independent of seabirds on the abundance or availability of myctophids (lantern fish) in the vicinity of the Pribilof Islands. Myctophids are the principal prey of the primarily nocturnally-foraging red-legged kittiwake, a relatively specialized forager. They are a secondary prey of the primarily diurnally-foraging black-legged kittiwake, a relatively generalized forager, which forages for myctophids at night (Hunt *et al.* 1981, Decker *et al.* 1995). Myctophids were less used by black-legged kittiwakes on St. George Island in 1984 and 1988 than in the 1970s, although there was no apparent change in their use by red-legged kittiwakes between the decades (Fig. 19.7). We interpret these data as an indication that the availability of myctophids

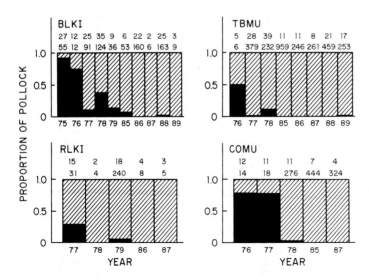

Fig. 19.4. Changes in the relative proportions of 0-group (diagonal fill) and 1-group walleye pollock (solid fill) in the diets of black-legged kittiwakes (BLKI), red-legged kittiwakes (RLKI), Brünnich's guillemots (thick-billed murres TBMU) and guillemots (common murres COMU) on the Pribilof Islands, 1975 to 1989. The top number above each column is the number of stomach samples examined; the number beneath is the number of pollock obtained from samples in a given year. From Hunt *et al.* (in press), with permission.

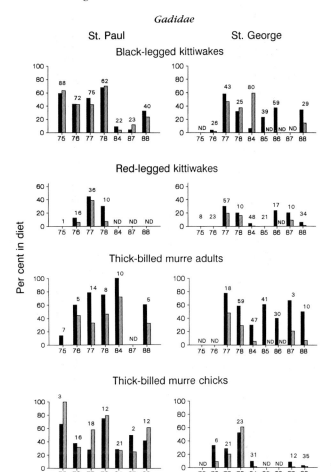

Fig. 19.5. Use of gadids, mostly walleye pollock, by kittiwakes and Brünnich's guillemots (thick-billed murres) on the Pribilof Islands. Solid filled bars represent percentage occurrence, diagonal filled bars percentage volume. Numbers at the tops of columns are the numbers of stomach samples or food loads examined. Redrawn from Decker *et al.* (1995), with permission.

decreased in the 1980s and that red-legged kittiwakes on St. George Island continued to seek myctophids, whereas black-legged kittiwakes there gave up foraging for them and turned to alternative prey in 1984 and 1988. The data from St. Paul Island are more difficult to interpret. We cannot determine, on the basis of these data alone, whether myctophids became less available to black-legged kittiwakes, or if some other prey, such as sandlance, became more available (see below). Guillemots did not use myctophids to an appreciable extent.

In contrast to the decreases in availability of some prey types, sandlance appeared to become more available in the 1980s than in the 1970s. Although there were no measures of their availability independent of seabirds, consumption of sandlance at both St. Paul and St. George Islands increased in the two seabird species,

black-legged kittiwake and Brünnich's guillemots, which preyed upon them (Fig. 19.8).

PREY USE AND SEABIRD REPRODUCTIVE PERFORMANCE

Springer (1992) found negative correlations between the average annual productivity of kittiwakes on the Pribilof Islands and both the total biomass of pollock and the abundance of 1-group pollock in the eastern Bering Sea between 1976 and 1990. At a smaller spatial scale, Decker *et al.* (1995) failed to find a significant positive correlation between the annual variability in numbers of 1-group pollock in the vicinity of the Pribilof Islands and the reproductive performance of seabirds nesting there. Further, they found no significant positive

Mallotus villosus

St. Paul   St. George

Black-legged kittiwakes

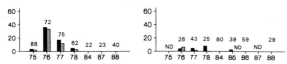

Red-legged kittiwakes

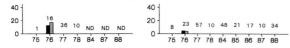

Thick-billed murre adults

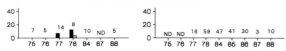

Thick-billed murre chicks

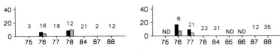

Year

Fig. 19.6. Use of capelin by kittiwakes and Brünich's guillemots (thick-billed murres) on the Pribilof Islands. Solid filled bars represent percentage occurrence, diagonal filled bars percentage volume. Numbers at the tops of the columns are the numbers of stomach samples or bill loads examined. Redrawn from Decker *et al.* (1995), with permission.

Myctophiidae

St. Paul   St. George

Black-legged kittiwakes

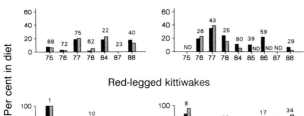

Red-legged kittiwakes

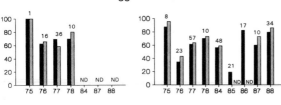

Year

Fig. 19.7. Use of myctophids (lantern fish) by kittiwakes on the Pribilof Islands. Solid filled bars represent percentage occurrence, diagonal filled bars percentage volume. Numbers at the tops of columns are the number of stomach samples examined. Redrawn from Decker *et al.* (1995), with permission.

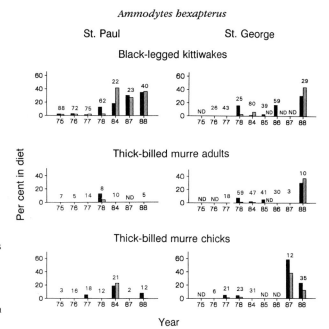

Fig. 19.8. Use of sandlance by black-legged kittiwakes and Brünnich's guillemots (thick-billed murres) on the Pribilof Islands. Solid filled bars represent percentage occurrence, diagonal filled bars percentage volume. Numbers at the tops of columns are the number of stomach samples or bill loads examined. Redrawn from Decker *et al.* (1995), with permission.

correlations between the percentage of pollock in seabird diets and reproductive success. There was, however, a significant negative relationship between the total amount of pollock in the diets of black-legged kittiwakes nesting on St. George Island and their production of young; separate examination of the role of 1-group pollock also showed no significant correlation with reproductive output (Hunt *et al.*, in press). On the basis of these findings, Hunt *et al.* concluded that variation in the abundance of walleye pollock in the vicinity of the Pribilof Islands was not the direct cause of interannual variation in seabird reproduction at the Pribilof Islands.

To examine the importance of fatty fish (myctophids, sandlance and capelin) to kittiwake reproductive performance, we compared inter-annual variation in the use of these fish to the number of chicks produced by each species of kittiwake on each island, but found no statistically significant relationships. We then examined the possibility that variations in the availability of fatty fish influenced the production of young by making the parents work harder to get these prey, even if the proportion of these fish in the diets might not correlate with reproductive output. We assumed that the proportion of fatty fish in the diets of generalist black-legged kittiwakes was an index of the availability of these fish to both species of kittiwakes at the Pribilof Islands. The proportion of fatty fish in the diets of black-legged kittiwakes on St. George Island was a

significant predictor of the reproductive performance of red-legged kittiwakes on St. George Island (r = 0.962, n = 5, p = 0.009) (Fig. 19.9A). The use of fatty fish by black-legged kittiwakes on St. George Island also appeared to have been positively correlated with the production of young by black-legged kittiwakes on St George Island (r = 0.639, n = 5, p = 0.246) (Fig. 19.9B) and the production of young by red-legged kittiwakes on St. Paul Island (r = 0.723, n = 5, p = 0.168) (Fig. 19.9C), although these correlations were not statistically significant due to the small sample sizes involved.

Use of fatty fish by black-legged kittiwakes on St. Paul Island was not a useful predictor of either red-legged kittiwake (r = 0.048, n = 7, p = 0.919) or black-legged kittiwake (r = −0.403, n = 6, p = 0.370) reproductive performance there (Fig. 19.9D). If in the latter test the value for 1984 is removed, the strength of the relationship between use of fatty fish and reproductive performance for black-legged kittiwakes on St. Paul Island is similar to those obtained on St. George Island (Fig. 19.9B).

Fatty fish consumption by black-legged kittiwakes on St. George Island was a more useful predictor of red-legged kittiwake reproductive performance on both St. Paul and St. George Islands than was fatty fish consumption by black-legged kittiwakes on St. Paul Island. This discrepancy is probably a result of differences in

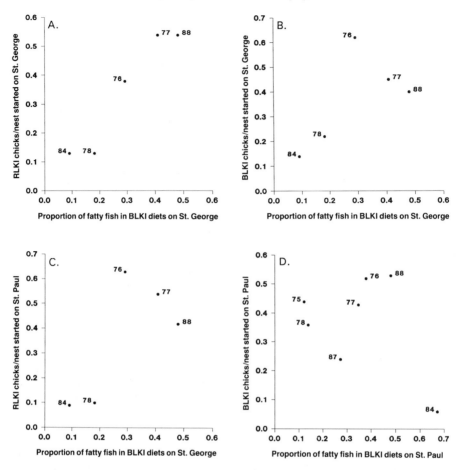

Fig. 19.9. Relationships between the proportion of fatty fish in the diets of black-legged kittiwakes (BLKI) and the production of chicks by black-legged and red-legged kittiwakes (RLKI) on the Pribilof Islands during the 1970s and 1980s. Sample sizes vary depending on the available data. The use of fatty fish by black-legged kittiwakes on St. Paul Island was not a useful predictor of the reproductive output of red-legged kittiwakes, and this relationship is not illustrated here.

the foraging areas used. Black-legged kittiwakes on St. Paul Island tend to forage north of the region used by red-legged kittiwakes from St. George Island, which forage primarily south and west of the island, often near the shelf edge (Hunt *et al.* 1981, Schneider and Hunt 1984). Red-legged kittiwakes fly south from St. Paul Island and forage in areas used by birds from St. George Island. Springer and Byrd (1989) have commented on the strong correlations between the reproductive performances of red-legged and black-legged kittiwakes on the two islands. In fact, the strongest correlation of chick production was between black-legged kittiwakes on St. George Island and red-legged kittiwakes on St. Paul Island (Pearson r = 0.95, n = 12, p < 0.001). We interpret these results as further

evidence that the two species of kittiwakes from St. George Island and red-legged kittiwakes from St. Paul Island share a common foraging ground.

The species composition of the fatty fish taken by black-legged kittiwakes changed over the course of the study (Table 19.1). The primary fatty fish used shifted from myctophids to sandlance. This shift is compatible with the interpretation that the availability of myctophids may have decreased over the course of the study period, and could account for the relatively low level of reproduction of red-legged kittiwakes in the 1980s as compared with the 1970s (Decker *et al.* 1995). The year 1988 was an exception to the rest of the 1980s; both red-legged and black-legged kittiwakes took myctophids (Fig. 19.7), and both kittiwake species reproduced

Table 19.1. Principal types of fatty fish used by black-legged kittiwakes breeding on the Pribilof Islands, by percent volume.

Year	St. Paul Island		St. George Island	
	Fatty fish	% by vol.	Fatty fish	% by vol.
1975	Myctophids	8.0	No data	
	Capelin	2.4		
1976	Capelin	33.4	Myctophids	22.7
	Myctophids	2.4	Capelin	6.3
1977	Myctophids	20.3	Myctophids	39.2
	Capelin	11.8	Capelin	1.6
1978	Sandlance	5.6	Myctophids	15.3
	Myctophids	5.2	Sandlance	2.6
1984	Sandlance	44.2	Sandlance	8.6
	Myctophids	20.9	Myctophids	0.4
1987	Sandlance	27.3	No data	
1988	Sandlance	36.3	Sandlance	43.8
	Myctophids	12.7	Myctophids	2.3

successfully (Decker *et al.* 1995) (Figs 19.5 and 19.6). Although the availability of sandlance may have increased, they were not taken by red-legged kittiwakes.

DISCUSSION

This paper presents evidence that changes in access to fatty fishes may have played a more important role in the decrease in the reproductive output of kittiwakes on the Pribilof Islands than did changes in the availability of juvenile pollock. Other authors have suggested that because fatty fish are higher in energy content than pollock, a decrease in the availability of fatty fish could have resulted in decreases in seabird populations (Springer 1992, Anon. 1993). Similar arguments have been proposed to explain the decreases in fur seals and northern sea lions, and limited data support this hypothesis (e.g. Alverson 1991).

It is not clear why changes in the use by black-legged kittiwakes on St. George Island of all species of fatty fish combined should be a better predictor of red-legged kittiwake reproductive output on the Pribilof Islands than black-legged kittiwake use of myctophids. The principal fatty fish used by black-legged kittiwakes was sandlance, whereas red-legged kittiwakes, that specialize on myctophids, did not use sandlance. The most likely explanation is that conditions in the upper water column that draw foraging myctophids to the surface also are favourable for sandlance foraging. For example, in years with cool surface waters, copepod development may be slowed (Vidal & Smith 1986), and

it is possible that larger numbers of copepods and other fish prey remain at the surface late into the summer as compared to years with warmer water. In the California Current, Roemmich and McGowan (1995) found a long-term negative relationship between sea surface warming and zooplankton biomass. They hypothesized that the 80% decline in zooplankton stocks was the result of decreased vertical flux of nutrients depressing primary production. Whatever the mechanism, it seems likely that the connection between the two species of kittiwakes involves not only their joint use of myctophids, but also a connection mediated by a broad variation in the foraging conditions experienced by their shared prey.

Although evidence from trawl surveys corroborates data from kittiwakes that 1-group pollock and capelin decreased in abundance near the Pribilof Islands between the 1970s and the 1980s, we have no data independent of the seabirds that indicate whether myctophids or sandlance changed in abundance or availability. Springer (1992) suggested that increased numbers of pollock may have preyed upon or competed with fish such as myctophids, capelin and sandlance, thereby reducing their availability to seabirds. Decker *et al.* (1995) provide limited data that argue against control of sandlance and capelin populations by pollock, but there is insufficient evidence to resolve the issue. Additionally, changes in sea surface temperatures and locations of the domains and fronts may have influenced the horizontal or vertical distribution of prey (Springer 1992, Decker *et al.* 1995, Hunt *et al.*, in press). A change in the vertical migration patterns of myctophids that resulted in their not approaching the sea surface at night would have had immediate effect on the diets of both species of kittiwakes, which are confined to foraging in the upper 25–50 cm of the water column.

Our ability to examine rigorously the importance of fatty fishes to the kittiwakes was severely diminished by the lack of a complete time series of seabird diets, and by the lack of independent data on the abundance and distribution of fatty fish. As efforts to develop multi-species models of fisheries interactions progress, our lack of knowledge about the ecology and population biology of these small forage fishes will become more critical. These fishes may be a key to the population regulation of marine bird and mammal populations. They may also play an important role in energy transfer to predatory fishes. The rapid decrease in some populations of marine birds and mammals in the northeast Pacific Ocean, and the potential or perceived threat to commercial fisheries if greater control measures are imposed when species of birds and mammals are declared threatened or endangered, should encourage investment in knowledge of forage fish biology.

ACKNOWLEDGEMENTS

We thank G.V. Byrd, P. Hunt, E. Woehler, and two anonymous referees for helpful comments on a previous draft of the manuscript. Research on the Pribilof Islands in the 1970s was supported by Bureau of Land Management, National Oceanographic and Atmospheric Administration Outer Continental Shelf Assessment Program contracts to G. Hunt. Studies in the 1980s were supported by a National Science Foundation Grant DPP-8521178 to G. Hunt. Data on seabird productivity in the 1980s and 1990s were provided by the Alaska Maritime National Wildlife Refuge. M.B. Decker was supported by a National Aeronautics and Space Administration Fellowship in Global Change Research.

REFERENCES

Alverson, D.L. (1991) Commercial fisheries and the Steller sea lion (*Eumetopias jubatus*): The conflict arena. *FRI-UW-9106*, 90 pp. Fisheries Research Institute, University of Washington School of Fisheries.

Anon. (1993) *Is it Food?: Addressing Marine Mammal and Seabird Declines: Workshop Summary.* Alaska Sea Grant Report 93-01.

Baird, P.H. (1990) Influence of abiotic factors and prey distribution on diet and reproductive success of three seabird species in Alaska. *Ornis Scandinavica*, **21**, 224–235.

Castellini, M. (1993) Report of the Marine Mammal Working Group. *Is it Food?: Addressing Marine Mammal and Seabird Declines: Workshop Summary*, pp. 4–13. Alaska Sea Grant Report 93-01.

Climo, L. (1993) The status of cliff-nesting seabirds at St. Paul Island, Alaska, in 1992. *US Fish and Wildlife Service Report AMNWR 93/15*, 53 pp. Homer, Alaska.

Cooney, R.T. & Coyle, K.O. (1982) Trophic implications of cross-shelf copepod distributions in the southeastern Bering Sea. *Marine Biology*, **70**, 187–196.

Decker, M.B., Hunt Jr, G.L. & Byrd Jr, G.V. (1995) The relationships among sea-surface temperature, the abundance of juvenile walleye pollock (*Theragra chalcogramma*), and the reproductive performance and diets of seabirds at the Pribilof Islands, southeastern Bering Sea. *Climate Change and Northern Fish Populations* (ed. R.J. Beamish), pp. 425–437. Canadian Special Publication in Fisheries and Aquatic Science, **121**.

Dragoo, B.K. & Sundseth, K. (1993) The status of northern fulmars, kittiwakes, and murres at St. George Island, Alaska, in 1992. *US Fish and Wildlife Service Report AMNWR 93/10*, 92 pp. Homer, Alaska.

Fritz, L.W., Wespestad, V.G. & Collie, J.S. (1993) Distribution and abundance trends of forage fishes in the Bering Sea and Gulf of Alaska. *Is it Food?: Addressing Marine Mammal and Seabird Declines: Workshop Summary*, pp. 30–44. Alaska Sea Grant Report 93-01.

Hatch, S.A. & Sanger, G.A. (1992) Puffins as samplers of juvenile pollock and other forage fish in the Gulf of Alaska. *Marine Ecology Progress Series*, **80**, 1–14.

Hunt Jr, G.L., Eppley, Z., Burgeson, B. & Squibb, R. (1981) Reproductive ecology, foods and foraging areas of seabirds nesting on the Pribilof Islands 1975–1979. *Environmental Assessment of the Alaska Continental Shelf; Final Reports of Principal Investigators*, 12, pp. 1–257. National Oceanic and Atmospheric Administration, Washington DC.

Hunt Jr, G.L., Kitaysky, A.S., Decker, M.B., Dragoo, B.K. & Springer, A.M. (in press) Changes in the distribution and size of juvenile walleye pollock as indicated by seabirds breeding on the Pribilof Islands, 1975 to 1993. *NOAA Technical Report Series*. National Oceanic and Atmospheric Administration, Washington DC.

Iverson, R.L., Coachman, L.K., Cooney, R.T., English, T.S., Goering, J.J., Hunt Jr, G.L., Macauley, M.C., McRoy, C.P., Reeburg, W.S. & Whitledge, T.E. (1979) Ecological significance of fronts in the southeastern Bering Sea. *Ecological Processes in Coastal and Marine Systems* (ed. R.L. Livingston), pp. 437–466. Plenum Press, New York.

Kinder, T.H. & Schumacher, J.D. (1981) Hydrographic structure over the continental shelf of the southeastern Bering Sea. *The Eastern Bering Sea Shelf: Its Oceanography and Resources* (eds D.W. Hood & J.A. Calder), pp. 31–51. US Department of Commerce/Department of the Interior, Washington DC.

Loughlin, T.R. (1987) *Report of the Workshop on the Status of Northern Sea Lions in Alaska*. Processed Report No. 87-04. Northwest and Alaska Fisheries Research Center, National Marine Fisheries Service, National Oceanic and Atmospheric Administration, Anchorage.

Miller, A.J., Cayan, D.R., Barnett, T.P., Graham, N.E. & Oberhuber, J.M. (1994) Interdecadal variability of the Pacific Ocean: model response to observed heat flux and wind stress anomalies. *Climate Dynamics*, **9**, 287–302.

Monaghan, P., Uttley, J.D., Burns, M.D., Thaine, C. & Blackwood, J. (1989) The relationship between food supply, reproductive effort and breeding success in Arctic terns *Sterna paradisaea*. *Journal of Animal Ecology*, **58**, 261–274.

Monaghan, P., Wright, P.J., Bailey, M.C., Uttley, J.D. & Walton, P. (in press) The influence of changes in food abundance on diving and surface feeding seabirds. *Studies of high latitude seabirds 4, Trophic Relationships of Marine Birds and Mammals* (ed. W.A. Montevecchi). Canadian Wildlife Service Occasional Paper, Ottawa.

Roemmich, D. & McGowan, J. (1995) Climatic warming and the decline of zooplankton in the California Current. *Science*, **267**, 1324–1326.

Royer, T.C. (1989) Upper ocean temperature variability in the Northeast Pacific Ocean: is it an indicator of global warming? *Journal of Geophysical Research*, **94 C12**, 18175–18183.

Royer, T.C. (1993) High-latitude oceanic variability associated with the 18.6-year nodal tide. *Journal of Geophysical Research*, **98 C3**, 4639–4644.

Springer, A.M. (1992) A review: walleye pollock in the North Pacific – how much difference do they really make? *Fisheries Oceanography*, **1**, 80–96.

Springer, A.M. & Byrd, G.V. (1989) Seabird dependence on

walleye pollock in the southeastern Bering Sea. *Proceedings of the International Symposium on the Biology and Management of Walleye Pollock, Anchorage 1988*, pp. 667–677. Alaska Sea Grant Report 89-1.

Schneider, D.C., & Hunt Jr, G.L. (1984) A comparison of seabird diets and foraging distribution around the Pribilof Islands. *Marine Birds: Their Feeding Ecology and Commercial Fisheries Relationships* (eds D.N. Nettleship, G.A. Sanger & P.F. Springer), pp. 86–95. Canadian Wildlife Service Special Publication, Ottawa.

Schneider, D.C., Hunt Jr, G.L. & Harrison, N.M. (1986) Mass and energy transfer to seabirds in the southeastern Bering Sea. *Continental Shelf Research*, **5**, 241–257.

Trenberth, K.E. & Hurrell, J.W. (1994) Decadal atmosphere–ocean variations in the Pacific. *Climate Dynamics*, **9**, 303–319.

Vidal, J. & Smith, S.L. (1986) Biomass, growth, and development of populations of herbivorous zooplankton in the southeastern Bering Sea during spring. *Deep Sea Research*, **33**, 525–556.

Walsh, J.J. & McRoy, C.P. (1986) Ecosystem analysis in the southeastern Bering Sea. *Continental Shelf Research*, **5**, 259–288.

Wright, P.J. & Bailey, M.C. (1993) Biology of sandeels in the vicinity of seabird colonies at Shetland. *Fisheries Research Services Report 14/93*. Marine Laboratory, Aberdeen.

CHAPTER 20

Is there a conflict between sandeel fisheries and seabirds? A case study at Shetland

P.J. Wright

SUMMARY

(1) There has been concern that fishing may affect the availability of sandeels to seabirds, either through effects on the total stock, stock recovery or through local depletions near seabird colonies. This problem has been addressed through a study of changes in sandeel availability around Shetland.

(2) Stock changes around Shetland largely appeared to reflect variability in pre-recruit immigration and survival rather than changes in the total stock. As a consequence it is unlikely that the Shetland fishery played a significant part in either the stock decline or recovery.

(3) The effect of changes in total stock abundance on sandeel density was not uniform throughout the region. As a consequence of this distributional response, seabirds feeding in areas preferred by sandeels probably experienced far less variability in food supply than total stock abundance would imply, whilst those in less preferred areas will have experienced much greater variation.

(4) In addition to changes in stock size, prey availability to seabirds was found to be influenced by the size of 0-group sandeels and the time when they become available to seabirds at chick rearing.

Key-words: *Ammodytes marinus*, distribution, prey, recruitment, sandeel

INTRODUCTION

The lesser sandeel *Ammodytes marinus* is one of the commonest fish species on the continental shelf of north-west Europe and accounts for somewhere between 10 and 15% of the total fish biomass of the North Sea (Sparholt 1990). This species is of considerable ecological and commercial importance, being prey to many fish, seabirds and marine mammals (Harwood & Croxall 1988, Daan 1989, Furness 1990) as well as supporting the largest single-species fishery in the North Sea (Anon. 1992). The importance of *A. marinus* in marine food-webs has led to concern over the potential impact of sandeel harvesting on prey availability to marine predators. In particular there has been con-

siderable speculation over the involvement of sandeel fisheries in regional declines in the breeding success of seabirds.

The northern UK coast supports the largest concentration of seabird colonies bordering the North Sea (Lloyd *et al.* 1991). Given that seabird predation on sandeels is highest during the summer breeding season it is therefore not surprising that around 80% of sandeels consumed by birds are from the western North Sea, north of 53°N (Anon. 1994). In contrast, the majority of the North Sea sandeel catch comes from other areas (Anon. 1993), so there is little overlap in the areas exploited by fisheries and seabirds at a North Sea scale (Anon. 1994). Up until 1990 the main fishery in the north-western North Sea (ICES areas IVa west and

IVb west) operated close to Shetland and had a peak in landings of just over 52 000 tonnes in 1982. The landings from the Shetland fishery fell during the 1980s due to a decline in both the stock and the economic value of sandeels (Goodlad 1989).

The decline in landings from the Shetland sandeel fishery in the 1980s and the parallel decline in seabird breeding success and sandeel abundance was well publicized and provides an excellent example of the perceived competition between seabirds and sandeel fisheries (Bailey 1991). Due to the proximity of the fishery grounds to areas where seabirds foraged, many argued that the fishery competed for the same resource as the seabirds, and was responsible for the decline in sandeel availability (Avery & Green 1989). The importance of sandeels to seabirds around Shetland can be explained by the scarcity of other small schooling prey in this region (Kunzlik 1989).

Annual assessments of sandeel numbers around Shetland have been carried out using virtual population analysis (VPA) since the fishery began in 1974. These assessments suggested that the decline in sandeel abundance was the result of a decline in recruitment to the Shetland stock (recruitment is defined here as the number of young-of-the-year (known as 0-group) alive on 1 July), which preceded any change in the spawning stock (Kunzlik 1989). It appeared that a change in pre-recruit mortality was the main cause of the decline in fishery landings and prey for seabird chicks. Nevertheless, it was still argued that the fishery may have accelerated the decline in recruitment and delayed the recovery of the sandeel stock (Monaghan 1992). Further it was suggested that any fishery-induced local reductions near seabird colonies could go undetected at the total stock level (Monaghan 1992). Bailey (1991) also highlighted that there is some uncertainty in assessments based on fisheries data alone because trends in the spawning stock biomass and recruitment derived from VPA are sensitive to the values of natural mortality used in analyses. It was also unclear whether stock changes alone could account for the changes in sandeel availability to seabirds. For example, Monaghan *et al.* (1989) showed that Arctic tern *Sterna paradisaea* chicks at Shetland died of starvation in 1987 as a result of inadequate provisioning by their parents despite a high frequency of foraging trips compared with a successful breeding colony in north-east England. This problem was linked with the low energy content of feeds per trip, because of the relatively small size of sandeels being brought to the Shetland colony.

Although seabirds prey on all age-classes, the species that appear most susceptible to declines in sandeel abundance are those that predominantly feed their chicks on young of the year (0-group). These include birds such as terns and kittiwakes *Rissa tridactyla* which feed close to the sea surface, and to a lesser extent the puffin *Fratercula arctica*, which is a shallow-diving species. It is unlikely that there is much direct competition between seabirds and fisheries for 0-group sandeels during the chick rearing period (June–July) because most fishing mortality on 0-group occurs after 1 July (Bailey 1991, Anon. 1992). Hence, the main impact on prey availability may be indirect, through reductions in spawning stock biomass, which are often implicitly assumed to affect recruitment (e.g. Serebryakov 1990). Therefore it is necessary to assess whether fishing is capable of reducing the spawning stock to a level at which recruitment is affected because of insufficient egg production (recruitment overfishing).

Due to the uncertainty surrounding the causes of changing sandeel availability and the international importance of the Shetland Isles as an area for breeding seabirds (see Tasker *et al.* 1987, Lloyd *et al.* 1991), a research programme was initiated in 1990 to investigate sandeel biology around Shetland. In this paper, data from this study, together with assessment and archival information is used to examine the key factors responsible for changes in sandeel availability at spatial and temporal scales relevant to both the fish stock and seabird colonies. From this appraisal the importance of fishery versus naturally-induced changes in sandeel stocks is considered.

METHODS

Sandeel distribution

Although it is recognized that sandeels are not solely confined to exploited areas, all fishery assessments are based on commercial and research survey catch and effort data from fished grounds. There are no published accounts of sandeel distribution outside these areas. Consequently, in order to consider sandeel availability, historical information on sandeel distribution was extracted from a variety of research survey sources. The data sources available provided information on sandeel distribution both for the pelagic 0-group and sandeels which have settled to sandy areas (mainly 1-group and older). Information on pre-settlement 0-group and ≥1-group sandeel density distribution by ICES statistical squares was obtained from the International Young Gadoid Surveys conducted in June (I0GS, 1971–1988). The age composition of catches from this survey were distinguished using modal analysis, the modal total length of 0-group sandeels generally being less than 100 mm. Data on the occurrence of post-settled sandeels were taken from Scottish trawl surveys (1922–

1990) and Marine Laboratory exploratory sandeel surveys (1967–1974). Together these two surveys provided good coverage of the North Sea and sandeel grounds on the east mainland coast. However, the coverage of the west coast was generally poor.

Variability in sandeel abundance

The relationship between year-class strength and spawning stock biomass was examined using the 1990 and 1993 VPA runs. This analysis assumes that natural mortality for a given age-class is constant from year to year. Since 1984, estimated fishing mortality has been considerably lower than natural mortality. The catch data is derived from landings at the local fishmeal plant and consequently no account is taken of the number of fishing grounds from which the data are derived. Different types of VPA were used for the periods 1974–1989 and 1984–1992 (see Anon. 1992 for further information). In order to consider year-class effects independently of the effect of spawning stock size, the mean number of recruits per spawning stock biomass ($r \cdot ssb^{-1}$) was determined from the 1974–1989 semi-annual VPA series. D-tests were then used to test for significant deviations from this mean recruit/spawning stock biomass value, and thereby infer the relative year-class strength.

Independent information on sandeel abundance at fished grounds (see Fig. 20.1) was obtained from Marine Laboratory recruitment surveys of all Shetland grounds conducted in August between 1984 and 1992, and trawl surveys conducted during June and July between 1990 and 1993 (18 June–26 July 1990, 24 June–15 July 1991, 1–2 June and 25 June–1 July 1992). The June surveys only covered grounds at Mousa Sound, Baas, Voe and Fair Isle. However, the number and period over which samples were taken were both greater in the June–July surveys than in the August survey (June N = 3–6 per ground, August N = 2–3 per ground). Samples were collected using a commercial bottom trawl and used to derive age-stratified catch per unit effort data. Inter-annual differences in 0-group sandeel abundance (determined as catch per unit swept area) between regions was examined from the August survey data. Abundance on each fishing ground could be estimated knowing the extent of each region (J.A. Gauld, unpublished data).

Variability in sandeel density distribution

Sandeel distribution outside the areas commonly fished was examined from acoustic surveys that were conducted in conjunction with the June surveys. The main

area surveyed included south Shetland and Fair Isle; 60°05′N–59°30′N and 000°00′–002°00′W. A zig-zag cruise track with either a 5 nautical mile spacing between tracks or 2–3 km spacing between tracks was used to assess large-scale distribution (5–15 km from the coast) and finer-scale distribution (within 5 km of the coast). Cruise tracks were divided into a series of track lengths 1, 2, 3 and 6 km long, for data analysis. Fish distribution was recorded using a Simrad EY-200 echosounder with a 200 kHz transducer in 1990 and 1992 and either an EK-400 or EK-500 echointegration system with a 120 kHz transducer in 1991 and 1993, respectively. The large vertical height of most sandeel shoals (over 80% of shoals were more than 5 m in height during the three years) suggests that, despite the problems of detection in the top and bottom 3 m of the water column, it is unlikely that many shoals would have gone undetected. The high working frequencies used were necessary to resolve the weak signals given by sandeel shoals. Since there are no reliable target strength measurements for sandeels at 120 kHz, relative fish density was expressed on a scale ranging from 0–1 units. Echotraces were recorded on a colour printer and assigned to species on the basis of samples taken from hauls with a pelagic trawl.

In order to compare the results from echointegration surveys in 1991 and 1993 with echosounder surveys carried out in 1990 and 1992, an echotrace area/echointegral relationship for sandeels was derived from a comparison of $\log_n$ transformed ($\ln x + 1$) data on shoal area and echointegrals (see Monaghan *et al.* 1994). Using the 95% confidence intervals for this regression, shoal area could be assigned to one of six echointegral values. This classification of acoustic data was then used for inter-annual comparisons of density distribution. In order to consider whether density changes in fished and unfished areas varied in a proportionate manner with respect to abundance, acoustic data were grouped into five regions, which included fishing grounds, other coastal areas and areas away (>5 km) from the coast (see Fig. 20.1).

The patchiness of shoals was considered for the intensively surveyed coastal area, using the relationship between Lloyd's (1967) mean crowding index and mean density. The slope of this relationship is called the density contagiousness coefficient and indicates how the basic components distribute themselves over the habitat (Iwao 1968).

Prey quality

The size of sandeels between 1990 and 1992 was examined from length–frequency data derived from pelagic and bottom trawl samples. The age composition

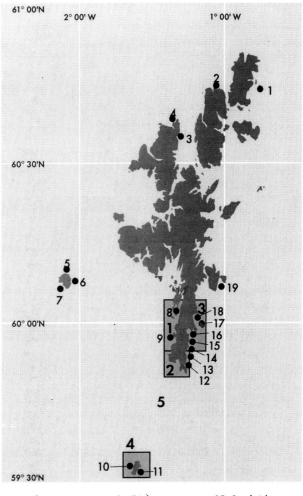

Fig. 20.1. Chart showing locations of sandeel fishing grounds around Shetland. Boxes show component regions of acoustic survey (1–5: 1 = Colsay, 2 = Sumburgh, 3 = Mousa, 4 = Fair Isle, 5 = Offshore).

1. Balta
2. Brekkin
3. Fethaland
4. Sand Voe
5. North Foula
6. Ham o'Foula
7. South Foula

8. Trink (Clift Sound)
9. Colsay
10. West Fair Isle
11. East Fair Isle
12. Grutness
13. Boddam Voe
14. Clumlie

15. Sandwick
16. Mousa Sound
17. Braeside
18. Helliness
19. Sound Sands

of 0-group fish was also examined from daily otolith increments. A detailed account of sandeel otolith microstructure is given by Wright (1993).

RESULTS

Sandeel distribution

The comparison between areas fished and the distribution of 1-group and older sandeels in the western North Sea shows that there were many unexploited sandeel concentrations in this region (Fig. 20.2). Included in the Shetland assessment area defined by the ICES working group were unexploited sandeel concentrations around Orkney. In addition, a concentration of 1-group and older sandeels was present just to the east of the assessment area. The distribution of 1-group and older sandeels around Orkney appeared to be continuous with concentrations off the Scottish east coast. With the exception of a few concentrations offshore, 1-group and older sandeel distribution was largely confined

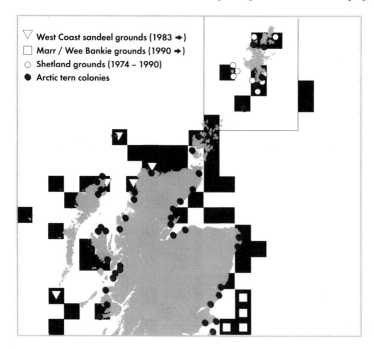

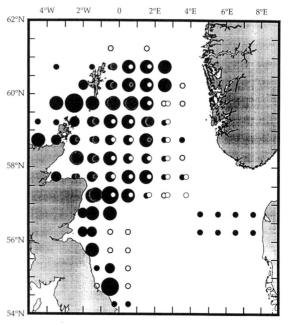

Fig. 20.2. Chart showing the distribution of ≥1+ sandeels. Data based on the presence of sandeels in demersal trawls from Scottish trawl surveys (1922–1990) and Marine Laboratory exploratory sandeel surveys (1967–1974). Fishing grounds and major Arctic tern breeding colonies (derived from Lloyd *et al.* 1991) are also shown. Box indicates Shetland sandeel assessment area.

to the coastal margins. Although the distribution of all age-classes overlapped, that of juvenile 0-group sandeels was found to be more extensive in some years. For example, in 1977 juveniles were caught in large numbers offshore, from 2°E to the extent of sampling north of Scotland at 4°W (Fig. 20.3).

Changes in sandeel abundance

Trends in spawning stock biomass and recruitment based on VPA are given in Figure 20.4. Spawning stock biomass did not account for a significant proportion of the variation in recruitment for either VPA run. However, recruitment accounted for a significant proportion of the variation in spawning stock biomass when compared at a lag of two years, based on a cross correlation analysis (1974–1989 $r^2 = 0.63$, $P < 0.01$). As sandeels are mature by two years old (Gauld & Hutcheon 1990) this result suggests that the spawning stock biomass is heavily influenced by the size of the maturing year-class at recruitment. Deviations from the mean recruits per spawning stock biomass relationship indicated that there were periods of unusually high or low recruitment between 1974 and 1989. Significant ($P < 0.05$) positive deviations occurred in the mid-1970s and 1982 whilst negative deviations occurred from 1984 to 1989.

Variability in relative recruitment and 0-group distri-

Fig. 20.3. 0-Group (45–95 mm TL) sandeel density (as number per hour tow) distribution in the North Sea in June 1971 (shaded) and 1977 (solid), as determined by the International Young Gadoid Surveys. Open circles represent stations where no sandeels were present. ●, $<10\,h^{-1}$; ●, $10–100\,h^{-1}$; ●, $100–1000\,h^{-1}$; ●, $1000–10\,000\,h^{-1}$; ●, $>10\,000\,h^{-1}$.

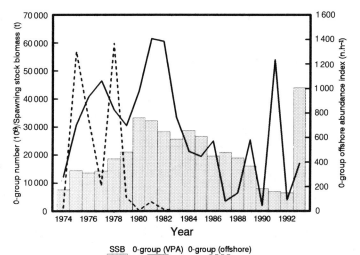

Fig. 20.4. Comparison between recruitment (0-group numbers) and spawning stock biomass (ssb) in the Shetland assessment area as assessed from virtual population analysis. Data for 1974–1989 is based on 1990 VPA run, data from 1990–1992 is based on 1994 VPA run (Anon. 1991, S. Reeves personal communication). Data on 0-group abundance in offshore trawl catches is also presented.

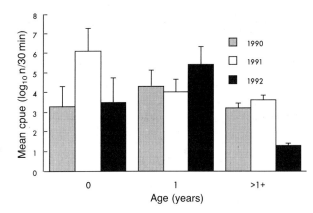

Fig. 20.5. Comparison of mean catch per unit effort ($\log_{10} n/30$ min) for catches from three Shetland grounds combined (Mousa, Baas, Voe). Data derived from 9 trawls in 1990, 18 trawls in 1991 and 9 trawls in 1992. Data for 0-group, 1-group and >1-group sandeels are given.

bution, evident from the IOGS, was compared to see if there was any relationship between large-scale changes in 0-group density-distribution and recruitment around Shetland. Values of $r \cdot ssb^{-1}$ were ranked as +1, 0 and −1 according to whether they significantly differed, either positively or negatively, from the mean. Comparison between these rankings and IOGS data indicated that there were significant differences in the density of 0-group sandeels between Orkney and Shetland with respect to relative recruitment (Kruskal–Wallis $\chi^2 = 9.98$; $P = 0.007$). The period of high $r \cdot ssb^{-1}$ coincided with a period when offshore concentrations of 0-group sandeels were significantly higher than the long-term mean, and their distribution extended from Orkney to Shetland. Conversely, the period of low $r \cdot ssb^{-1}$ and low recruitment coincided with a period when virtually no sandeels were caught at offshore stations. These changes in the relative abundance of

0-group sandeels inshore (from VPA) and offshore (from trawl indices given by Bailey *et al.* 1991) are shown in Figure 20.4.

Trends in abundance indices from research trawl and acoustic surveys around Shetland during June and July agreed with those derived from the VPA for the period 1990–1992. Mean catch per 30 min trawl of 0-group sandeels from south Shetland grounds increased from just over 1500 in 1990 to 1.26×10^6 in 1991 (Fig. 20.5). The 1991 year-class formed 99% of catches in 1991, 97% of catches in 1992 and 40% of catches by June 1993. Acoustic estimates of overall density within the area 60°00'N and 59°30'N indicated a similar pattern of change to that indicated by trawl catch per unit effort (cpue) at grounds. Median and range in density of sandeels, based on 5' latitude × 5' longitude rectangles, were 0.0005, 0–0.015 units·km^{-1} in 1990, 0.024, 0–1.00 units·km^{-1} in 1991 and 0.005, 0–0.27

units$\cdot$km^{-1} in 1992, 0.008, 0–0.1 units$\cdot$km^{-1} in 1993. Pelagic trawl sampling indicated that 0-group sandeels comprised over 95% of concentrations found away from recognized fishing grounds in 1991.

Density related changes in distribution

The geographical extent of sandeel distribution within the Shetland–Fair Isle study area was found to vary in proportion to the overall abundance between 1990 and 1993. However, the rate of change in local density with respect to median density differed among regions (Friedman test statistic = 10.6, P = 0.03, df = 4). From differences in the slope of local density on area density it would appear that the rate of change in Shetland coastal regions was far lower than in offshore regions (Fig. 20.6). The highest density of sandeels (mostly 0-group) was found in the offshore region during the year of highest overall abundance (1991) but sandeels were absent from this region in the year of lowest abundance (1990). In all years a relatively high density of sandeels was found around the south-east Shetland coast and variation in local density was lower than in other coastal regions (coefficient of variation 1990–1993: south-east 15.9%, Sumburgh 32.8%, south-west 71%). The importance of the south-east coast, near Mousa, was also evident during the 1980s from a comparison between 0-group variability in this and other regions to the north west and south west of the mainland (Fig. 20.7). The relation between local

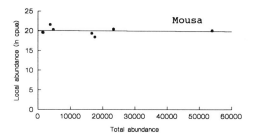

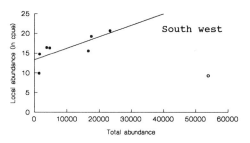

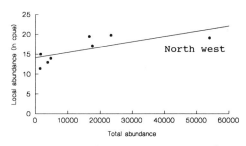

Fig. 20.7. Relation between local and total abundance (derived from VPA) of 0-group sandeels in three regions around Shetland between 1984 and 1992. Linear regressions were fitted to the data (excluding the 1991 south-west region datum represented by an open circle). The Mousa region refers to grounds at Mousa Sound and Braeside, the south-west region refers to grounds at Colsay and Trink and the north-west region refers to grounds at Sandsvoe and Fethaland (see Fig. 20.1 for locations of these grounds).

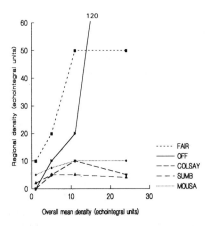

Fig. 20.6. Comparison of the changes in regional density with respect to overall mean density for different regions in the Shetland study area (60°05′N–59°30′N and 000°00′–1002°00′W). The regions are shown in Fig. 20.1. Data based on relative echointegral values × 10^3 per km^{-1} travelled for June–July surveys conducted between 1990 and 1993.

and total abundance of 0-group sandeels differed between the three regions examined (Kruskal–Wallis χ^2 = 9.04; P = 0.01). Local abundance did not vary significantly in relation to total 0-group abundance at grounds near Mousa (r^2 = 0.05, P = 0.87) but did in the north-west region (r^2 = 0.54, P = 0.04). Local abundance also varied with respect to total abundance in the south-west region, although local abundance was unexpectedly low in this region in 1991. Estimates of 1-group abundance in the south-west region in 1992 were the highest recorded, suggesting that the local abundance of 0-group must have increased shortly after sampling in 1991. As a result of the differences in recruitment rates to the different regions, 0-group sandeels from Mousa grounds accounted for 98% of the total catch in the 1990 August survey compared with 25% in 1984.

Changes in the abundance and geographic extent of sandeel distribution were paralleled by the dispersion of sandeel shoals. Shoals became more widely and randomly dispersed within the coastal waters around Shetland at high densities. For example, based on the occurrence of sandeels per 2 km track lengths, 94.5% of these (N = 146) contained sandeels in 1991 compared with only 21.3% (N = 145) in 1990. The regression of mean crowding (m*) on mean density (m) for the series of surveys in both 1990 and 1991 was linear (1990 m* = 2.14m, r^2 = 0.80, P < 0.001; 1991 m* = 1.09m, r^2 = 0.59, P < 0.01) suggesting that sandeels were dispersed in a density independent pattern over the range of mean densities. In 1990, the dispersion of shoals was non-random as the slope of the regression of m*/m was significantly different from 1.0 (t_{16} = 94, P < 0.05). In contrast the slope m*/m in 1991 was not significantly different from 1.0 (t_{11} = 1.08, P = 0.3).

Temporal variation in abundance

The abundance of 0-group sandeels in coastal waters was very variable during the chick rearing period (late June–July). For example, the coefficient of variation in bottom trawl catch per unit effort (cpue) for five south Shetland grounds based on $\log_n$ transformed data was 165–173% in 1990 and 66–131% in 1991. In 1990, relatively high numbers of 0-group shoals did not appear until late July, after the chick rearing season of most seabirds (Fig. 20.8). The sudden increase in 0-group sandeels in late July 1990 was related to the

appearance of 0-group with a significantly later mean hatch date to those found earlier in the season (mean hatch date of 0-group on 25.6.90: x̄ = 77 ± 1.76, N = 51; hatch date of 0-group on 26.7.90: x̄ = 56 ± 0.96, N = 44; t_{93} = 32.4, P < 0.001). The later hatched cohort was not detected in earlier samples from any of the south Shetland or Fair Isle grounds.

Fish size

The range in 0-group length present in June differed between years, ranging from 35–95, 55–120 and 55–95 mm TL, in 1990–1992, respectively (Fig. 20.9). A comparison between the length composition of 0-group sandeels in late June 1990 and 1991 indicated that the more abundant of the two 0-group size-classes in 1991 was significantly larger than 0-group in 1990 and 1992, with a median length of 95 mm, compared with 70 mm in 1990 (Wilcoxon test W = 15 022, P < 0.001). Hence, not only were 0-group sandeels more abundant in 1991, they also tended to be larger than in other years. Studies using daily increments in otoliths (see Wright 1993) showed that there were significant differences in mean specific growth rates between years (ANOVA, P < 0.001) which could account for the differences in length at chick rearing.

DISCUSSION

It is not surprising that the Shetland fishery was implicated in the decline in prey availability to seabirds. Sandeel abundance generally declined following the peak in landings until the closure of the fishery in 1991. However, both fisheries' assessments and the research survey data presented in this study indicate that variability in the survival and movement of pre-recruit

Fig. 20.8. Changes in the median frequency and size of sandeel shoals in Mousa Sound between 16.6.90 and 26.7.90, based on acoustic surveys. Numbers above histograms refer to mean hatch-date of respective trawl samples in julian days.

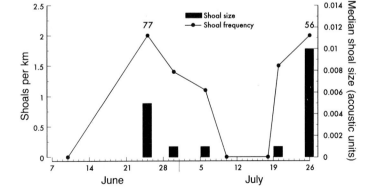

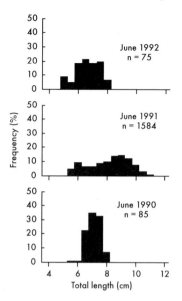

Fig. 20.9. Length frequency distributions for 0-group sandeels at the south-east Shetland grounds in late June 1990, 1991 and 1992.

stages can explain the observed changes in sandeel abundance. As Kunzlik (1989) described from fishery assessments, the initial decline in recruitment preceded any change in the size of the spawning stock, suggesting that the Shetland sandeel fishery was not responsible for the reduction in the total stock. This apparent lack of a relationship between recruitment and spawning stock is supported by more recent changes in recruitment and spawning stock size following the closure of the fishery. The large year-class in 1991 had a profound effect on the overall abundance and distribution of sandeels within the Shetland–Fair Isle region and resulted in a marked increase in the spawning stock biomass by 1993. Indices of 0-group sandeel abundance derived from recruit surveys in August support this view, and suggest that whilst the year-classes in 1990 and 1992 were two of the lowest since surveys began in 1984, the 1991 year-class was the highest (Anon. 1992). Comparisons of mean cpue for 2-group sandeels in June between years did not suggest any marked change in the size of the potential spawning stock between 1990 and 1992, although due to seasonal changes in adult availability to fishing gear (see Reeves 1994) it is difficult to assess spawning stock size accurately from survey data. Consequently, the marked increase in abundance of 0-group sandeels around Shetland in 1991 and the decrease in 1992 could only be explained by changes in the survivorship of larvae and/or immi-

gration of 0-group sandeels from other regions into the study area. Such rapid changes in sandeel populations are not unprecedented (see Sherman *et al.* 1981) and comparative studies of teleost life-history characteristics suggest that sandeel populations tend to fluctuate at shorter and more irregular intervals than most other marine species (Kawasaki 1980).

Whilst variability in pre-recruit survivorship is often the primary cause of fish stock declines, fishing is often believed to exacerbate the effect of such changes (Beverton 1990). One reason for this concerns the relationship between recruitment and spawning stock. Recruitment is generally believed to be related to spawning stock biomass below some, unknown, low level of spawning stock, although clearly this relationship is rarely a linear one. Consequently, fishing mortality on the spawning stock is presumed to have a greater effect on recruitment, when the stock has already been depressed by a series of low year-classes. Various studies have attempted to predict a minimum spawning stock size necessary for average recruitment, by examining long-term variability in recruitment and spawning stock biomass (e.g. Serebryakov 1990). Indeed, the management decision to close the Shetland sandeel fishery in 1991 was partly based on a prediction that the spawning stock size had declined to a level where above average recruitment was unlikely. In the event, there was a large year-class in 1991, despite the apparently low spawning stock size. However, the assumption that Shetland sandeels could be regarded as a unit stock, in such a comparison of spawning biomass and recruitment, may have been invalid.

The treatment of sandeel concentrations around Shetland as a single stock was based on the isolation of these fishing grounds from those of other sandeel fisheries, the comparatively low growth rates of sandeels from Shetland in relation to those in the north-east North Sea and the absence of any known migration of post-settled stages (Anon. 1979). However, pre-settlement stages (i.e. larvae and juvenile 0-group) can be widely dispersed (Langham 1971, Hart 1974, Wright & Bailey 1993) (Fig. 20.3) and so there is no reason to suppose that the Shetland fishing grounds are isolated from nearby unexploited sandeel concentrations found to the south and east of Shetland. Indeed, inter-annual comparisons of 0-group distribution in June and July show that there is often a continuous distribution of 0-group sandeels between Orkney, Shetland and eastwards to 0°. Further, periods of high or low recruitment per spawning stock biomass at Shetland have coincided respectively with an abundance or scarcity of 0-group sandeels in offshore waters. Surveys of recently hatched larvae carried out during this century indicate that the major areas of spawning in the north-west North Sea are centred around Orkney, rather than Shetland (Bowman 1914, Langham 1971, P.J. Wright & M. Bailey unpublished data). Given these findings, it would seem reasonable to infer that the variability in 0-group abundance in the Shetland assessment area was linked to the rate of influx of 0-group sandeels from outside the assessment area. This raises the possibility that the sudden changes in sandeel availability to both Shetland seabirds and the fishery were largely unrelated to local conditions. As such the situation at Shetland may therefore be inappropriate for assessing whether sandeel fisheries exacerbate the effects of natural stock collapses. However, the changes in sandeel abundance at Shetland do draw attention to the need for a fuller understanding of the population structure of this species.

Monaghan (1992) postulated that direct local competition for sandeels between the fishery and seabirds near colonies could occur, even though fishing may not affect sandeel stocks as a whole. The close proximity of Shetland fishing grounds to seabird colonies and the energetic importance of feeding close to the colony (<10 km) (Monaghan *et al.* 1994) would imply there was overlap in the areas exploited by seabirds and the fishery. However, it is important to note that fishing grounds comprise a very small part (total ground area = $32 \, km^2$, J.A. Gauld, unpublished data) of the area where seabirds forage (Tasker *et al.* 1987, Wright & Bailey 1993) and many colonies at which seabird breeding success declined were several kilometres away from fished grounds (see Heubeck 1989). Nevertheless, given this overlap in resource utilization, it is possible that local competition may have occurred under two conditions. Firstly, had local sandeel density been largely governed by the carrying capacity of the grounds, high fishing pressure could have lead to a localized depletion. However, because sandeels are a low-value species fishermen may stop fishing when catch rates become economically non-viable, even though the fish are still rather abundant. Further, the present study provides evidence for a continuous recruitment to certain grounds even in the years of low recruitment.

Another potential form of local competition could arise if sandeel densities at fished grounds were maintained by an influx from neighbouring areas, which were important to foraging seabirds. The disproportionate changes in local sandeel density demonstrated by the present study does suggest that densities at certain fishing grounds were maintained by sandeel movements away from unfished areas and a more limited settlement distribution. This pattern of geographic change is consistent with density-dependent habitat selection theory and implies that fishing grounds

are areas of high 'suitability' (Fretwell & Lucas 1970, see also MacCall 1990). The suitable sediment characteristics and relatively large size of fishing grounds ($0.6-7.5 \, km^2$, J.A. Gauld, unpublished data, see also Reay 1970) may partly explain why sandeels prefer these areas. However, the differences in recruitment among grounds suggests that there are also other factors affecting habitat suitability. In terms of sandeel availability to fisheries, the nature of this population response could enable fishermen to fish economically from certain grounds even when the overall stock density was low. This might affect marginal sandeel concentrations, when overall stock levels are low and when there is little immigration of 0-group sandeels into Shetland waters. Unfortunately, there is insufficient information on sandeel concentrations within and outside the fishing grounds during the period of fishing to test this view. It is also important to note that year-class strength is a very important factor to local abundance, as all grounds and marginal areas were quickly re-colonized in the year of high recruitment. This recolonization of all areas leads to a reduction in the patchiness of shoal distribution which is likely to considerably enhance the birds' chances of encountering sandeel shoals.

Although much attention has been placed on stock changes that can be influenced by fishing pressure, there are other aspects of the population dynamics of sandeels that influence their availability to seabirds. Seabirds rely on suitably sized prey being available near their colony by the onset of chick rearing. The present study has demonstrated that both the timing and size of 0-group available to seabirds at chick rearing can vary. In 1990, appreciable numbers of 0-group sandeels did not appear in coastal waters until late July, i.e. after most birds had reared chicks. This resulted in kittiwakes having to spend a longer time foraging and as a result chicks were left unattended at times, which would have increased the chicks' thermoregulatory costs, risk of predation and of attack by other kittiwakes (Hamer *et al.* 1993). This was probably a major reason why kittiwakes suffered a total breeding failure in many Shetland colonies in 1990. What determines 0-group movements into coastal waters is not known, although given the widespread distribution of 0-group in some years it seems likely that shoals of 0-group sandeels will be actively seeking suitable areas for settlement around the time of chick rearing.

Changes in the size of 0-group fish available to seabirds at the onset of chick rearing have been implicated as a key factor affecting seabird breeding success (Martin, 1989, Monaghan *et al.* 1989). From a comparison with the energy content/length relationships for sandeels given by Hislop *et al.* (1991) it would appear that the differences in 0-group length composition found in the present study represent marked differences in the energy content of prey available to small seabirds. For example, based on median lengths, the calorific value of 1991 0-group was approximately twice that of 1990 0-group. Given that the smallest sandeels occurred in the year of lowest overall abundance, this would be expected to have exacerbated the prey shortage experienced by Arctic terns and kittiwakes that year. These differences in size of prey may be linked to the same processes affecting pre-recruit mortality, as there was a positive association between pre-recruit growth rate and year-class size.

Many studies in which a fishery-induced effect on prey availability has been inferred have been based solely on a coincidence between declines in fishery landings and seabird populations (see Camphuysen 1990 for review). However, as this study demonstrates, such an approach is flawed in that it ignores the complexity of factors that influence prey availability to seabirds. Understanding local prey availability clearly requires a consideration of the prey's population dynamics and the predator's location in relation to the habitat preference of the prey. Consequently, information on the nature of population structure and the location of seabird foraging areas with respect to the preferred areas of prey habitat is needed. As seabird colonies present at the margin of a stock's range could be adversely affected by small changes in total stock, due to either natural or fishery-induced causes, it is necessary to examine seabird prey availability at a smaller scale than that required for fisheries assessments. Apart from Shetland, the lack of information on small-scale changes in sandeel abundance in the North Sea makes it impossible to evaluate whether competition between sandeel fisheries and seabirds occurs.

ACKNOWLEDGEMENTS

I am grateful to Martin Bailey for assistance in the Shetland sandeel project and to Robin Cook, John Hislop, Steve Hall and Stuart Reeves for constructive comments on a draft of this paper. The work was supported by The Scottish Office Agriculture and Fisheries Department, Department of Environment (CUE and ERA), Scottish Natural Heritage, The Scottish Office Environment Department, Worldwide Fund for Nature, UK, and the Royal Society for the Protection of Birds. Prof. A.D. Hawkins provided facilities at the Marine Laboratory, Aberdeen.

REFERENCES

Anon. (1979) *Report of the working group on Norway pout and sandeels in the North Sea*, ICES CM 1979/G:26.

Anon. (1991) *Report of the Industrial Fisheries Working Group*. ICES CM 1991/Assess:14.

Anon. (1992) *Report of the Industrial Fisheries Working Group*. ICES CM 1992/Assess:14.

Anon. (1993) *Report of the working group on Norway pout and sandeels*. ICES CM 1993/Assess:14.

Anon. (1994) *Report of the study group on seabird/fish interactions*. ICES CM 1994/L:3.

Avery, M. & Green, R. (1989) Not enough fish in the sea. *New Scientist*, **1674**, 28–29.

Bailey, R.S. (1991) *The interaction between sandeels and seabirds – a case study at Shetland*. ICES CM 1991/L:41.

Bailey, R.S., Furness, R.W., Gauld, J.A. & Kunzlik, P.A. (1991) Recent changes in the population of the sandeel (*Ammodytes marinus* Raitt) at Shetland in relation to estimates of seabird predation. *ICES Marine Science Symposia*, **193**, 209–216.

Beverton, R.J.H. (1990) Small marine pelagic fish and the threat of fish; are they endangered? *Journal of Fish Biology*, **37** (Suppl. A), 5–16.

Bowman, A. (1914) The spawning areas of sandeels in the North Sea. *Scientific Investigations of the Fisheries Board Scotland*, 1913(3), 13 pp.

Camphuysen, C.J. (1990) Fish stocks, fisheries and seabirds in the North Sea. *Technisch rapport Vogelbescherming*, **5**, 119 pp.

Daan, N. (ed.) (1989) Data base report of the stomach sampling project 1981. *Co-operative Research Report Conseil Internationale Exploration de la Mer* 164, 144 pp.

Fretwell, S. & Lucas, H (1970) On the territorial behaviour and other factors influencing habitat distribution in birds. *Acta Biotheoretica*, **19**, 16–36.

Furness, R.W. (1990) A preliminary assessment of the quantities of Shetland sandeels taken by seabirds, seals, predatory fish and the industrial fishery in 1981–83. *Ibis*, **132**, 205–217.

Gauld, J.A. & Hutcheon, J.R. (1990). Spawning and fecundity in the lesser sandeel, *Ammodytes marinus* Raitt, in the north-western North Sea. *Journal of Fish Biology*, **36**, 611–613.

Goodlad, J. (1989) Industrial fishing in Shetland waters. *Seabirds and Sandeels: Proceedings of a Seminar held in Lerwick, Shetland, 15–16 October 1988* (ed. M. Heubeck), pp. 50–59. Shetland Bird Club, Lerwick.

Hamer, K., Monaghan, P., Uttley, J.D., Walton, P. & Burns, M.D. (1993). The influence of food supply on the breeding ecology of kittiwakes *Rissa tridactyla* in Shetland. *Ibis*, **135**, 255–263.

Hart, P.J.B. (1974) The distribution and long-term changes in abundance of larval *Ammodytes marinus* (Raitt) in the North Sea. *The Early Life History of Fish* (ed. J.H.S. Blaxter), pp. 172–182. Springer-Verlag, Berlin.

Harwood, J. & Croxall, J.P. (1988) The assessment of competition between seals and commercial fisheries in the North Sea and the Antarctic. *Marine Mammal Science*, **4**, 13–33.

Heubeck, M. (1989) Breeding success of Shetland's seabirds:

Arctic skua, kittiwake, guillemot, razorbill and puffin. *Seabirds and Sandeels: Proceedings of a Seminar held in Lerwick, Shetland, 15–16 October 1988* (ed. M. Heubeck), pp. 11–18. Shetland Bird Club, Lerwick.

Hislop, J.R.G., Harris, M.P. & Smith, J.G.M. (1991) Variation in the calorific value and total energy content of the lesser sandeel (*Ammodytes marinus*) and other fish preyed on by seabirds. *Journal of Zoology*, **224**, 501–517.

Iwao, S. (1968) A new regression method for analyzing the aggregation patttern of animal populations. *Researches on Population Ecology*, **14**, 97–128.

Kawasaki, T. (1980) Fundamental relations among the selections of life history in the marine teleosts. *Bulletin of the Japanese Society Scientific Fisheries*, **46**, 289–293.

Kunzlik, P.A. (1989) Small fish around Shetland. *Seabirds and Sandeels: Proceedings of a Seminar held in Lerwick, Shetland, 15–16 October 1988* (ed. M. Heubeck), pp. 38–49. Shetland Bird Club, Lerwick.

Langham, L.P.E. (1971) The distribution and abundance of larval sand-eels (*Ammodytidae*) in Scottish waters. *Journal of the Marine Biological Association, UK*, **51**, 697–707.

Lloyd (1967) Mean crowding. *Journal of Animal Ecology*, **36**, 1–30.

Lloyd, C.S., Tasker, M.L. & Partridge, K. (1991) *The Status of Seabirds in Britain and Ireland.* T and A.D. Poyser, London.

MacCall, A.D. (1990) *Dynamic Geography of Marine Fish Populations*, 153 pp. University of Washington Press, Seattle and London.

Martin, A.R. (1989) The diet of Atlantic puffin (*Fratercula arctica*) and northern gannet (*Sula bassana*) chicks at Shetland colony during a period of changing prey availability. *Bird Study*, **36**, 170–180.

Monaghan, P. (1992) Seabirds and sandeels: The conflict between exploitation and conservation in the northern North Sea. *Biodiversity and Conservation*, **1**, 98–111.

Monaghan, P., Uttley, J.D., Burns, M.D., Thaine, C. & Blackwood, J. (1989) The relationship between food supply, reproductive effort and breeding success in Arctic terns *Sterna paradisaea. Journal of Animal Ecology*, **58**, 261–274.

Monaghan, P., Wright, P.J., Bailey, M.C., Uttley, J.D., & Walton, P. (1994) The influence of changes in food abundance on diving and surface feeding seabirds. *Canadian Wildlife Services Occasional Series*. Canadian Wildlife Service, Ottawa.

Reay, P.J. (1970) Synopsis of biological data on north Atlantic sandeels of the genus *Ammodytes*. FAO Fisheries Synopsis No. 82.

Reeves, S.A. (1994) *Seasonal and annual variation in catchability of sandeels at Shetland*, 8 pp. ICES CM 1994/D:19.

Serebryakov, V.P. (1990) Population fecundity and reproductive capacity of some food fishes in relation to year class strength fluctuations. *Journal de Conseil Internationale Exploration de la Mer*, **47**(2), 267–272.

Sherman, K., Jones, C., Sullivan, L., Smith, W., Berrien, P. & Ejsymont, L. (1981) Congruent shifts on sand eel abundance in western and eastern North Atlantic ecosystems. *Nature*, **291**, 486–489.

Sparholt, H. (1990) An estimate of the total biomass of fish in the North Sea. *Journal du Conseil Internationale Exploration de la Mer*, **46**, 200–210.

Tasker, M.L., Webb, A., Hall, A.J., Pienkowski, M.W. & Langslow, D.R. (1987) *Seabirds in the North Sea*. Nature Conservancy Council, Peterborough.

Wright, P.J. (1993) Otolith microstructure of the lesser sandeel, *Ammodytes marinus. Journal of the Marine Biological Association, UK*, **73**, 245–248.

Wright, P.J. & Bailey, M. (1993) Biology of sandeels in the vicinity of seabird colonies at Shetland. *Fisheries Research Report No. 15/93.* SOAFD Marine Laboratory, Aberdeen.

CHAPTER 21

A review of seabird responses to natural or fisheries-induced changes in food supply

R.W. Furness

SUMMARY

(1) Seabirds are generally long-lived and breed colonially, mostly producing less than one fledgling per year that will not recruit until several years old.

(2) Seabird prey, such as small pelagic fish, cephalopods or zooplankton, generally exhibit short lifespans, early and highly variable recruitment.

(3) Seabird population sizes do not track short-term (annual) changes in prey population size, so that seabirds can be expected to have behavioural buffering mechanisms to cope with natural fluctuations in food supply. These may include prey switching, selecting high-density patches of prey, seasonal migration, flexible time budgets, brood reduction, flexible chick growth rates and periodic non-breeding.

(4) Fisheries may alter food supply to seabirds by reducing stock biomass, increasing variability of recruitment, by altering food-web structure and by making available to scavenging seabirds foods that they could not otherwise reach.

(5) Empirical evidence suggests that seabird numbers may be limited by food-related winter mortality and by food-related reproductive success. There is evidence for buffering against effects of food shortage, but buffering ability and behavioural responses of seabirds to food shortage vary among species.

(6) Although sensitive indicators of food shortage can be identified, it is extremely difficult to distinguish between fishery-induced and natural changes in terms of their effects on seabirds. Monitoring of seabird buffering capacity may be more useful than monitoring of seabird numbers or breeding success.

Key-words: ecosystem ecology, fisheries, monitoring, seabirds

INTRODUCTION

Seabirds might interact with fisheries in several ways that can be beneficial or detrimental to either party. In some parts of the world artesanal fisheries exploit seabirds as guides to where they should fish. In the Azores, tuna fishermen seek flocks of terns as these birds plunge-dive on shoals of small fish that are driven to the surface by foraging tuna and so indicate good places to fish tuna by baited line (Batty 1989). Fishermen broadcast a steady trickle of dead bait fish to attract tuna, and a considerable number of these fish are scavenged by terns, which tend to follow working tuna boats; both terns and fishermen appear to derive some benefit from this association.

Several fisheries may cause mortality of seabirds as a result of entanglement in nets, especially monofilament nylon drift nets (Strann *et al.* 1991), and on baited hooks, such as in some tuna or squid fisheries. Although seabirds generally consume fish that are smaller than those harvested in most fisheries, seabird predation may affect yield in some situations (Barrett *et al.* 1990). In this review I shall concentrate on two effects of fisheries on food supply to seabirds, especially in the context of the North Sea; the possible reduction in supply of small shoaling fish as a result of industrial fishing or changes in food-web structure and dynamics, and the provision to scavenging seabirds of offal and discards of large demersal fish that would not be available as a natural food supply (because large demersal fish live too deep to be accessible or are too big for deep-diving seabirds to swallow).

CHARACTERISTICS OF SEABIRD ECOLOGY

Seabirds occupy positions in the upper trophic levels of marine food webs. Details of the diet vary regionally, seasonally, among seabird species and according to prey abundance and availability (Hislop & Harris 1985, Martin 1989, Barrett & Furness 1990, Bailey *et al.* 1991). During the breeding season many seabirds feed predominantly on fish, especially small shoaling species of planktivores such as sandeels *Ammodytes* spp., sprats *Sprattus sprattus*, young herring *Clupea harengus*, pilchards *Sardinops* spp., or anchovies *Anchoveta* spp. (Harris & Hislop 1978, Furness & Monaghan 1987). These fish are especially favoured as food for chicks, since they are energy-rich and permit rapid chick growth (Furness & Hislop 1981, Anker-Nielsen & Lorentsen 1990, Klaassen *et al.* 1994). Species adapted to a diet of squid, such as many pelagic Procellariiformes, tend to have slower chick growth rates than species adapted to fish diets, and while species adapted to a diet of squid can raise chicks on fish, the reverse is not necessarily true (Prince & Morgan 1987).

Outside the breeding season, seabird diets appear to be more diverse than during chick-rearing, and often involve foraging at lower trophic levels (e.g. on zooplankton) not usually exploited during breeding (Furness & Monaghan 1987). While some species show a high degree of dietary specialization (for example, common guillemots *Uria aalge* and shags *Phalacrocorax aristotelis* feed predominantly on one to three species of small shoaling fish throughout the year (Blake 1983, Blake *et al.* 1985, Bradstreet & Brown 1985, Vader *et al.* 1990a,b, Durinck *et al.* 1991, Harris & Wanless 1991)), others may switch between different diets. For example, great skuas *Catharacta skua* feed their chicks in Shetland predominantly on sandeels, but switch to a diet of

discarded haddock *Melanogrammus aeglefinus*, whiting *Merlangius merlangus* and Norway pout *Trisopterus esmarkii* in late summer when sandeel availability decreases (Furness & Hislop 1981). However, the same species in the north-west of Scotland and in the Hebrides feeds predominantly on seabirds (auks, gulls, storm petrels) and on goose-barnacles *Lepas* spp. (Furness 1987). The diet of most seabird species has been studied during the breeding season, often by sampling food brought to chicks (Duffy & Jackson 1986), but diets of seabirds in winter are only rather poorly documented (Dunnet *et al.* 1990).

Although the biomass of seabirds in marine ecosystems may be lower than that of predatory fish or marine mammals, seabirds have much higher metabolic rates and so their consumption of marine foods is much higher on a mass-specific basis (Bennett & Harvey 1987, Gabrielsen *et al.* 1987, Nagy 1987, Birt-Friesen *et al.* 1989, Bryant & Furness 1995). The field metabolic rate of a seabird is approximately 50 times that of a benthic fish in the North Sea, for example. Furthermore, almost all seabirds are colonial breeders and so their consumption of marine prey is strongly concentrated around major breeding areas (Furness 1978, 1990, Cairns *et al.* 1986, Diamond *et al.* 1993). Even in winter, when most seabirds are dispersed from colonies, in many regions the bulk of the seabird community feeds in a coastal zone. Many potential food supplies are unavailable to seabirds as a consequence of distance offshore or depth distribution.

Seabirds are generally long-lived. The annual survival rate of adults is typically over 85% (Coulson & Horobin 1976, Dunnet *et al.* 1990). They usually start to breed only after several years of immaturity and produce few eggs per year, with single egg clutches in all Procellariiformes (Warham 1990) and most alcids, one or two eggs in most penguins, skuas and boobies, two or three eggs in most gulls and terns (Furness & Monaghan 1987). Seabirds also show very strong fidelity to their established nest site and generally show strong natal philopatry. Thus numbers breeding at colonies can usually change only slowly over time. Rapid changes in breeding numbers can occur in some species as a consequence of low natal and breeding site philopatry, as in pomarine skuas *Stercorarius pomarinus*, high rates of emigration under conditions of food shortage or disturbance, as seen in some terns, and as a consequence of high rates of non-breeding (Aebischer & Wanless 1992).

PREDICTED EFFECTS ON SEABIRDS OF FLUCTUATIONS IN FOOD SUPPLY

The life history characteristics of seabirds are important since they contrast with the population dynamics of their main prey. For example, small shoaling fish tend to be short-lived and fluctuate in abundance from year to year with highly variable recruitment (Sherman *et al.* 1981, Daan *et al.* 1990, Wright & Bailey 1993). Thus seabird food supply naturally fluctuates far more than seabird numbers. Seabird population sizes do not track annual changes in prey population sizes (Heubeck *et al.* 1991, Klomp & Furness 1992, Montevecchi 1993). Seabirds therefore require behavioural buffering mechanisms to cope with natural fluctuations in food supply. Fluctuations in food supply caused by fisheries are likely simply to exaggerate these natural variations.

Based on ecological theory, Cairns (1987, 1992a,b) suggested that only an extreme reduction in food supply would affect seabird population size, whereas moderate reductions would result in reduced breeding success and slower chick growth. Small reductions in food supply might only result in an increased foraging effort by adults, reallocating time spent 'off-duty' when food supply is good to foraging in order to maintain the level of chick provisioning and to maintain adult body condition (but see Drent & Daan 1980). Buffering mechanisms are likely to be many and diverse, including prey switching, selecting high-density patches of prey, seasonal migrations, flexible time budgets, brood reduction, flexible chick growth rates, and periodic non-breeding and emigration according to food abundance.

EMPIRICAL EVIDENCE FOR SEABIRD POPULATION LIMITATION BY FOOD

Birkhead & Furness (1985) reviewed evidence for population regulation in seabirds and concluded that food may limit numbers through effects on winter mortality or through effects on breeding production. Support for regulation through winter mortality in relation to prey abundance comes from observations that winter mortality of auks may correlate with prey fish stock biomass (Blake 1984, Vader *et al.* 1990a,b, Harris & Bailey 1992). Support for regulation through effects of food supply on breeding performance comes from empirical observations of increases in prey fish densities at increasing distance from colonies (Birt *et al.* 1987), observed reduced chick growth or breeding success in larger seabird colonies (Gaston *et al.* 1983, Birkhead & Furness 1985, Hunt *et al.* 1986), patterns

of colony sizes and distributions suggesting competition for food during breeding (Furness & Birkhead 1984, Cairns 1992b), and evidence not only of reduced breeding productivity when food is scarce (Anderson *et al.* 1982, Anker-Nielsen 1987, Monaghan *et al.* 1989, Baird 1990, Harris & Wanless 1990, Hatch & Sanger 1992, Noordhuis & Spaans 1992, Paterson *et al.* 1992) but also of increased adult mortality during the breeding season and as a function of food supply (Coulson *et al.* 1983, Hamer, *et al.* 1991).

EMPIRICAL EVIDENCE FOR BUFFERING AGAINST VARIABLE FOOD SUPPLY

Many studies of breeding seabirds have shown that populations (Klomp & Furness 1992) and individuals (Burger & Piatt 1990, Cairns *et al.* 1990, Hamer *et al.* 1991, Wanless & Harris 1992, Uttley *et al.* 1994) are behaviourally buffered against short-term reductions in food supply. However, theory (Furness & Ainley 1984) predicts, and observations concur, that buffering capabilities vary among species (Anderson & Gress 1984, Springer *et al.* 1986, Montevecchi, *et al.* 1988, Anderson 1989, Monaghan 1992). In particular, small seabirds seem to have to work close to their maximum capacity even when food supply is good, whereas large seabirds tend to have more flexibility in their time budget (Furness & Ainley 1984, Monaghan *et al.* 1989, 1994).

EMPIRICAL EVIDENCE FOR FISHERIES AFFECTING SEABIRDS

Seabirds in the North Sea may consume around 600 000 t of food per annum, including about 200 000 t of sandeels, 30 000 t of sprats and small herring, 109 000 t of discards and 71 000 t of offal (Anon. 1994). Consumption of sandeels is predominantly in summer and in the northwest North Sea so is spatially separated from the main industrial fishery (Anon. 1994). However, this large food requirement raises the possibility of competitive interactions with fisheries. Although it is clear that a reduced abundance of prey fish can reduce seabird breeding success, it is impossible, given the present state of knowledge, to suggest how much a stock must be reduced in order to exceed the buffering capacity of seabirds. Nor is it possible in most cases to discriminate between fisheries and natural phenomena in causing a short-term reduction in fish stock biomass. Recent research reviewed above suggests that winter survival rates of auks and breeding success of many seabirds, especially smaller surface-feeding species, may be reduced if fisheries cause a reduction in sandeel or sprat stocks.

The key question here is whether industrial fishing reduces food availability to seabirds. It has been suggested that sandeel recruitment is independent of sandeel spawning stock biomass, because recruitment is highly variable according to environmental conditions, being a function of larval survival and not of egg production (Wright, this volume). If so, then the removal of adult sandeels by the fishery would not affect abundance of 0-group sandeels on which seabirds could feed. However, the fact that a significant correlation between sandeel spawning stock biomass and recruitment cannot be demonstrated does not mean that no such relationship exists. It may simply indicate the poor quality of sandeel stock assessment data, and lack of knowledge of stock identity.

Even if industrial fishing did not affect sandeel recruitment, food supply to seabirds would be affected. Although sandeel fisheries do not remove 0-group fish until late summer, by which time seabird breeding seasons are nearly complete, seabirds that feed on sandeels in winter would have a reduced food supply, while the food supply for seabirds feeding on older sandeels would be more severely depleted. Consider a hypothetical stock of sandeels with (typical) natural mortality rates of 50% on the 0-group and 25% on all older age classes. To this we can add a new industrial fishery, causing fishing mortality of 0% on the 0-group but 25% on all older age classes. If recruitment to this stock is independent of spawning stock biomass and is held constant at ten million fish per year, then the numbers of fish in each age class before and after the establishment of the fishery are as in Fig. 21.1. The unfished stock has large numbers of fish more than six years old, but hardly any survive to this age in the fished stock. Thus for a seabird feeding on the stock, the effect of the fishery on food supply depends very much on the ages of sandeels taken. For a predator taking sandeels one year old and older the fishery reduces food supply by 50%, whereas a seabird taking only five year old and older sandeels would suffer a 90% reduction in food supply as a result of the fishery (Fig. 21.2). In practice, many seabirds feed predominantly on 1-group or older sandeels rather than on 0-group fish (Anon. 1994). Since the model illustrated in Figs 21.1 and 21.2 considers only fish numbers and not biomass, it is worth emphasizing that the differences in food supply to seabirds would be very much greater when expressed in terms of fish biomass, which is likely to be a more important measure for seabirds than numbers of

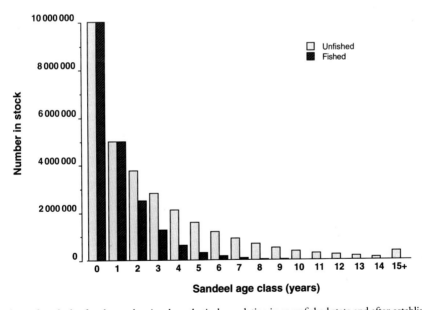

Fig. 21.1. Numbers of sandeels of each age class in a hypothetical population in an unfished state and after establishment of an industrial fishery. The population parameters are set (unfished) at: 10 000 recruits per year, natural mortality of 50% of 0-group and 25% of each older age group. When fished the natural mortality and recruitment remain unchanged, with a fishery mortality of 0% on the 0-group and 25% on all older age groups.

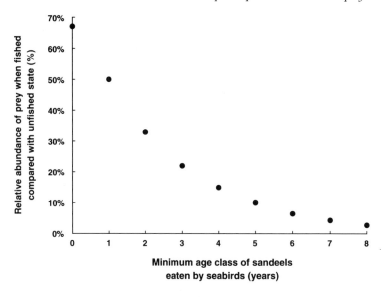

Fig. 21.2. Relative abundance of sandeels (expressed as numerical abundance in fished state as a percentage of numerical abundance in the unfished state) for seabirds feeding on sandeels of particular minimum ages. Sandeel stock parameters as set in model in Figure 21.1.

prey. Even a sustainable industrial fishery may therefore reduce food supply for seabirds, especially those depending on larger, so older, fish. Unfortunately, we know too little of the functional relationship between food abundance and seabird population dynamics to be able to assess how important changes in food availability of this magnitude may be.

Effects of providing additional food in the form of offal and discards are easy to see because this supply would be entirely unavailable under all forseeable natural conditions. Although about 1.2 million pairs of scavenging seabirds (fulmars, gannets, great skuas, great black-backed gulls, herring gulls, lesser black-backed gulls, kittiwakes) breed on North Sea coasts, many species increase in numbers in the North Sea in winter as migrants arrive from breeding areas further north (Furness 1992). Breeding numbers of scavenging seabirds on North Sea coasts have increased at least tenfold from 1900–1990 (Lloyd *et al.* 1991). Studies at sea show that there is strong competition among species for discards and offal provided by fishing vessels (Hudson & Furness 1988, 1989, Furness *et al.* 1992, Camphuysen *et al.* 1993, Garthe 1993, Hüppop & Garthe 1993) and that a very high proportion of all discharged offal and roundfish discards is consumed (Hudson & Furness 1989, Berghahn & Rösner 1992, Furness *et al.* 1992, Camphuysen *et al.* 1993). Almost all offal discharged experimentally was consumed, predominantly by fulmars in the northern North Sea and by gulls in the southern North Sea. Smallest discards were taken by kittiwakes, and largest by gannets. Fulmars had a low success rate in consuming discards, because they are not able to swallow

large fish easily, and smaller gulls lost many fish to kleptoparasitism by larger gulls, great skuas and gannets. Furness (1992) suggested that numbers of scavenging seabirds in the North Sea may now be limited by amounts of offal and discards made available, and that changes in fisheries' practices in future to reduce discarding could alter the competitive relationships among seabirds and lead to decreases in numbers. Similar important effects of discarding on seabird ecology in other areas have been documented by Blaber & Wassenberg (1989) in Australia, and by Thompson (1992) in the Falklands, where discards have become a major contribution to the diet of some seabirds.

Although reducing amounts discarded may be desirable both for better management of demersal fish stocks and to reduce impact on seabirds, the short-term effect of any reduction may be to cause declines in numbers of several seabird species, and not necessarily to return to the same population levels as before discarding began. There is, therefore, a need to monitor effects of changes in discarding practices on seabird communities, as well as to investigate the effects of fisheries on the abundance of prey fish taken by seabirds and the impact of these fisheries on seabird populations as a consequence of this altered food supply. Furness & Ainley (1984) identified terns and other small seabirds feeding several chicks as likely to be most sensitive to changes in food supply. Cairns (1987) suggested that behavioural buffering of adult seabirds would respond first to reduced food before effects became evident in chick growth or breeding success, with adult survival being rather insensitive to food supply. These predictions have generally been supported in case studies (Monaghan *et al.* 1989, Burger & Piatt 1990, Hamer *et al.* 1991, Klomp & Furness 1992, Monaghan *et al.* 1994, Uttley *et al.* 1994). Thus the monitoring of buffering capacity in populations of small seabirds such as terns and kittiwakes is likely to be more informative and provide an earlier warning of ecological change than the monitoring of seabird numbers. Nevertheless, there is a clear exception to this rule. Common guillemot numbers at colonies and winter survival seem to be especially closely related to the status of their fish prey (Vader *et al.* 1990a,b, Heubeck *et al.* 1991, Harris & Bailey 1992), perhaps because the common guillemot remains largely dependent on small shoaling fish in winter when many other seabirds expoit a wider range of foods.

REFERENCES

Aebischer, N.J. & Wanless, S. (1992) Relationships between colony size, adult non-breeding and environmental conditions for shags *Phalacrocorax aristotelis* on the Isle of May, Scotland. *Bird Study*, **39**, 43–52.

Anderson, D.J. (1989) Differential responses of boobies and other seabirds in the Galapagos to the 1986–87 El Niño–Southern Oscillation event. *Marine Ecology Progress Series*, **52**, 209–216.

Anderson, D.W. & Gress, F. (1984) Brown pelicans and the anchovy fishery off southern California. *Marine Birds, their Feeding Ecology and Commercial Fisheries Relationships* (eds, D.N. Nettleship, G.A. Sanger & P.F. Springer), pp. 128–135. Canadian Wildlife Service, Ottawa.

Anderson, D.W., Gress, F. & Mais, K.F. (1982) Brown pelicans: influence of food on reproduction. *Oikos*, **39**, 23–31.

Anker-Nielsen, T. (1987) The breeding performance of puffins *Fratercula arctica* on Rost, northern Norway in 1979–1985. *Fauna Norvegica Series C, Cinclus*, **10**, 21–38.

Anker-Nielsen, T. & Lorentsen, S.H. (1990) Distribution of puffins *Fratercula arctica* feeding off Rost, northern Norway, during the breeding season, in relation to chick growth, prey and oceanographic parameters. *Polar Research*, **8**, 67–76.

Anon. (1994) *Report of the Study Group on Seabird/Fish Interactions*. ICES CM 1994/L:3, 119 pp.

Bailey, R.S., Furness, R.W., Gauld, J.A. & Kunzlik, P.A. (1991) Recent changes in the population of the sandeel (*Ammodytes marinus* Raitt) at Shetland in relation to estimates of seabird predation. *ICES Marine Science Symposium*, **193**, 209–216.

Baird, P.H. (1990) Influence of abiotic factors and prey distribution on diet and reproductive success of three seabird species in Alaska. *Ornis Scandinavica*, **21**, 224–235.

Barrett, R.T. & Furness, R.W. (1990) The prey and diving depths of seabirds on Hornoy, north Norway after a decrease in Barents Sea capelin stocks. *Ornis Scandinavica*, **21**, 179–186.

Barrett, R.T., Rov, N., Loen, J. & Montevecchi, W.A. (1990) Diets of shags *Phalacrocorax aristotelis* and cormorants *P. carbo* in Norway and implications for gadoid stock recruitment. *Marine Ecology Progress Series*, **66**, 205–218.

Batty, L. (1989) Birds as monitors of marine environments. *Biologist*, **36**, 151–154.

Bennett, P.M. & Harvey, P.H. (1987) Active and resting metabolism in birds: allometry, phylogeny and ecology. *Journal of Zoology, London*, **213**, 327–363.

Berghahn, R. & Rösner, H.-U. (1992) A method to quantify feeding on discard from the shrimp fishery in the North Sea. *Netherlands Journal of Sea Research*, **28**, 347–350.

Birkhead, T.R. & Furness, R.W. (1985) Regulation of seabird populations. *Behavioural Ecology* (eds R.M. Sibly and R.H. Smith), pp. 145–167. Blackwell, Oxford.

Birt, V.L., Birt, T.P., Goulet, D., Cairns, D.K. & Montevecchi, W.A. (1987) Ashmole's halo: direct evidence for prey depletion by a seabird. *Marine Ecology Progress Series*, **40**, 205–208.

Birt-Friesen, V.L., Montevecchi, W.A., Cairns, D.K. & Macko, S.A. (1989) Activity-specific metabolic rates of free-living northern gannets and other seabirds. *Ecology*, **70**, 357–367.

Blaber, S.J.M. & Wassenberg, T.J. (1989) Feeding ecology of the piscivorous birds *Phalacrocorax varius*, *P. melanoleucos* and *Sterna bergii* in Moreton Bay, Australia: diets and dependence on trawler discards. *Marine Biology*, **101**, 1–10.

Blake, B.F. (1983) A comparative study of the diet of auks killed during an oil incident in the Skagerrak in January 1981. *Journal of Zoology, London*, **201**, 1–12.

Blake, B.F. (1984) Diet and fish stock availability as possible factors in the mass death of auks in the North Sea. *Journal of Experimental Marine Biology and Ecology*, **76**, 89–103.

Blake, B.F., Dixon, T.J., Jones, P.H. & Tasker, M.L. (1985) Seasonal changes in the feeding ecology of guillemots (*Uria aalge*) off north and east Scotland. *Estuarine Coastal and Shelf Science*, **20**, 559–568.

Bradstreet, M.S.W. & Brown, R.G.B. (1985) Feeding ecology of the Atlantic Alcidae. *The Atlantic Alcidae* (eds D.N. Nettleship and T.R. Birkhead), pp. 264–318. Academic Press, London.

Bryant, D.M. & Furness, R.W. (1995) Basal metabolic rates of North Atlantic seabirds. *Ibis*, **137**, 219–226.

Burger, A.E. & Piatt, J. (1990) Flexible time budgets in breeding common murres: buffers against variable prey abundance. *Studies in Avian Biology*, **14**, 71–83.

Cairns, D.K. (1987) Seabirds as indicators of marine food supplies. *Biological Oceanography*, **5**, 261–271.

Cairns, D.K. (1992a) Bridging the gap between ornithology and fisheries science: use of seabird data in stock assessment models. *Condor*, **94**, 811–824.

Cairns, D.K. (1992b) Population regulation of seabird colonies. *Current Ornithology*, **9**, 37–61.

Cairns, D.K., Montevecchi, W.A. & Birt, V.L. (1986) Energetics and prey consumption by seabirds breeding in Newfoundland. *Pacific Seabird Group Bulletin*, **13**, 102.

Cairns, D.K., Montevecchi, W.A., Birt-Friesen, V.L. & Macko, S.A. (1990) Energy expenditures, activity budgets and prey harvest of breeding common murres. *Studies in Avian Biology*, **14**, 84–92.

Camphuysen, C.J., Ensor, K., Furness, R.W., Garthe, S., Hüppop, O., Leaper, G., Offringa, H. & Tasker, M.L. (1993) *Seabirds Feeding on Discards in Winter in the North Sea*. NIOZ Rapport 1993–8. Netherlands Institute of Sea Research, Texel.

Coulson, J.C. & Horobin, J. (1976) The influence of age on the breeding biology and survival of the Arctic tern *Sterna paradisaea*. *Journal of Zoology, London*, **178**, 247–260.

Coulson, J.C., Monaghan, P., Butterfield, J., Duncan, N., Thomas, C. & Shedden, C. (1983) Seasonal changes in the herring gull in Britain: weight, moult and mortality. *Ardea*, **71**, 235–244.

Daan, N., Bromley, P.J., Hislop, J.R.G. & Nielsen, N.A. (1990) Ecology of North Sea fish. *Netherlands Journal of Sea Research*, **26**, 343–386.

Diamond, A.W., Gaston, A.J. & Brown, R.G.B. (1993) Studies of high latitude seabirds 3. A model of the energy demands of the seabirds of eastern and Arctic Canada. *Canadian Wildlife Services Occasional Paper No. 77*, pp. 1–39. Canadian Wildlife Services, Ottawa.

Drent, R.H. & Daan, S. (1980) The prudent parent: energetic adjustments in avian breeding. *Ardea*, **68**, 225–252.

Duffy, D.C. & Jackson, S. (1986) Diet studies of seabirds: a review of methods. *Colonial Waterbirds*, **9**, 1–17.

Dunnet, G.M., Furness, R.W., Tasker, M.L. & Becker, P.H. (1990) Seabird ecology in the North Sea. *Netherlands Journal of Sea Research*, **26**, 387–425.

Durinck, J., Skov, H. & Danielsen, F. (1991) Winter food of guillemots *Uria aalge* in the Skagerrak. *Dansk Ornitologisk Forening Tidsskrift*, **85**, 145–150.

Furness, R.W. (1978) Energy requirements of seabird communities: a bioenergetics model. *Journal of Animal Ecology*, **47**, 39–53.

Furness, R.W. (1987) *The Skuas*. T. and A.D. Poyser, Calton.

Furness, R.W. (1990) A preliminary assessment of the quantities of Shetland sandeels taken by seabirds, seals, predatory fish and the industrial fishery in 1981–83. *Ibis*, **132**, 205–217.

Furness, R.W. (1992) Implications of changes in net mesh size, fishing effort and minimum landing size regulations in the North Sea for seabird populations. *JNCC Report No. 133*, 72 pp. Joint Nature Conservation Committee, Peterborough.

Furness, R.W. & Ainley, D.G. (1984) *Threats to Seabirds Presented by Commercial Fisheries. Status and Conservation of the World's Seabirds* (eds J.P. Croxall, P.G.H. Evans & R.W. Schreiber), pp 701–708. International Council for Bird Preservation, Cambridge.

Furness, R.W. & Birkhead, T.R. (1984) Seabird colony distributions suggest competition for food supplies during the breeding season. *Nature*, **311**, 655–656.

Furness, R.W. & Hislop, J.R.G. (1981) Diets and feeding ecology of great skuas *Catharacta skua* during the breeding season in Shetland. *Journal of Zoology, London*, **195**, 1–23.

Furness, R.W. & Monaghan, P. (1987) *Seabird Ecology*. Blackie, Glasgow.

Furness, R.W., Ensor, K. & Hudson, A.V. (1992) The use of

fishery waste by gull populations around the British Isles. *Ardea*, **80**, 105–113.

Gabrielsen, G.W., Mehlum, F. & Nagy, K.A. (1987) Daily energy expenditure and energy utilization of free-ranging black-legged kittiwakes. *Condor*, **89**, 126–132.

Garthe, S. (1993) Quantifizierung von Abfall und Beifang der Fischerei in der südöstlichen Nordsee und deren Nutzung durch Seevögel. *Hamburger Avifauna Beitrung*, **25**, 125–237.

Gaston, A.J., Chapdelaine, G. & Noble, D.G. (1983) The growth of thick-billed murre chicks at colonies in Hudson Strait: inter- and intra-colony variation. *Canadian Journal of Zoology*, **61**, 2465–2475.

Hamer, K.C., Furness, R.W. & Caldow, R.W.G. (1991) The effects of changes in food availability on the breeding ecology of great skuas *Catharacta skua* in Shetland. *Journal of Zoology, London*, **223**, 175–188.

Harris, M.P. & Bailey, R.S. (1992) Mortality rates of puffin *Fratercula arctica* and guillemot *Uria aalge* and fish abundance in the North Sea. *Biological Conservation*, **60**, 39–46.

Harris, M.P. & Hislop, J.R.G. (1978) The food of young puffins (*Fratercula arctica*). *Journal of Zoology, London*, **185**, 213–236.

Harris, M.P. & Wanless, S. (1990) Breeding success of British kittiwakes *Rissa tridactyla in* 1986–88: evidence for changing conditions in the northern North Sea. *Journal of Applied Ecology*, **27**, 172–187.

Harris, M.P. & Wanless, S. (1991) The importance of the lesser sandeel *Ammodytes marinus* in the diet of the shag *Phalacrocorax aristotelis*. *Ornis Scandinavica*, **22**, 375–382.

Hatch, S.A. & Sanger, G.A. (1992) Puffins as predators on juvenile pollack and other forage fish in the Gulf of Alaska. *Marine Ecology Progress Series*, **80**, 1–14.

Heubeck, M., Harvey, P.W. & Okill, J.D. (1991) Changes in the Shetland guillemot *Uria aalge* population and the pattern of recoveries of ringed birds, 1959–1989. *Seabird*, **13**, 3–21.

Hislop, J.R.G. & Harris, M.P. (1985) Recent changes in the food of puffins *Fratercula arctica* on the Isle of May in relation to fish stocks. *Ibis*, **127**, 234–239.

Hudson, A.V. & Furness, R.W. (1988) Utilization of discarded fish by scavenging seabirds behind whitefish trawlers in Shetland. *Journal of Zoology, London*, **215**, 151–166.

Hudson, A.V. & Furness, R.W. (1989) The behaviour of seabirds foraging at fishing boats around Shetland. *Ibis*, **131**, 225–237.

Hunt, G.L., Eppley, Z.A. & Schneider, D.C. (1986) Reproductive performance of seabirds: the importance of population and colony size. *Auk*, **103**, 306–317.

Hüppop, O. & Garthe, S. (1993) Seabirds and fisheries in the southeastern North Sea. *Sula*, **7**, 9–14.

Klaassen, M., Habekotte, B., Schinkelshoek, P., Stienen, E. & Van Tienen, P. (1994) Influence of growth retardation on time budgets and energetics of Arctic tern *Sterna paradisaea* and common tern *S. hirundo* chicks. *Ibis*, **136**, 197–204.

Klomp, N.I. & Furness, R.W. (1992) Non-breeders as a buffer against environmental stress: declines in numbers of great skuas on Foula, Shetland, and prediction of future recruitment. *Journal of Applied Ecology*, **29**, 341–348.

Lloyd, C.S., Tasker, M.L. & Partridge, K. (1991) *The Status of*

Seabirds in Britain and Ireland. T. and A.D. Poyser, London.

Martin, A.R. (1989) The diet of Atlantic puffin (*Fratercula arctica*) and northern gannet (*Sula bassana*) chicks at a Shetland colony during a period of changing prey availability. *Bird Study*, **36**, 170–180.

Monaghan, P. (1992) Seabirds and sandeels: the conflict between exploitation and conservation in the northern North Sea. *Biodiversity and Conservation*, **1**, 98–111.

Monaghan, P., Uttley, J.D., Burns, M.D., Thaine, C. & Blackwood, J. (1989) The relationship between food supply, reproductive effort and breeding success in Arctic terns *Sterna paradisaea*, *Journal of Animal Ecology*, **58**, 261–274

Monaghan, P., Uttley, J.D. & Burns, M.D. (1992) Effects of changes in food availability on reproductive effort in Arctic terns. *Ardea*, **80**, 71–81.

Monaghan, P., Walton, P., Wanless, S., Uttley, J.D. & Burns, M.D. (1994) Effects of prey abundance on the foraging behaviour, diving efficiency and time allocation of breeding guillemots *Uria aalge*. *Ibis*, **136**, 214–222.

Montevecchi, W.A. (1993) Seabirds as monitors of fish stocks. *Birds as Monitors of Environmental Change* (eds R.W. Furness & J.J.D. Greenwood), pp. 217–266. Chapman and Hall, London.

Montevecchi, W.A., Birt, V.L. & Cairns, D.K. (1988) Dietary changes of seabirds associated with local fisheries failures. *Biological Oceanography*, **5**, 153–161.

Nagy, K.A. (1987) Field metabolic rate and food requirement scaling in mammals and birds. *Ecological Monographs*, **57**, 111–128.

Noordhuis, R. & Spaans, A.L. (1992) Interspecific competition for food between herring *Larus argentatus* and lesser black-backed gulls *L. fuscus* in the Dutch Wadden Sea area. *Ardea*, **80**, 115-132.

Paterson, A.M., Martinez, V.A. & Dies, J.I. (1992). Partial breeding failure of Audouin's gull in two Spanish colonies in 1991. *British Birds*, **85**, 97–100.

Prince, P.A. & Morgan, R.A. (1987) Diet and feeding ecology of Procellariiformes. *Seabirds: Feeding Ecology and Role in Marine Ecosystems* (ed. J.P. Croxall), pp. 135–172.

Cambridge University Press, Cambridge.

Sherman, K., Jones, C., Sullivan, L., Smith, W., Berrien, P. & Ejsymont, L. (1981) Congruent shifts in sandeel abundance in western and eastern North Atlantic ecosystems. *Nature*, **291**, 486–489.

Springer, A.M., Roseneau, D.G., Lloyd, D.S., McRoy, C.P. & Murphy, E.C. (1986) Seabird responses to fluctuating prey availability in the eastern Bering Sea. *Marine Ecology Progress Series*, **32**, 1–12.

Strann, K.-B., Vader, W. & Barrett, R.T. (1991) Auk mortality in fishing nets in north Norway. *Seabird*, **13**, 22–29.

Thompson, K.R. (1992) Quantitative analysis of the use of discards from squid trawlers by black-browed albatrosses *Diomedea melanophris* in the vicinity of the Falkland Islands. *Ibis*, **134**, 11–21.

Uttley, J.D., Walton, P., Monaghan, P. & Austin, G. (1994) The effects of food abundance on breeding performance and adult time budgets of guillemots *Uria aalge*. *Ibis*, **136**, 205–213.

Vader, W., Barrett, R.T., Erikstad, K.E. & Strann, K.-B. (1990a) Differential responses of common and thick-billed murres to a crash in the capelin stock in the southern Barents Sea. *Studies in Avian Biology*, **14**, 175–180.

Vader, W., Anker-Nilssen, T., Bakken, V., Barrett, R.T. & Strann, K.-B. (1990b) Regional & temporal differences in breeding success and population development of fisheating seabirds in Norway after collapses of herring and capelin stocks. *Transactions 19th International Union of Game Biologists Congress (Trondheim) 1989*, pp. 143–150. International Union of Game Biologists, Oslo.

Wanless, S. & Harris, M.P. (1992) Activity budgets, diet and breeding success of kittiwakes *Rissa tridactyla* on the Isle of May. *Bird Study*, **39**, 145–154.

Warham, J. (1990) *The Petrels: their Ecology and Breeding Systems*. Academic Press, London.

Wright, P.J. & Bailey, R.S. (1993) Biology of Sandeels in the Vicinity of Seabird Colonies at Shetland. *Fisheries Research Services Report 14/93*. SOAFD Marine Laboratory, Aberdeen.

CHAPTER 22

Manipulating the fish–zooplankton interaction in shallow lakes: a tool for restoration

G.L. Phillips, M.R. Perrow and J. Stansfield

SUMMARY

(1) The restoration of eutrophic lakes remains a major problem and there is a growing awareness that 'top-down' processes such as fish–zooplankton–phytoplankton interactions are an important factor that need to be taken into consideration.

(2) Grazing by large cladoceran zooplankton such as *Daphnia hyalina* L. can reduce chlorophyll *a* concentrations. However predation by planktivorous fish during the summer prevents such populations from developing in most of the Norfolk Broads.

(3) These fish populations have been artificially removed from some of the broads and the resulting increase in zooplankton grazing has reduced the chlorophyll *a* concentration, creating clear water and enabling aquatic vegetation to establish in these shallow lakes.

(4) In unmanipulated broads there is a significant positive correlation between large grazing zooplankton and macrophyte biomass. In addition the least number of fish were found associated with dense macrophytes, corresponding to extremely low predation pressure, and it is clear that macrophytes provide a refuge for grazing zooplankton and may thus assist in the maintenance of low algal standing crops.

(5) Macrophyte dominated lakes also have a different fish population structure. Although little evidence is as yet available from the Broads, initial observations confirm this view with macrophyte dominated broads having a greater proportion of pike to prey species than plankton dominated lakes.

(6) Data from the Broads are similar to published relationships between total phosphorus concentrations and chlorophyll *a*, although macrophyte and plankton dominated broads have distinct regression lines. Biomanipulated lakes fall between these two groups, and suggest that within a range of phosphorus concentrations either plankton or macrophyte states can form stable communities. This factor needs to be included in the planning of restoration programmes.

Key-words: biomanipulation, eutrophication, lake restoration, Norfolk Broads, plankton

INTRODUCTION

The control of eutrophication and its undesirable effects such as excessive phytoplankton growth or the production of toxic cyanobacteria blooms has occupied the minds of limnologists for more than four decades. Despite this, many standing water bodies still exhibit excessive growth of algae causing problems to a variety of water users. Eutrophication is generally accepted as resulting from the enrichment of water by plant nutrients, particularly nitrogen and phosphorus and many quantitative relationships between various surrogates of nutrient supply and algal biomass have been published (Lund 1970, Dillon & Rigler 1974, Vollenweider & Kerekes 1980). These have reinforced the traditional concept that nutrient supply ultimately controls the biomass of higher trophic levels and restoration has thus focused on reducing nutrient, particularly phosphorus, inputs. Although there have been several successes, many lakes show considerable resilience to change (Sas 1989, Jeppesen *et al.* 1990) and there is growing evidence that the biological community structure of the lake must be taken into account when attempting to manage these lakes (Edmondson 1991, Carpenter & Kitchell 1993, Phillips & Moss 1994).

These ideas emerge from the early work of Brooks & Dodson (1965) on size selective predation by fish on zooplankton and the seminal observations of Hrbáček *et al.* (1961) who noted that the greatest concentration of algae was found in fish ponds where abundant fish stocks had eliminated large-bodied daphnids. A large number of observations and experimental work have subsequently investigated the importance of these 'top-down' influences on lakes, and the relative importance of this, in contrast to the more traditional bottom-up approach, has been widely discussed (McQueen 1990, Carpenter & Kitchell 1992, DeMelo *et al.* 1992).

The usefulness of this to lake managers is the potential of reducing phytoplankton by removing planktivorous fish to create large populations of large-bodied grazing cladoceran zooplankton (Lammens *et al.* 1990). This approach is particularly important in shallow lakes, such as the Norfolk Broads, where interactions between the plankton and benthic or littoral biotic communities are much more significant than in deeper lakes (Mortensen *et al.* 1994).

The Norfolk Broads

The Norfolk Broads are a series of small, very shallow lakes formed from flooded medieval peat diggings. They are interconnected by tidal freshwater rivers and the whole system forms one of the United Kingdom's most extensive and important wetland areas. Until the mid-1960s these lakes contained relatively clear water, a variety of submerged aquatic plants and a productive and well renowned recreational fishery (George 1992). However, in recent years the lakes have become dominated by phytoplankton, the aquatic vegetation has largely vanished and the fishery has become characterized by numerous small fish, typically roach *Rutilus*

rutilus (L.) supplemented by large bream *Abramis brama* (L.) (Moss *et al.* 1979).

Restoration of the Norfolk Broads began in 1980 with the removal of phosphorus from sewage effluents discharging to the River Ant and ultimately to Barton Broad. Despite a 90% reduction in the amount of phosphorus discharged to the River Ant upstream of Barton Broad the lake remained dominated by phytoplankton (Phillips 1984) and it has become clear that to restore these lakes additional steps, such as enhancing zooplankton grazing, will be required (Madgwick & Phillips 1992).

Grazing by large bodied cladoceran zooplankton can significantly reduce the phytoplankton standing crop in the Broads (Phillips & Kerrison 1991) and the effect of zooplanktivorous fish on reducing these zooplankton is well documented (Cryer *et al.* 1986, Townsend & Perrow 1989, Perrow & Irvine 1992, Perrow *et al.* 1994), suggesting a potential direct 'top-down' control in these lakes. The effect was demonstrated using small mesh cages, which protected zooplankton from fish predation, placed in Barton Broad (Phillips & Kerrison 1991). Inside the cages substantial populations of large cladocerans developed and despite free water movement into the cage clear water was produced allowing aquatic vegetation to flourish. These observations led to larger scale experimental manipulations (Moss *et al.* 1986, in press) and a need to understand how the changes in the fish–zooplankton–phytoplankton community might be made stable. In this paper we show how manipulating fish populations in a series of very shallow lakes in Eastern England has been used as a restoration tool and demonstrate how submerged macrophytic vegetation may affect the zooplankton community structure of these lakes.

MATERIALS AND METHODS

Details of lakes mentioned, together with any experimental manipulations, are given in Table 22.1. Water for chemical analysis and zooplankton samples from three sites, Upton Broad, Hoveton Little Broad and a fish exclosure in an isolated bay of Hoveton Little Broad (Pound End), (and sites for which data are shown in Figure 22.6) were collected at two weekly intervals from March to October and monthly during the winter. Details of chemical methods are given in Phillips & Kerrison (1991). Routine zooplankton samples were collected using a plastic tube which sampled the entire water column. Five seven litre samples were filtered through a 64 µm mesh net to produce a single bulked sample.

Parameters of the fish population in Alderfen, Upton,

Table 22.1. Details of experimental and reference Broads.

Site	Area (ha)	Status	Restoration measures	Further details
Cockshoot Broad	5.5	Biomanipulated, some macrophytes since 1992	Isolated from river 1980, sediment removed 1980, annual fish removal since 1989	Moss *et al.* (1986) Moss *et al.* in press
Cromes Broad	2.0	Unstable macrophytes	Partial sediment removal prior 1982, isolated from river 1992	
Hoverton Little Broad	12.0	Algal dominated	Sediment removal	
Hoverton Little Broad Fish exclosure (Pound End)	5.0	Biomanipulated	Sediment removal 1989/90, fish barrier 1990, annual fish removal since 1990	
Upton Broad	6.9	Macrophyte dominated	None (pristine site)	
Alderfen Broad	4.7	Unstable macrophytes	Inflow stream diverted 1979, fish kill 1990, annual fish removal since 1993	Moss *et al.* (1986) Perrow *et al.* (1994)
Barton Broad	77.3	Algal dominated	Phosphorus reduction from effluents since 1980	Phillips (1984)
Wroxham Broad	34.4	Algal dominated	Phosphorus reduction from effluents since 1986	
Salhouse Broad	12.7	Algal dominated	Phosphorus reduction from effluents since 1986	

Cromes and Cockshoot Broads and the fish exclosure in Hoveton Little Broad (Pound End), were derived from electro-fishing surveys. Descriptions of the techniques used are given in Perrow *et al.* (1990) and Perrow (1994a). Fish removals from Cockshoot and the fish exclosure in Hoveton Little Broad were carried out using non-destructive techniques, mainly electro-fishing carried out during the winter and spring spawning period. Full details of the techniques used are provided by Perrow (1994b).

On one occasion during August 1993 fish, zooplankton and macrophytes were sampled in Cromes Broad to determine their distribution in relation to each other. One hundred sample points were selected in the Broad covering four habitat types, open water, dense weeds, sparse weeds and the edge of the lake. At each sample point the fish population was determined using point-sample electro-fishing (Perrow 1994a). A sample of fish collected was returned to the laboratory and dietary items were determined by dissection of their stomach or foregut. At a sub-set of 40 sites a single seven litre zooplankton sample was obtained and the biomass of aquatic vegetation (mainly *Ceratophyllum demersum* L.) estimated from samples collected using a 50 × 50 cm box quadrat.

RESULTS

Seasonal pattern of zooplankton and phytoplankton

The seasonal patterns of zooplankton and chlorophyll *a* in three contrasting sites are shown in Figure 22.1. Hoveton Little Broad, a eutrophic Broad with no aquatic vegetation, is directly connected to the River Bure by a navigable channel and is likely to contain a fish population similar to the river, which is dominated by small roach and bream with a biomass of between 100 and $200 \, \text{kg} \cdot \text{ha}^{-1}$ (National Rivers Authority, unpublished data). The seasonal pattern of zooplankton in the lake is typical of many of the eutrophic broads, with a substantial late spring peak of the small-bodied cladoceran *Bosmina longirostris* Müller, together with a smaller, but significant peak of the larger *D. hyalina*. The *Daphnia* population then declined leaving the zooplankton dominated by *B. longirostris* for the remainder of the summer. Chlorophyll *a* concentration in Hoveton Little Broad increased during the spring, showed a marked dip in May during the cladoceran peak, and then rose again to a maximum in summer when the grazing pressure of large bodied cladocerans was low.

Pound End is a small bay of Hoveton Little Broad which was isolated from the main basin of the lake by a

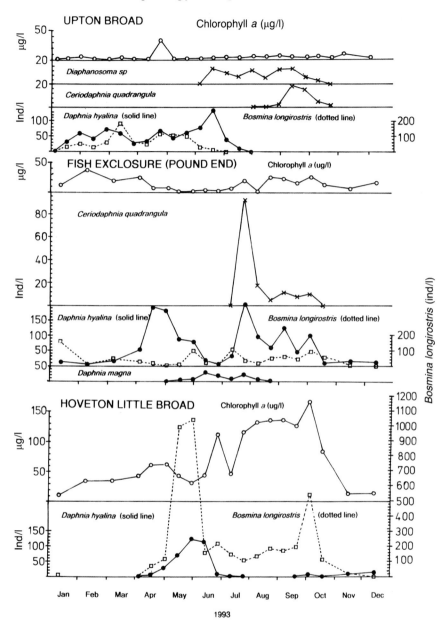

Fig. 22.1. Chlorophyll *a* concentration (open circles) and number of cladoceran zooplankton (*Daphnia* spp. solid circles, solid line; *Bosmina longirostris* open squares, dotted line; other species crosses, solid line) in three contrasting sites, Upton Broad (macrophyte dominated), a fish exclosure in a bay of Hoveton Little Broad (Pound End) and the main basin of Hoveton Little Broad.

fish barrier (constructed from plastic boards with 1 mm steel mesh windows) in 1990. Problems with construction of the barrier allowed fish to enter the lake, but by 1993 the barrier was improved and 719 kg of fish were removed ($175 \, kg \cdot ha^{-1}$) from the bay between April and June, mostly large bream. Fish removal was thought to have reduced the fish population by at least 90% and point electro-fishing surveys conducted the following

May (1994) revealed that the biomass of planktivorous
fish (mainly roach and bream) were approximately
12.5 kg·ha^{-1}, confirming that the fish removal carried
out in 1993 had reduced the populations of these
species substantially. At this site *D. hyalina* dominated
the zooplankton for most of the growing season. *B.
longirostris* was present, but at much lower numbers
and between June and August the much larger cladoceran *Daphnia magna* Straus was found, together with
Ceriodaphnia quadrangula Müller. In contrast to Hoveton
Little Broad, chlorophyll *a* concentration in the isolated
bay remained relatively low, despite tidal water ex-
change with the main lake basin through the mesh
windows of the fish barrier.

Upton Broad is an isolated lake with a stable
macrophyte population (dominated by *Najas marina* L.)
and is generally considered to have been unaffected by
enrichment. Electro-fishing surveys carried out during
1993 and 1994 did not catch any roach or bream and
showed the fish population was dominated by pike
Esox lucius (L.) (51.6 kg·ha^{-1}), tench *Tinca tinca* (L.)
(15.2 kg·ha^{-1}) and perch *Perca fluviatilis* (L.) (1.5 kg·
ha^{-1}). Here the cladoceran zooplankton consisted of
D. hyalina and *B. longirostris* from early February until
July, when macrophytes became abundant. The *D.
hyalina* was then replaced by *Diaphanosoma* sp. and
C. quadrangula, both species typical of macrophyte-
dominated lakes. With the exception of a very small
peak in late April the chlorophyll *a* concentration in
Upton Broad remained low throughout the year.

Biomanipulation of Cockshoot Broad

Further evidence for the role of fish–zooplankton–
algal interactions came from observations made at
Cockshoot Broad. This lake was isolated from the
River Bure in 1981 to reduce the input of nutrient rich
water and had the top 0.5 m of its sediment removed
by suction dredging. Prior to 1982 typical summer
chlorophyll *a* concentrations were 100 μg·l^{-1} (Moss *et
al.* 1986), but following isolation these were reduced to
between 15 and 30 μg·l^{-1} (Fig. 22.2). This reduction
was initially thought to have been a direct response to
the reduced phosphorus concentration in the lake, as
enriched river water no longer entered the lake (Moss *et
al.* 1986). However, between 1984 and 1987 chlorophyll
a increased again and was mirrored by a decline in the
D. hyalina population and a shift in the cladoceran
community from dominance by *Daphnia pulex* L. in
1988 through *D. hyalina* to *B. longirostris* by 1987
(Perrow 1992, Perrow & Irvine 1992). The explanation
for these events was that the isolation and mud-pumping
operation in this lake had effectively removed the fish
population in 1981 allowing *Daphnia* to develop and
graze the phytoplankton. The combined effect of
reduced phosphorus supply and grazing reduced the
phytoplankton standing crop. Subsequently, successful
annual recruitment of fish re-established this population, the *Daphnia* declined and chlorophyll *a* began to
increase again (Moss *et al.*, in press). As a consequence
in the winter of 1988/89 a regular fish removal operation
was initiated in the lake with the result that *Daphnia*
numbers began to increase and chlorophyll *a* concentration decreased. The only exception to this was in

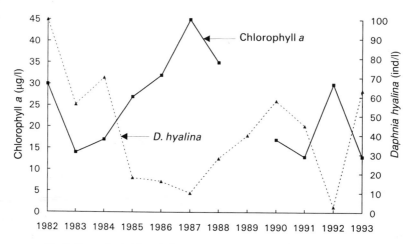

Fig. 22.2. Mean summer (April–September) chlorophyll *a* concentration (square symbols, solid line) and *Daphnia hyalina*
population (triangular symbols, dotted line) in Cockshoot Broad 1982–1993.

1992 when a mysid shrimp *Neomysis integer* (Leach) was found in significant numbers in the broad. This shrimp which can feed on *Daphnia* (Irvine *et al.* 1993) is normally associated with brackish water and apparently occurred in Cockshoot as a result of abnormal surge tides which enabled brackish estuary water to enter the lake during periods of low fluvial flow in the tidal River Bure. Since 1993 very few *N. integer* have been found and the lake has retained very low phytoplankton populations. The result of this has been that by the summer of 1994 a substantial and diverse submerged aquatic plant population had become established.

Importance of macrophytes and fish populations

Clearly the long-term maintenance of low fish populations in lakes is ultimately not desirable. However, once aquatic vegetation is established the habitat structure that this provides may help in restructuring the aquatic community by providing shelter for zooplankton (Timms & Moss 1984). Unfortunately there are few sites in the Broads where the relationships between natural fish populations, zooplankton and macrophytes can be determined. However, data collected from Cromes Broad, one of the few lakes in the Broads containing significant macrophyte beds, demonstrate that most of the large grazing cladocerans were significantly correlated with plant biomass (Table 22.2), suggesting that in a lake where the fish population has not been reduced the vegetation provides some advantage to their survival. In addition the fish population in this lake was not distributed evenly amongst the various habitats (Fig. 22.3). Very few fish were found in dense weed beds, with the majority of roach associated with either the open water or sparse weed beds. In contrast pike were clearly associated with the edge of the lake, a habitat where no roach were caught.

To estimate the relative fish predation pressure

Table 22.2. Relationship between the numbers of zooplankton (num $\cdot$ l^{-1}) and biomass (g $\cdot$ m^{-2}) of macrophytes. Spearman Rank Correlation coefficients (r_s) are shown with associated probabilities: *, p < 0.05; **, p < 0.01; ***, p < 0.001; ns, no significant correlation.

Prey taxon	r_s	Probability
Grazers		
Daphnia hyalina	+0.51	**
Ceriodaphnia quadrangula	+0.40	*
Simocephalus vetulus	+0.67	***
Sida crystallina	+0.56	**
Scrapers		
Eurycercus lamellatus	+0.57	***
Alona spp.	+0.37	*
Pleuroxus/Leydigia spp.	+0.65	***
Chydorus sphaericus	+0.54	**
Other Cladocera		
Bosmina longirostris	−0.50	**
Polyphemus pediculatus	+0.31	ns
Copepoda		
Adult cyclopoids	+0.72	***
Adult calanoids	+0.41	*
Cyclopoid copepodites	+0.48	**
Calanoid copepodites	+0.17	ns

on the zooplankton in each of these habitats a point predation pressure was derived, by combining fish density with the mean number of zooplankters consumed per species and size class of fish, estimated from gut contents. Maximum predation pressure occurred among the sparse weed beds where the greater density of both predator and prey combine, while the lowest predation pressure is seen in the dense weed beds, where zooplanktivorous fish density is low despite high prey numbers (Fig. 22.4).

The dominance of macrophytes in clear water lakes

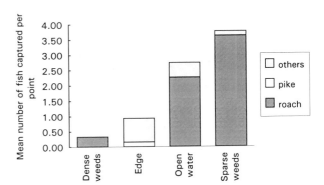

Fig. 22.3. Fish community structure in the various habitats of Cromes Broad (1993), shown by the mean number of fish captured per point.

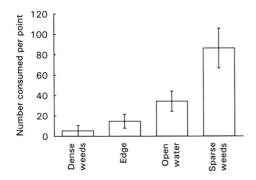

Fig. 22.4. Index of point predation pressure (mean ± SE) on zooplankton in different habitats in Cromes Broad (1993).

also directly influences fish community structure. In particular the number of pike, a visual ambush piscivore, may be much greater in these lakes. Actual predation pressure upon small fish, combined with the greater invertebrate food diversity associated with the macrophytes available to larger fish, may reduce the predation pressure on the zooplankton. The relative proportions of pike and prey fish in a number of broads are shown in Figure 22.5. Very few pike appear to be present in the eutrophic algal dominated broads. Their populations are much greater in Cromes and Alderfen Broads, two sites where macrophytes are returning, while at Upton Broad pike clearly dominate the lake.

DISCUSSION

Hoveton Little and Upton Broads provide good examples of the contrasts seen in the plankton of the Broads. In Upton Broad moderate populations of

D. hyalina occur from early January until July when macrophytes become well established in the lake and weed associated cladoceran species are found, while in Hoveton Little Broad *D. hyalina* only occurs for a short period when it is associated with a spring clear water phase. It is clear that planktivorous fish in open water environments such as Hoveton Little Broad rapidly reduce the populations of these larger cladocerans, leaving the zooplankton dominated by smaller species such as *B. longirostris* during the summer (Cryer *et al.* 1986, Townsend & Perrow 1989, Perrow & Irvine 1992). In contrast, in the absence of planktivorous fish, as at Upton Broad or in the fish exclosure, large cladoceran species predominate throughout the year.

Large cladocerans, such as *Daphnia* species, graze more efficiently than small species (Burns 1969, Mourelatos & Lacroix 1990) and there is little doubt that cladocerans such as *D. hyalina* can exert a considerable grazing pressure on the phytoplankton in the Broads. The importance of these processes relative to the nutrient supply in determining the standing crop of phytoplankton is still open to debate (Carpenter & Kitchell 1992, DeMelo *et al.* 1992). It is often argued that factors such as differential population turnover times will prevent top-down control of phytoplankton growth from being more than a temporary event, typically seen during the spring (Reynolds 1994). Such a clear water phase was seen in Hoveton Little Broad, but when planktivorous fish are removed from these lakes, as at Cockshoot Broad and in the fish exclosure at Hoveton Little Broad, it is clear that despite zooplankton population cycles low chlorophyll concentrations can be maintained. This may in part be a direct result of the very shallow nature of these lakes, as some larger cladocerans such as *D. magna* may be able to switch to sediment detritus as an alternative food source when phytoplankton populations are reduced (Gulati 1990).

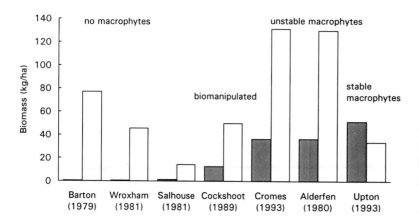

Fig. 22.5. Biomass of pike *Esox lucius* (hatched bar) and biomass of prey (open bar) populations at Broadland sites. Data from Upton, Cromes and Cockshoot (Perrow 1994a), Alderfen (Peirson 1983), all others from NRA reports.

Whatever the explanation, this form of biomanipulation appears to be a practical way of creating clear water conditions provided the control of the fish population is continued.

Although possible, it is not desirable to have to continually maintain artificially low planktivorous fish populations. In many other European lakes biomanipulation carried out on only one or two occasions has resulted in a shift in the community structure of the lake that has been maintained for a number of years (Shapiro 1990, Phillips & Moss 1994). In almost all of these cases this was associated with the rapid development of aquatic vegetation. The mechanism through which this acts is still unclear, but two factors are thought to play a major role. The vegetation may reduce the availability of nutrients for algal growth (Ozimek *et al.* 1990), either through a temporary nutrient sink by uptake into plant tissue or perhaps by the promotion of denitrification (Meijer *et al.* 1990). Alternatively, the physical nature of the plants may provide a refuge for grazing zooplankton. Timms & Moss (1984) demonstrated the effectiveness of dense plant beds for sheltering zooplankton and our data support this view. In broads containing unmanipulated fish populations, large cladocerans are positively correlated with weed density. Fish appear to be associated with sparse rather than dense weed beds, perhaps because the dominant cyprinids forage less effectively among dense plants (Engel 1988) and we conclude that macrophytes provide a refuge for zooplankton.

The importance of piscivorous fish in determining lake structure is still very unclear. It is an extremely difficult issue to test experimentally due to effects of scale. In shallow lakes in the Netherlands, Lammens (1986) has shown that bream and roach dominate lakes without vegetation, while those with rich macrophyte stands contain a more diverse community including tench, rudd *Scardinius erythrophthalmus* (L.), perch and

pike. Jeppesen *et al.* (1990) observed a similar shift, from a system dominated by pike and perch in low nutrient lakes containing macrophytes, to one exclusively dominated by planktivorous–benthivorous fish at higher nutrient levels, where phytoplankton were abundant, adding further evidence to the link between fish populations and lake community structure. Grimm & Backx (1990) demonstrated a relationship between plant cover and pike populations and suggested that in macrophyte dominated lakes pike populations greater than $100\,kg \cdot ha^{-1}$ could control cyprinid fish populations provided nutrient enrichment was less than $400\,\mu g \cdot l^{-1}$. There is insufficient data at present to derive firm conclusions for the Broads, but our initial observations support the view that high pike biomass in relation to planktivorous fish is clearly linked to macrophyte dominated lakes and the restructuring of the fish community is likely to be an essential final step in the permanent restoration of these lakes.

The published relationships (e.g. Dillon & Rigler 1974) between algal biomass and nutrient supply tend to suggest that all lakes follow the same general curve. The relationship between summer chlorophyll *a* and total phosphorus for a number of broads is shown in Figure 22.6. Taken as a group these data are similar to those of Dillon & Rigler (1974). However, when different lakes are identified it is clear that they fall into two groups. Those that are still dominated by phytoplankton are quite distinct from the macrophyte dominated Upton Broad. The slopes of the fitted regression lines are very similar although the intercepts are distinct. It is interesting to note that many of the biomanipulated sites lie close to an extrapolated regression line fitted to the Upton Broad data, suggesting that in the longer term, once macrophytes are well established, these lakes may become stable systems which will not require further artificial intervention. If this is the case then within a range of total phosphorus concentrations more

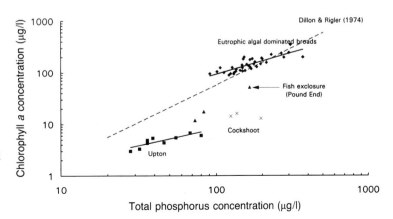

Fig. 22.6. Relationship between mean summer (April–September) chlorophyll *a* and total phosphorus concentration in a variety of broads, contrasted with regression line published by Dillon & Rigler (1974).

than one stable ecosystem state may be able to exist. These ideas have their roots in ecological theory (May 1974) and have recently been applied to the concept of biomanipulation (Scheffer *et al.* 1993). Our data support this view and imply that the simple reduction of nutrient supply to a lake will not always bring about the desired reduction in algal growth unless the food webs of the lake are able to restructure. Biomanipulation may be a way to assist this process and we suggest that for shallow lakes it may be an essential step in their restoration.

ACKNOWLEDGEMENTS

This work was carried out as part of an investigation into the restoration of the Norfolk Broads funded by the European Commission (LIFE 92-3/UK/031), National Rivers Authority, Broads Authority, Soap and Detergent Industry Association, and English Nature. The views expressed are those of the authors and not necessarily those of the above organizations.

REFERENCES

Brooks, J.L. & Dodson, S.I. (1965) Predation, body size and composition of the phytoplankton. *Science*, **150**, 28–35.

Burns, C.W. (1969) Relation between filtering rate, temperature and body size in four species of *Daphnia*. *Limnology and Oceanography*, **14**, 696–700.

Carpenter, S.R. & Kitchell, J.F. (1992) Trophic cascade and biomanipulation: interface of research and management – a reply to the comment by DeMelo *et al. Limnology and Oceanography*, **37**, 208–213.

Carpenter, S.R. & Kitchell, J.F. (1993) *The Trophic Cascade in Lakes*. Cambridge University Press, Cambridge.

Cryer, M., Peirson, G. & Townsend, C.R. (1986) Reciprocal interactions between roach, *Rutilus rutilus*, and zooplankton in a small lake: prey dynamics and fish growth and recruitment. *Limnology and Oceanography*, **31**, 1022–1038.

DeMelo, R., France, R. & McQueen, D.J. (1992) Biomanipulation: hit or myth? *Limnology and Oceanography*, **37**, 192–207.

Dillon, P.J. & Rigler, F.H. (1974) The phosphorus–chlorophyll relationship in lakes. *Limnology and Oceanography*, **19**, 767–773.

Edmondson, W.T. (1991) *The Uses of Ecology. Lake Washington and Beyond*. University of Washington Press, Seattle.

Engel, S. (1988) The role and interactions of submerged macrophytes in a shallow Wisconsin Lake. *Journal of Freshwater Ecology*, **4**, 229–341.

George, M. (1992) *The Land Use, Ecology and Conservation of Broadland*. Packard Publishing, Chichester.

Grimm, M.P. & Backx, J.J.G.M. (1990) The restoration of shallow eutrophic lakes, and the role of northern pike, aquatic vegetation and nutrient concentration. *Hydrobiologia*, **200/201**, 557–566.

Gulati, R.D. (1990) Structural and grazing responses of zooplankton community to biomanipulation of some Dutch water bodies. *Hydrobiologia*, **200/201**, 99–118.

Hrbáček, J., Dvorakova, M., Korinek, V. & Prochazkova, L. (1961) Demonstration of the effect of the fish stock on the species composition of zooplankton and the intensity of metabolism of the whole plankton assemblage. *Verhandlungen der Internationalen Vereinigung für Theoretische und Angewandte Limnologie*, **14**, 192–195.

Irvine, K., Bales, M.T., Moss, B., Stansfield, J.H. & Snook, D. (1993) The changing ecosystem of a shallow brackish lake, Hickling Broad, Norfolk, U.K. I. Trophic relationships with special reference to the role of *Neomysis integer* (Leach). *Freshwater Biology*, **29**, 119–139.

Jeppesen, E., Sondergaard, M., Sortkjaer, O., Mortensen, E. & Kristensen, P. (1990) Interactions between phytoplankton, zooplankton and fish in a shallow, hypertrophic lake: a study of phytoplankton collapses in Lake Sobygard, Denmark. *Hydrobiologia*, **191**, 149–164.

Lammens, E.H.R.R. (1986) *Interactions Between Fishes and the Structure of Fish Communities in Dutch Shallow Eutrophic Lakes*, PhD Thesis, Agricultural University of Wageningen.

Lammens, E.H.R.R., Gulati, R.D., Meijer, M.L. & van Donk, E. (1990) The first biomanipulation conference: a synthesis. *Hydrobiologia*, **200/201**, 619–627.

Lund, J.W.G. (1970) Primary production. *Water Treatment and Examination*, **19**, 332–358.

McQueen, D.J. (1990) Manipulating lake community structure: where do we go from here? *Freshwater Biology*, **23**, 613–620.

Madgwick, J. & Phillips, G.L. (1992) The role of research and experimental management in restoring the Norfolk Broads. *Assessing and Monitoring Changes in Wetland Parks and Protected Areas* (ed. J. Madgwick), pp. 70–74. Broads Authority, Norwich.

May, R.M. (1974) Biological populations with nonoverlapping generations, stable points, stable cycles and chaos. *Science*, **186**, 645–667.

Meijer, M.L., de Haan, M.W., Breuelaar, A.W. & Buiteveld, H. (1990) Is reduction of the benthivorous fish an important cause of high transparency following biomanipulation in shallow lakes? *Hydrobiologia*, **200/201**, 303–315.

Mortensen, E., Jeppesen, E., Søndergaard, M. & Kamp Nielsen, L. (1994) Nutrient dynamics and biological structure in shallow freshwater and brackish lakes. *Developments in Hydrobiologia 94*. Kluwer, Dordrecht.

Moss, B., Leah, R.T. & Clough, B. (1979) Problems of the Norfolk Broads and their impact on freshwater fisheries. *Proceedings of the First British Freshwater Fisheries Conference. University of Liverpool*, pp. 67–85.

Moss, B., Balls, H., Irvine, K. & Stansfield, J. (1986) Restoration of two lowland lakes by isolation from nutrient-rich water sources with and without the removal of sediment. *Journal of Applied Ecology*, **23**, 391–414.

Moss, B., Stansfield, J., Irvine, K., Perrow, M. & Phillips, G. (in press) Progressive restoration of a shallow lake – A twelve year experiment in isolation, sediment removal and biomanipulation. *Journal of Applied Ecology*.

Mourelatos, S. & Lacrois, G. (1990) In situ filtering rates of Cladocera: effect of body length, temperature, and food concentration. *Limnology and Oceanography*, 35, 1101–1111.

Ozimek, T., Gulati, R.D. & van Donk, E. (1990) Can macrophytes be useful in biomanipulation of lakes? The Lake Zwelmlust example. *Hydrobiologia*, **200/201**, 399–407.

Peirson, G. (1983) The diet of roach and bream during the first year of life in a small Norfolk Broad. *Proceedings of the Third British Freshwater Fish Conference, University of Liverpool*, pp. 229–235.

Perrow, M.R. (1992) Biomanipulation in Broadland. *Fisheries in the Year 2000, Proceedings of the 21st Anniversary Conference of the Institute of Fisheries Management* (eds K.T. O'Grady, A.J.B. Butterworth, P.B. Spillet & J.C.J. Domaniewski), pp. 335–337. Institute of Fisheries Management, Nottingham.

Perrow, M.R. (1994a) The nature of the fish communities in broads dominated by macrophytes. *The Development of Biomanipulation Techniques and Control of Phosphorus Release from Sediments* (eds J. Pitt & G.L. Phillips). Interim report 475/1/A LIFE project – Restoration of the Norfolk Broads. NRA, Bristol.

Perrow, M.R. (1994b) Practical aspects of the biomanipulation of fish populations. *The Development of Biomanipulation Techniques and Control of Phosphorus Release from Sediments* (eds J. Pitt & G.L. Phillips). Interim report 475/1/A LIFE project – Restoration of the Norfolk Broads. NRA, Bristol.

Perrow, M.R. & Irvine, K. (1992) The relationship between cladoceran body size and the growth of underyearling roach (*Rutilus rutilus* (L.)) in two shallow lakes: a mechanism for density-dependent reductions in growth. *Hydrobiologia*, 241, 155–161.

Perrow, M.R., Peirson, G. & Townsend, C.R. (1990) The dynamics of a population of roach (*Rutilus rutilus*) in a shallow lake: is there a 2-year cycle in recruitment? *Hydrobiologia*, 191, 67–73.

Perrow, M.R., Moss, B. & Stansfield, J. (1994) Trophic interactions in a shallow lake following a reduction in nutrient loading: a long-term study. *Hydrobiologia*, **275/276**, 43–52.

Phillips, G.L. (1984) A large scale field experiment in the control of eutrophication in the Norfolk Broads. *Journal of Institute of Water Pollution Control*, 83, 400–408.

Phillips, G.L. & Kerrison, P. (1991) The restoration of the Norfolk Broads: the role of biomanipulation. *Memorie del l'Istituto Italiano di Idrobiologia*, 48, 75–97.

Phillips, G.L. & Moss, B. (1994) Is Biomanipulation a Useful Technique in Lake Management? *R&D Note* 276. NRA, Bristol.

Reynolds, C.S. (1994) The ecological basis for the successful biomanipulation of degraded aquatic communities. *Archiv für Hydrobiologie*, 130, 1–33.

Sas, H. (1989) *Lake Restoration by Reduction of Nutrient Loadings: Expectations, Experiences, Extrapolations*. Academia Verlag Richarz, Sankt Augustin.

Scheffer, M., Hosper, S.H., Meijer, M.L., Moss, B. & Jeppesen, E. (1993) Alternative equilibria in shallow lakes. *Trends in Ecology and Evolution*, 8, 275–279.

Shapiro, J. (1990) Biomanipulation: the next phase – making it stable. *Hydrobiologia*, **200/201**, 13–27.

Timms, R.M. & Moss, B. (1984) Prevention of growth of potentially dense phytoplankton populations by zooplankton grazing in the prescence of zooplanktivorous fish in a shallow wetland ecosystem. *Limnology and Oceanography*, 29, 472–486.

Townsend, C.R. & Perrow, M.R. (1989) Eutrophication may produce population cycles in roach, *Rutilus rutilus* (L.), by two contrasting mechanisms. *Journal of Fish Biology*, 34, 161–164.

Vollenweider, R.A. & Kerekes, J. (1980) The loading concept as a basis for controlling eutrophication philosophy and preliminary results of the OECD programme on eutrophication. *Progress in Water Technology*, 12, 5–38.

CHAPTER 23

Aquatic predators and their prey: the take-home messages

G.M. Dunnet

I shall begin by congratulating the organizers of this conference for producing such a wide-ranging and well balanced programme on aspects of aquatic predators and their prey. The papers have been of a high quality and there is much in them to help us to understand many of the important relationships and processes involving predation in aquatic communities. I will not refer specifically to any individual paper, but see my task as having to identify what appear to me to be important general points which have emerged which could influence future research and the development of techniques for management. My comments therefore will be entirely personal and are not intended, or required, to be views which are representative of the conference as a whole. I have, however, found very valuable the comments of session chairmen who have highlighted general points which have emerged during their sessions. I believe many of these have been incorporated into my own comments which follow.

The title of the conference is 'Aquatic Predators and Their Prey', but though the concept of the conference included comparison with terrestrial systems, none of the papers addressed this directly. As pointed out in the Introduction, such an approach could be both revealing and rewarding.

The conference was arranged in three separate sections, the first entitled 'The Behavioural Strategies of Predators and Prey', the second 'The Role of Predators in Ecosystem Structure' and the third 'Predators, Prey and Man'. While my comments will not be restricted specifically to these sections, I will nonetheless present them within that framework.

BEHAVIOURAL STRATEGIES OF PREDATORS AND PREY

Predators and prey are integral components of biological communities and the process of predation is ubiquitous in ecological systems. It is only when predators compete

with ourselves, another predator, that they may be regarded unfavourably by man, but more of that later. To a zoologist, predators, like all animal species, merit detailed study in their own right. Their adaptations in their way of life, structure, physiology, behaviour and life-history excite scientific curiosity. Prey species are equally interesting to study. When we think of predation we are usually directing our attention at the interaction between predators and their prey, and one of the major elements in such interactions is the selection of prey by predators. A great deal of attention has been paid to the diet of predators which in many cases can be described in detail, both qualitatively and quantitatively. Less is known about the range of predators that take any particular prey species. The availability now of high technology equipment greatly adds to our ability to investigate the natural history of predators and their prey.

Natural history gives way to ecology when we become quantitative rather than qualitative in our approach to the biology of predators and their prey. This is particularly clear when we consider the ways in which predators select and obtain their food. All of the papers which have presented data on this aspect have shown that diets vary tremendously over time and from place to place; normally include a wide range of different prey species, prey sizes and life-cycle stages; and may change in relation to the age and experience of the predator. It is often assumed that the predators are *opportunists* selecting their prey from a number of *alternative species* according to their *availability*. These terms trip off our tongues very readily, but are in fact shorthand for very complex biological interactions about which we know only a little in quantitative terms. I think much research, including experimental work, if necessary in the laboratory, may yet be needed for us to fully understand these concepts. It is clear that such concepts are highly relevant to the issues discussed in the next section of the conference.

THE ROLE OF PREDATORS IN ECOSYSTEM STRUCTURE

If I may be a little pedantic, in some situations the term 'biological community' might be a more appropriate one than 'ecosystem': at least that would serve to simplify the concept! Numerous studies presented here have demonstrated that the biological communities, which include predators and their prey, are very complex. The long-established concept of predation as links within food webs seems inescapable. The prey components of communities may be composed of many species, and have been shown to vary greatly both from place to place and from time to time. Often ignored or unknown is whether or not the predator component of the community is equally variable. There have not been very many comprehensive long-term studies of whole systems (or extensive parts of systems) but where these have been reported this dynamism and variability have been strong features. The causes of major changes over long timescales are not known, but changes, for example in marine communities, have been associated with major changes in the distribution and quality of water masses possibly related to major climatological events. In such situations the concept of the ecosystem is highly appropriate!

For obvious reasons scientific investigations have concentrated on small areas over a relatively short term. Detailed studies have revealed many interacting variables involving daily and seasonal patterns of behaviour, effects of different stages of life history, effects of population age- and size-structure, patterns of dispersion and so on. Many studies explore these in ever greater detail, and as new techniques become available more and more high-quality data accumulate. Such empirical data from real world situations are generally regarded as essential both for increasing knowledge and understanding, and for the building of conceptual and mathematical models and for testing and validating them.

We have seen that these systems are naturally highly variable and dynamic over both time and geographical scales. The detailed biological studies referred to above are site-specific and time-specific. We recognize the dangers of extrapolating from models into areas beyond their immediate relevance. I believe, therefore, that there is perhaps a need for thought about the best approaches to gaining understanding of the real world as it is, and for ways of combining these approaches.

The development and use of models is at the heart of trying to find general principles and generalized processes by simplifying the complexity in order to see the wood from the trees (sorry for the terrestrial example). Such an approach seems to me to be essential, but I feel that the conclusion of some narrow short-term detailed studies being expressed in the form of a model is not always helpful. Indeed the simplifications inherent in the construction of the model may well exclude some of the interesting real-life interactions which are going on. Apart from their ability to generalize and predict, models enable the generation of hypotheses to be tested and also the robustness and sensitivity of systems to different influences to be investigated. An appeal has been made for field ecologists to be more numerate so that they are able to follow the developments of models and understand their value and limitations. While I applaud this I also feel that those who are capable of thinking in abstract terms and developing theoretical ecology in mathematical language also need to have more empathy with and understanding of natural history. After all, many sensible and critical hypotheses have been generated by the intuition of naturalists without the aid of models or even of sophisticated theory, and long may that continue in parallel with more modern practices.

I sometimes have difficulty in knowing why particular models have been developed and I think it important that, in every case, the objectives and limitations behind the production of any model should be made absolutely clear to potential users.

PREDATORS, PREY AND MAN

Numerous ways in which predators, prey and people interact have been illustrated during the conference and no doubt there are many others. I shall select only one for comment here – the situation where man as a predator is competing with other predators for a desired prey (resource). The biological problems are not particularly difficult in situations where the resource is more or less domesticated, for example in fish farming, but where we are dealing with wild populations in the sea or in rivers or lakes the situation is much more complex. We need to know whether it is worthwhile attempting to reduce competition between man and other top-predators, and if we believe it is worthwhile, how can it be achieved. The assessment of the impact of the predator (often equivalent to the damage done to man's interest by the predator) on the populations of its prey and its availability to us, presents very complex problems to be addressed. Commonly the approach is, in effect, to select a single link in the food web and to attempt to treat it in isolation. From all we know about interactions and processes within biological communities, this approach is quite inadequate. Though we may know, even in quantitative terms, the proportion of the diet of the predator which comprises our desired prey, it is extremely difficult to determine precisely the impact of the predation on the prey stocks as a whole. In most cases field experiments are very difficult, if not impossible, to mount, and confident and precise scientific interpretations and predictions are extremely difficult to achieve.

Even if it is established that there is a major impact, the second question of what to do about it raises similar scientific problems and also economic and social problems. It is possible that the best recommendation that scientists might make would be that the population of predators should be actively managed, which might include culling. Such a decision would be bound to provoke powerful opposition from conservationists and from animal rights activists. They might be placated, to some extent at least, if there was incontrovertible scientific evidence that such an approach would bring about the desired effect, and that the looked for effect would be measurable at the end of the day.

I believe that questions such as these are largely unanswerable in the short and medium term. The question therefore remains: should any action be taken to address perceived problems in the absence of rigorous scientific prediction and proof? Some say that scientists should be more prepared to say that they do not know the answers, thereby clearing the way for others to get on with the job. It is not as simple as that. It is not a matter of 'yes' or 'no' being the only alternatives. The fact is that scientists do know a great deal about these situations and indeed they know enough to be sure that the obvious *ad hoc* approach to solving the problem, i.e. controlling predator numbers, may well be inappropriate or ineffective. There is great pressure for action in some circumstances and we need to recognize the limitations of scientists to provide what we would all regard as the necessary information and understanding on which to base such action. The problem for us as ecological scientists is to find ways of addressing these problems, to make predictions on which management plans may be based, and to arrive at suitable mechanisms for monitoring the effectiveness of the programme. Our ability to predict from existing knowledge and understanding is better founded than that of many others, and needs to be recognized. The dilemma for managers is whether or not to proceed in the absence of complete scientific information, or to wait, possibly for decades, or even for ever, for such data to be available.

Index

Books published by **Fishing News Books**

Free catalogue available on request from Fishing News Books, Blackwell Science, Osney Mead, Oxford OX2 0EL, England

Abalone farming
Abalone of the world
Advances in fish science and technology
Aquaculture and the environment
Aquaculture & water resource management
Aquaculture development – progress and prospects
Aquaculture: principles and practices
Aquaculture in Taiwan
Aquaculture systems
Aquaculture training manual
Aquatic ecology
Aquatic microbiology
Aquatic weed control
Atlantic salmon: its future
The Atlantic salmon: natural history etc.
Bacterial diseases of fish
Better angling with simple science
Bioeconomic analysis of fisheries
British freshwater fishes
Broodstock management and egg and larval quality
Business management in fisheries and aquaculture
Cage aquaculture
Calculations for fishing gear designs
Carp farming
Carp and pond fish culture
Catch effort sampling strategies
Commercial fishing methods
Common fisheries policy
Control of fish quality
Crab and lobster fishing
The crayfish
Crustacean farming
Culture of bivalve molluscs
Design of small fishing vessels
Developments in electric fishing
Developments in fisheries research in Scotland
Dynamics of marine ecosystems
Ecology of fresh waters
The economics of salmon aquaculture
The edible crab and its fishery in British waters
Eel culture
Engineering, economics and fisheries management
The European fishing handbook 1993–94
FAO catalogue of fishing gear designs
FAO catalogue of small scale fishing gear
Fibre ropes for fishing gear
Fish catching methods of the world
Fisheries biology, assessment and management
Fisheries oceanography and ecology
Fisheries of Australia
Fisheries sonar
Fishermen's handbook
Fisherman's workbook
Fishery development experiences
Fishery products and processing
Fishing and stock fluctuations
Fishing boats and their equipment
Fishing boats of the world 1
Fishing boats of the world 2
Fishing boats of the world 3
Fishing ports and markets
Fishing with electricity
Fishing with light
Freshwater fisheries management
Fundamentals of aquatic ecology
Glossary of UK fishing gear terms
Handbook of trout and salmon diseases
A history of marine fish culture in Europe and North America

How to make and set nets
The Icelandic fisheries
Inland aquaculture development handbook
Intensive fish farming
Introduction to fishery by-products
The law of aquaculture: the law relating to the farming of fish and shellfish in Great Britain
A living from lobsters
Longlining
Making and managing a trout lake
Managerial effectiveness in fisheries and aquaculture
Marine climate, weather and fisheries
Marine fish behaviour in capture and abundance estimation
Marine ecosystems behaviour & management
Marketing: a practical guide for fish farmers
Marketing in fisheries and aquaculture
Mending of fishing nets
Modern deep sea trawling gear
More Scottish fishing craft and their work
Multilingual dictionary of fish and fish products
Multilingual dictionary of fishing vessels/safety on board
Multilingual dictionary of fishing gear
Multilingual illustrated dictionary of aquatic animals & plants
Multilingual illustrated guide to the world's commercial coldwater fish
Multilingual illustrated guide to the world's commercial warmwater fish
Navigation primer for fishermen
Netting materials for fishing gear
Net work exercises
Ocean forum
Pair trawling and pair seining
Pelagic and semi-pelagic trawling gear
Pelagic fish: the resource and its exploitation
Penaeid shrimps — their biology and management
Planning of aquaculture development
Pollution and freshwater fish
Purse seining manual
Recent advances in aquaculture IV
Recent advances in aquaculture V
Refrigeration on fishing vessels
Rehabilitation of freshwater fisheries
The rivers handbook, volume 1
The rivers handbook, volume 2
Salmon and trout farming in Norway
Salmon aquaculture
Salmon farming handbook
Salmon in the sea/new enhancement strategies
Scallop and queen fisheries in the British Isles
Scallop farming
Seafood science and technology
Seine fishing
Shrimp capture and culture fisheries of the US
Spiny lobster management
Squid jigging from small boats
Stability and trim of fishing vessels and other small ships
The state of the marine environment
Stock assessment in inland fisheries
Study of the sea
Sublethal and chronic toxic effects of pollution on freshwater fish
Textbook of fish culture
Trends in fish utilization
Trends in ichthyology
Trout farming handbook
Tuna fishing with pole and line